Earth Science

A Reference Guide

Staff for this Book

Dan Franck *Content Specialist*
Jane Willan *Content Specialist*
Suzanne Montazer *Art Director*
Lisa Dimaio Iekel *Production Manager*
Jayoung Cho *Designer*
Carol Leigh *Designer*
Christopher Yates *Cover Designer*
Jim Miller *Visual Designer*
Annette Scarpitta *Illustrations Editor*
Mary Beck Desmond *Text Editor*
Bud Knecht *Text Editor*
Susan Raley *Text Editor, Quality Control Specialist*
Jeff Pitcher *Instructional Designer*
John Hornyak *Instructional Designer*
Craig Ruskin *Project Manager*
Adam Rohner *Project Manager*
Connie Moy *Quality Control Manager*

Bror Saxberg *Chief Learning Officer*
John Holdren *Senior Vice President for Content and Curriculum*
Maria Szalay *Senior Vice President for Product Development*
Jennifer Thompson *Director of Product Delivery*
Tom DiGiovanni *Senior Director of Instructional Design*
Kim Barcas *Creative Director*
John G. Agnone *Director of Publications*
Charles Kogod *Director of Media and IP Management*
Jeff Burridge *Managing Editor*
Steve Watson *Product Manager*

About K12 Inc.

Founded in 1999, K12 Inc. is an elementary and secondary school service combining rich academic content with powerful technology. K[12] serves students in a variety of education settings, both public and private, including school classrooms, virtual charter schools, home schools, and tutoring centers. K[12] currently provides comprehensive curricular offerings in the following subjects: Language Arts/English, History, Math, Science, Visual Arts, and Music. The K[12] curriculum blends high quality offline materials with innovative online resources, including interactive lessons, teacher guides, and tools for planning and assessment. For more information, call 1-888-YOUR K12 or visit www.K12.com.

Published by K12 Inc., Herndon, Virginia

Printed by RR Donnelley, Hong Kong, March 2011, Lot 032011,

Earth Science

A Reference Guide

Contents

How to Use This Book **x**
Navigating a Page **xi**

Unit 1 Earth Science and Systems

Contributions to Science **2**
Our Place in the Universe **4**
Measurements **6**
Topographic Maps **8**
Geologic Maps **10**
Seismographs **12**
Scientific Method **14**
Landforms of the Contiguous United States **16**
Latitude and Longitude **18**
The Spheres of Earth **20**
Earth's Elements **22**
Atoms **24**
Types of Chemical Bonds **26**

Unit 2 Geology

Groups of Minerals **28**
Silicate Minerals **30**
Properties of Minerals **32**
Igneous Rocks **34**
Sedimentary Rocks **36**
Metamorphic Rocks **38**
The Rock Cycle **40**

Folds and Faults **42**

Soils of the World **44**

Changes in Earth's Surface **46**

Measuring Earthquake Severity **48**

Major Earthquakes **50**

Seismic Waves **52**

Earth's Layers **54**

Pangaea and Tectonic Activity **56**

Earth's Plates **58**

Types of Plate Boundaries **60**

Volcanoes **62**

Formation of the Himalayas **64**

Plate Tectonics: Development of a Theory **66**

Volcanic Processes **68**

Kinds of Volcanoes **70**

Geologic Time Scale **72**

History of Life on Earth **74**

Radioactive Decay **76**

Radiometric Dating **78**

The Grand Canyon **80**

Interpreting Geologic Cross Sections **82**

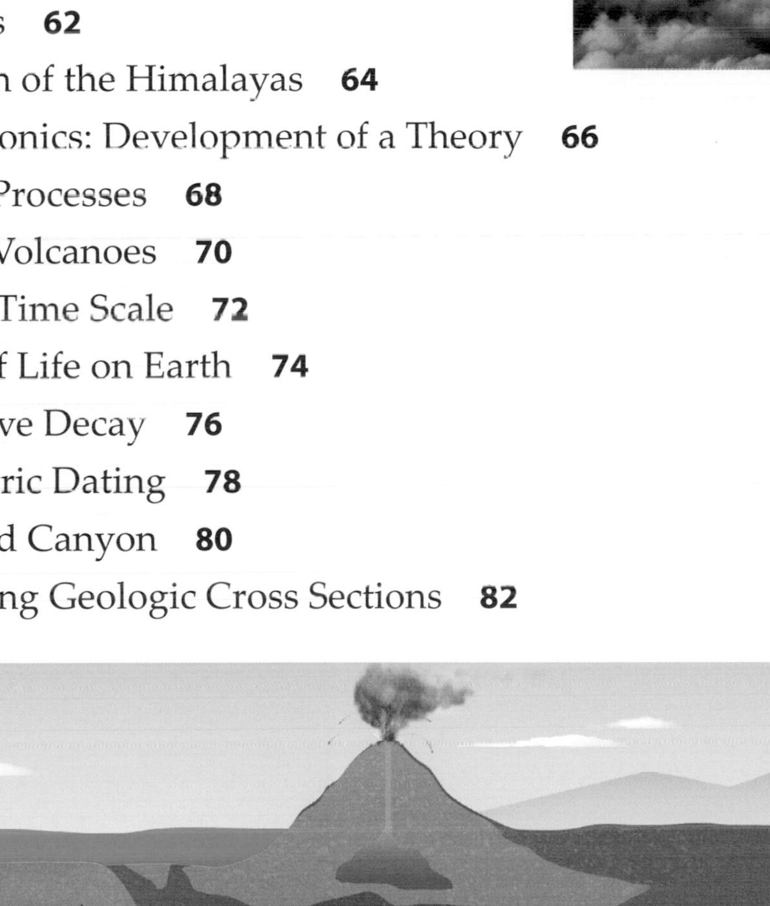

Unit 3 Earth's Atmosphere

History of Earth's Atmosphere **84**

Precambrian Life and Rocks **86**

The Layers of Earth's Atmosphere **88**

Air Pollution **90**

Solar Energy **92**

Composition of Atmospheres **94**

Sea-Level Temperatures **96**

Wind Patterns **98**

Atmospheric Circulation **100**

Unit 4 Climate and Weather

Precipitation and Climate **102**

Local Influences on the Weather **104**

Terrestrial Biomes **106**

Characteristics of Biomes **108**

El Niño–Southern Oscillation **110**

Air Masses in the United States and Canada **112**

Weather Instruments **114**

Air Masses and Fronts **116**

Weather Maps **118**

Clouds **120**

Tornadoes and Hurricanes **122**

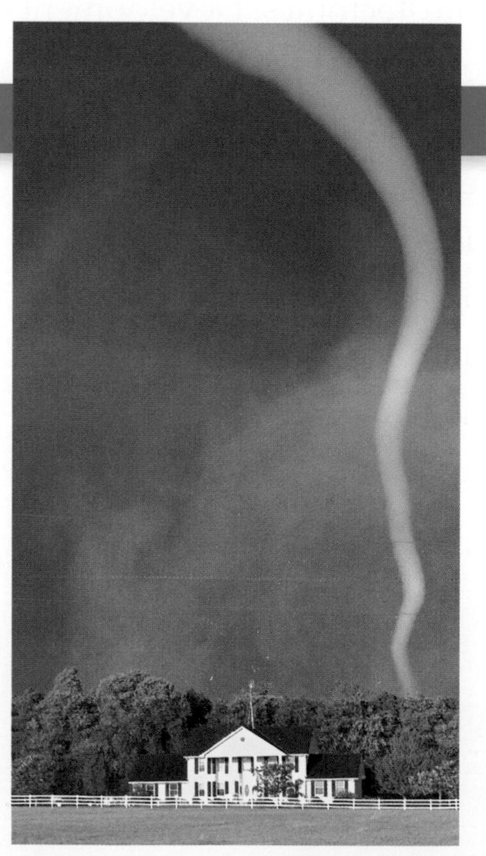

Unit 5 Oceans

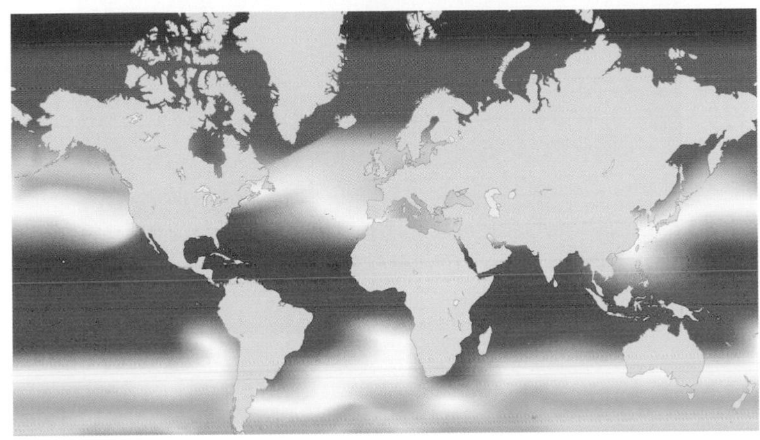

Earth's Oceans **124**

Features of the Ocean Floor **126**

Topography of the Ocean Floor **128**

Ocean Temperature **130**

Ocean Salinity **132**

Density of Ocean Water **134**

Ocean Surface Currents **136**

Ocean Waves **138**

Tides **140**

Zones of the Ocean **142**

Marine Organisms **144**

Unit 6 Cycles on Earth

The Nitrogen Cycle **146**

The Carbon Cycle **148**

The Phosphorus Cycle **150**

The Water Cycle **152**

Mineral Formation **154**

Ores **156**

Unit 7 Astronomy

Our Sun **158**
Phases of the Moon **160**
Eclipses **162**
Seasons **164**
Our Solar System **166**
Venus: A Terrestrial Planet **168**
Jupiter: A Gas Giant **170**
The Electromagnetic Spectrum **172**
Star Magnitudes **174**
Star Life Cycles **176**
Telescopes **178**

The Hertzsprung-Russell Diagram **180**
Element Formation in Stars **182**
Types of Galaxies **184**
The Scale of the Universe **186**
The Structure of the Universe **188**
Unique Objects in the Universe **190**
History of the Universe **192**
Red Shift **194**

Unit 8 Earth's Resources

Mineral Resources **196**

The Formation of Coal **198**

Coal Resources **200**

Iron Resources **202**

World Electricity Use **204**

Aquifers **206**

Water Treatment **208**

Human Population **210**

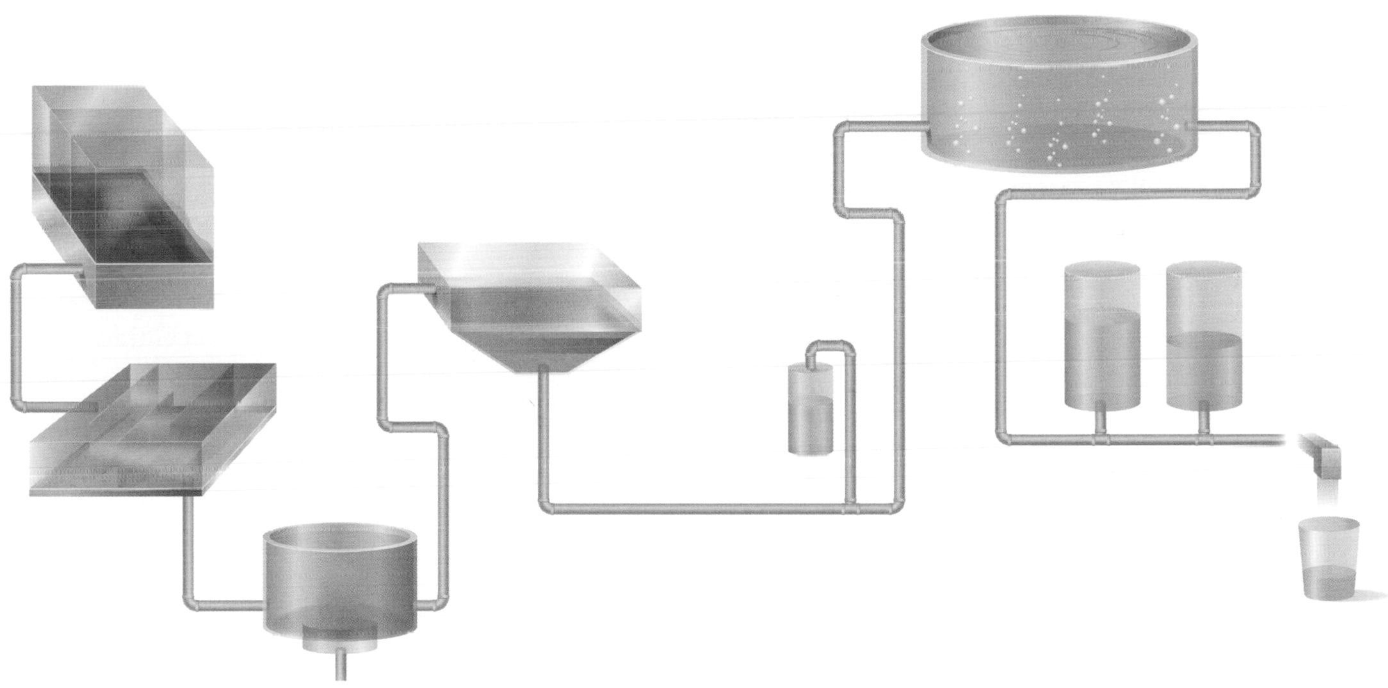

How to Use This Book

Welcome to *Earth Science: A Reference Guide*

This book serves as a reference guide for any student of earth science. It was developed as a companion to the online portion of K12 Inc.'s High School Earth Science program.

Each two-page spread of this book uses words and pictures to present an overview of a Key Idea. You can use this book to familiarize yourself with some aspects of earth science, or to review materials you're studying in other books or online sources.

How This Book Is Organized

Units of Study

This book is organized in the following units of study:

- Unit 1 – Earth Science and Systems
- Unit 2 – Geology
- Unit 3 – Earth's Atmosphere
- Unit 4 – Climate and Weather
- Unit 5 – Oceans
- Unit 6 – Cycles on Earth
- Unit 7 – Astronomy
- Unit 8 – Earth's Resources

Pronunciation Guide

To learn how to pronounce scientific terms, see the Pronunciation Guide on page 214.

Glossary

See pages 215–220 for a Glossary with brief definitions of some key terms.

Index

An Index is provided on pages 221–232.

Navigating a Page

Subject Each two-page spread explores a Key Idea in earth science.

Further Explanation This text provides more detailed information about the Key Idea.

Key Idea What's the Key Idea on these pages? Start reading here to find out.

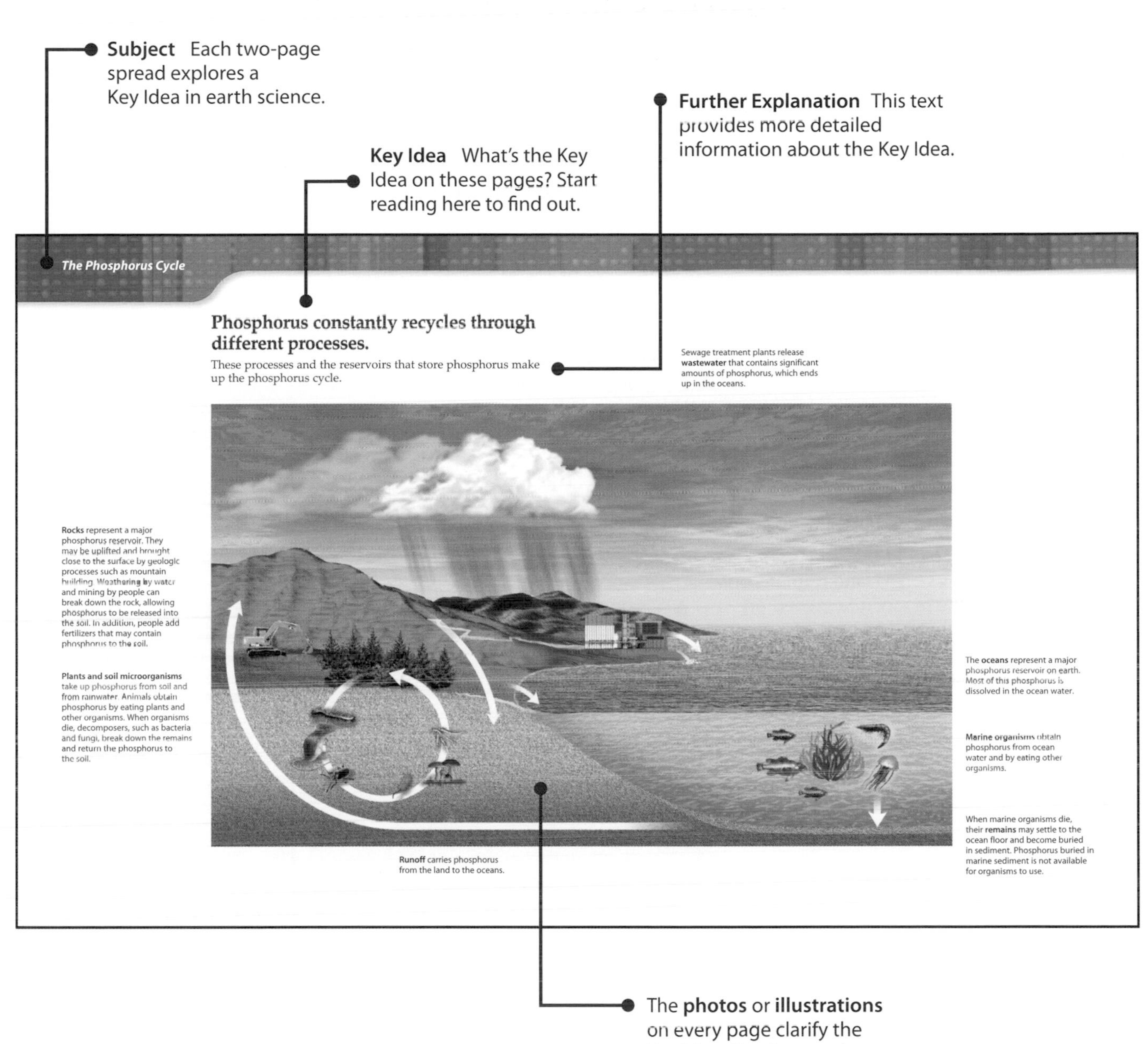

The **photos** or **illustrations** on every page clarify the Key Idea. Captions and labels help explain details in the images.

Throughout time, many people and cultures have contributed to our understanding of earth.

From the Polynesians, who sailed the Pacific Ocean 30,000 years ago, to the modern scientists who explore outer space, ancient and modern civilizations have done much to advance human understanding of earth and the solar system.

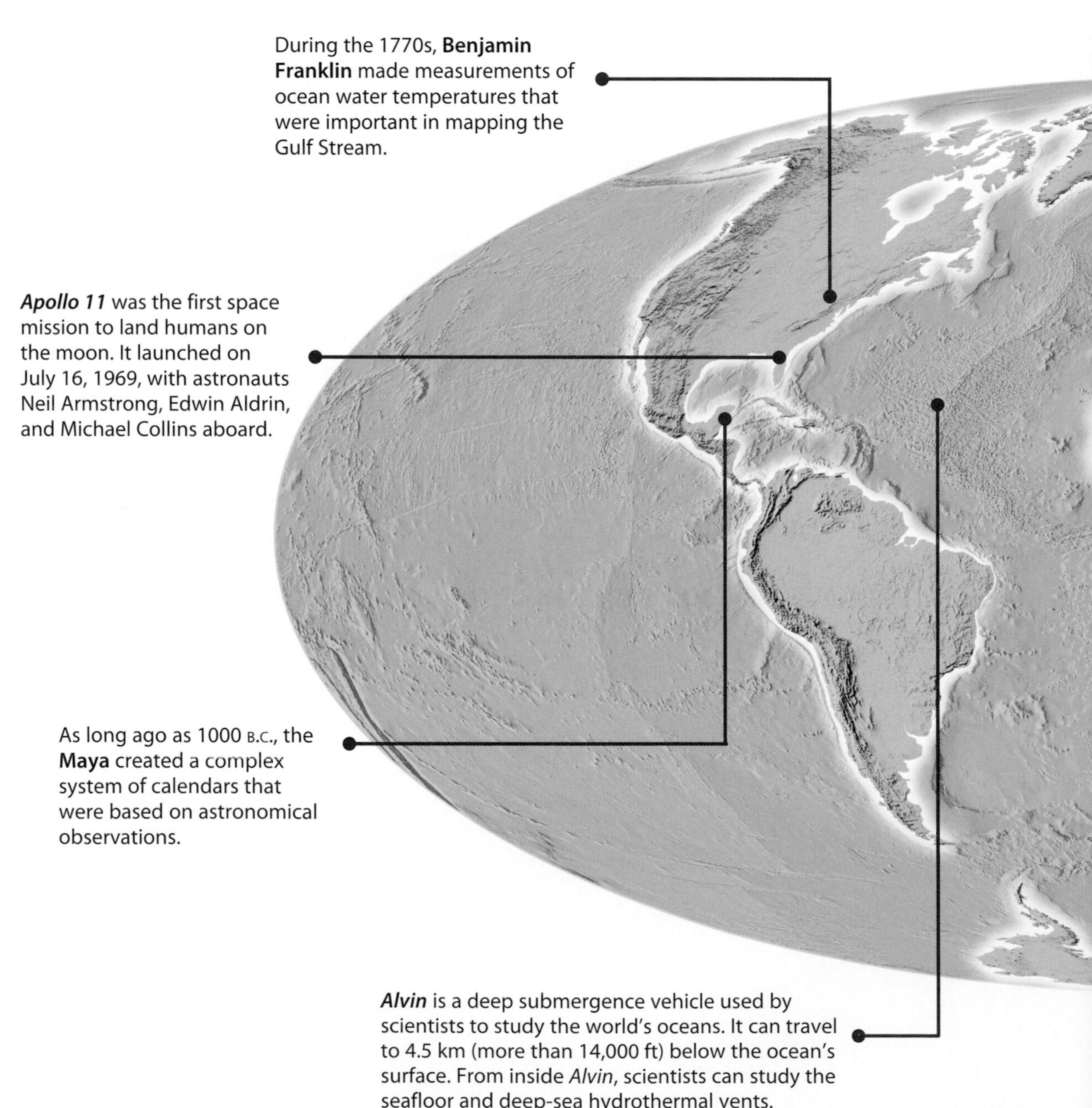

During the 1770s, **Benjamin Franklin** made measurements of ocean water temperatures that were important in mapping the Gulf Stream.

Apollo 11 was the first space mission to land humans on the moon. It launched on July 16, 1969, with astronauts Neil Armstrong, Edwin Aldrin, and Michael Collins aboard.

As long ago as 1000 B.C., the **Maya** created a complex system of calendars that were based on astronomical observations.

Alvin is a deep submergence vehicle used by scientists to study the world's oceans. It can travel to 4.5 km (more than 14,000 ft) below the ocean's surface. From inside *Alvin*, scientists can study the seafloor and deep-sea hydrothermal vents.

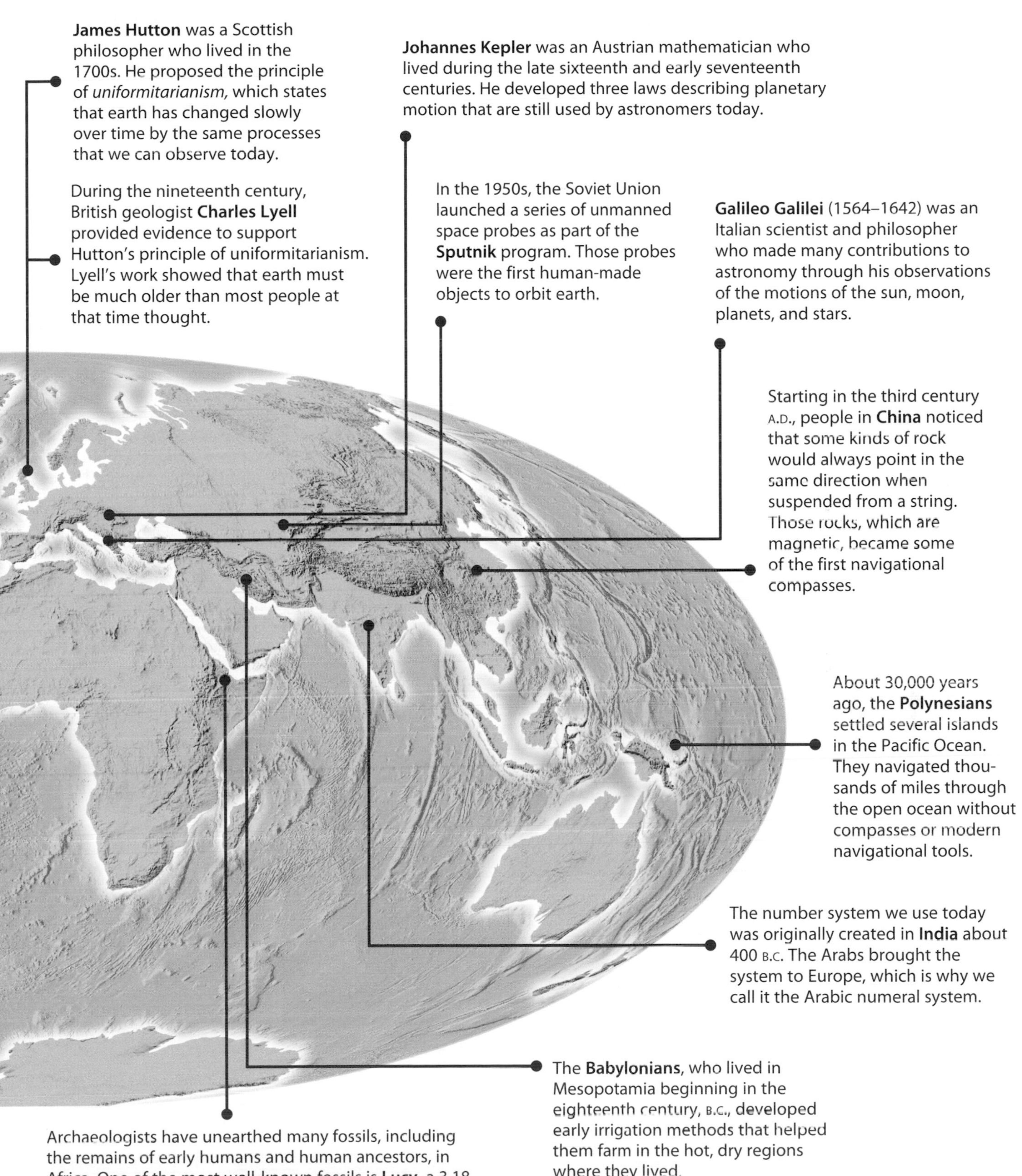

James Hutton was a Scottish philosopher who lived in the 1700s. He proposed the principle of *uniformitarianism,* which states that earth has changed slowly over time by the same processes that we can observe today.

During the nineteenth century, British geologist **Charles Lyell** provided evidence to support Hutton's principle of uniformitarianism. Lyell's work showed that earth must be much older than most people at that time thought.

Johannes Kepler was an Austrian mathematician who lived during the late sixteenth and early seventeenth centuries. He developed three laws describing planetary motion that are still used by astronomers today.

In the 1950s, the Soviet Union launched a series of unmanned space probes as part of the **Sputnik** program. Those probes were the first human-made objects to orbit earth.

Galileo Galilei (1564–1642) was an Italian scientist and philosopher who made many contributions to astronomy through his observations of the motions of the sun, moon, planets, and stars.

Starting in the third century A.D., people in **China** noticed that some kinds of rock would always point in the same direction when suspended from a string. Those rocks, which are magnetic, became some of the first navigational compasses.

About 30,000 years ago, the **Polynesians** settled several islands in the Pacific Ocean. They navigated thousands of miles through the open ocean without compasses or modern navigational tools.

The number system we use today was originally created in **India** about 400 B.C. The Arabs brought the system to Europe, which is why we call it the Arabic numeral system.

The **Babylonians**, who lived in Mesopotamia beginning in the eighteenth century, B.C., developed early irrigation methods that helped them farm in the hot, dry regions where they lived.

Archaeologists have unearthed many fossils, including the remains of early humans and human ancestors, in Africa. One of the most well-known fossils is **Lucy**, a 3.18 million-year-old skeleton discovered in 1973 in Ethiopia.

Knowledge of our place in the universe has changed over time due to advances in scientific understanding of the world around us.

At one time, for example, most people believed that earth was at the center of the solar system. Over time, technological innovations and new discoveries from cultures all over the world have changed our scientific perspective.

Chinese astronomers in 2137 B.C. were the first to record a solar eclipse.

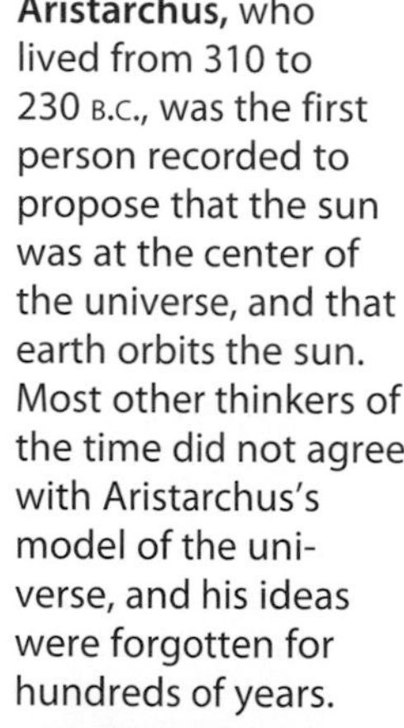

Aristarchus, who lived from 310 to 230 B.C., was the first person recorded to propose that the sun was at the center of the universe, and that earth orbits the sun. Most other thinkers of the time did not agree with Aristarchus's model of the universe, and his ideas were forgotten for hundreds of years.

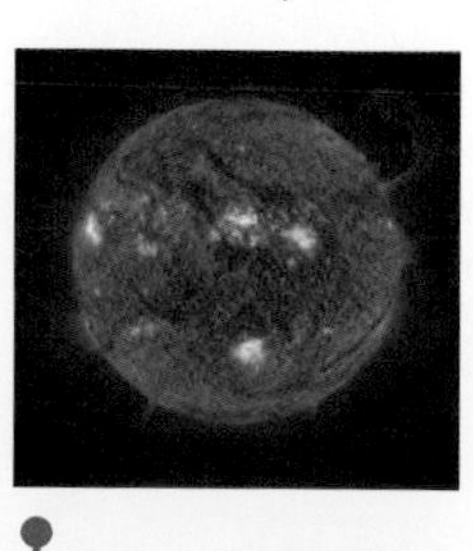

Between the first and second centuries A.D., **Ptolemy** put forth the idea that earth was at the center of the solar system. Technology was not advanced enough to disprove Ptolemy's claims, and many people believed his idea.

Chinese astronomers in 687 were the first to document a meteor shower.

Copernicus (1473–1543) proposed that the sun is at the center of the solar system (the heliocentric model) and that the planets travel in circular orbits around the sun. Although Copernicus's model of the solar system was close to being correct, it was not accepted until more than 100 years after his death.

2137 B.C.	310 B.C.	A.D. 100	687	1500

This time line is not to scale.

After years of observing the planets, **Johannes Kepler** concluded that the planets move in elliptical orbits around the sun. At the beginning of the seventeenth century, Kepler developed several laws of planetary motion. These laws are still used by astronomers today to describe the motions of the planets.

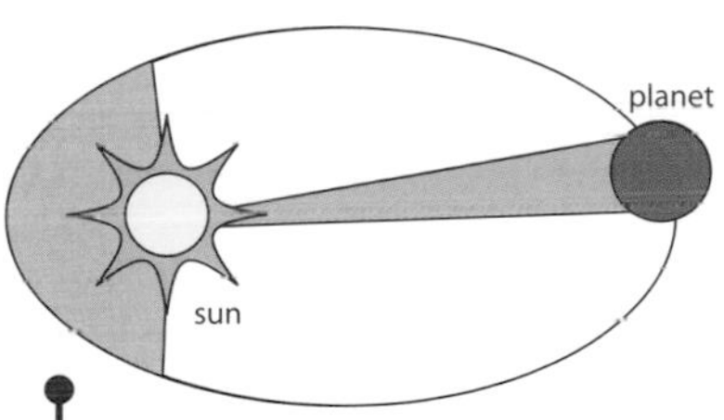

Galileo Galilei (1564–1642) improved the telescope and used it to observe the sun, moon, planets, and stars. He discovered four of Jupiter's moons and made many other important observations. Many of his observations provided support for a sun-centered solar system.

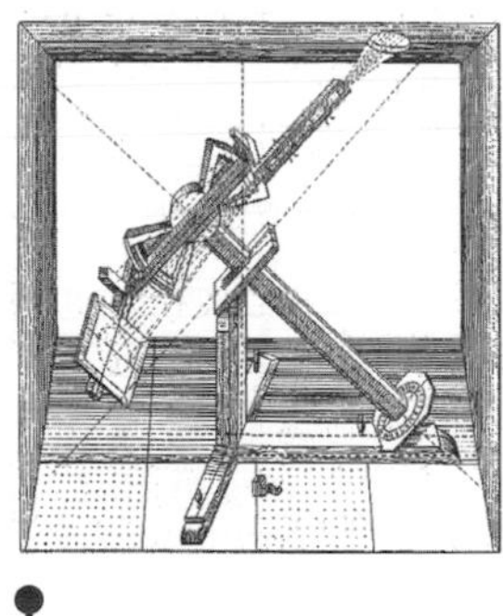

Near the end of the seventeenth century, **Sir Isaac Newton** described the force of gravity and developed three laws of motion. His ideas helped explain how the planets stay in orbit around the sun and how moons stay in orbit around planets. He also invented new forms of mathematics that can accurately describe the motions of objects.

Edwin Hubble's observations through a 100-in. Hooker telescope proved the existence of galaxies beyond the Milky Way. Evidence gained from his studies in the early 1900s led Hubble to theorize that the universe is continually expanding.

In 1927, Belgian priest **Georges-Henri Lemaître** proposed the big bang theory that the universe formed nearly 15 billion years ago when all matter and energy in the universe rapidly expanded from a single, infinitely small point. Since that time, scientists have found growing evidence to support the big bang theory.

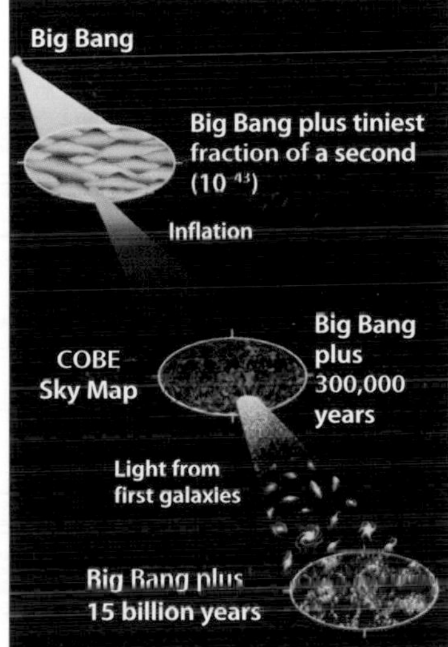

Cosmic microwave background (CMB) radiation is radiation thought to be left over from the big bang. **George Gamow** was the first to detect CMB radiation in 1948.

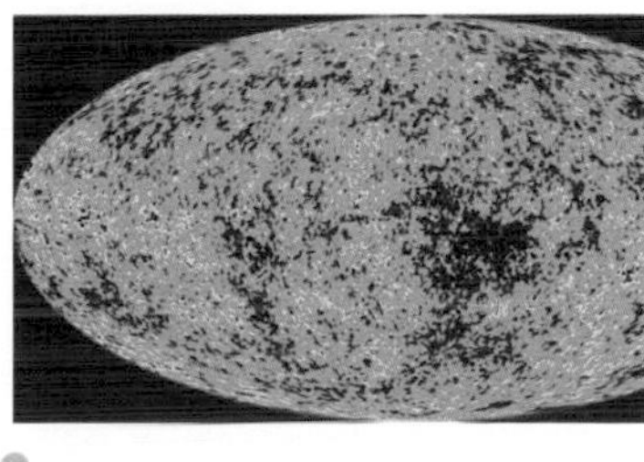

1600 | **1700** | **1900** | **1927** | **1948**

Although many people in the United States use inches, feet, or pounds to express units of measurement, scientists use a different system.

That system is the *Système International d'Unités* (International System of Units), and it is metric. Scientists often refer to this measurement system as SI.

SI BASE UNITS

QUANTITY	BASE UNIT	SYMBOL
time	second	s
mass	kilogram	kg
length	meter	m
electric current	ampere	A
amount of a substance	mole	mol
temperature	kelvin	K
luminous intensity	candela	cd

Seven units are fundamental to the SI. Other, more complex units are combinations of these seven base units.

SI PREFIXES

PREFIX	ABBREVIATION	FACTOR
pico-	p	0.000000000001
nano-	n	0.000000001
micro-	µ	0.000001
milli-	m	0.001
centi-	c	0.01
deci-	d	0.1
deka-	da	10
hecto-	h	100
kilo-	k	1,000
mega-	M	1,000,000
giga-	G	1,000,000,000

The prefix of a unit indicates its relationship to the base unit.

The SI organizes units by **powers of 10**. This organization makes the system easy for all scientists to use.

COMMON MEASUREMENTS

LENGTH		*AREA*	
1 kilometer (km)	1,000 m	1 square kilometer (km^2)	100 ha
1 meter (m)	standard unit	1 hectare (ha)	10,000 m^2
1 centimeter (cm)	0.01 m	1 square meter (m^2)	standard unit
1 millimeter (mm)	0.001 m	1 square centimeter (cm^2)	0.0001 m^2
1 micrometer (µm)	0.000001 m		
MASS		***LIQUID VOLUME***	
1 kilogram (kg)	1,000 g	1 kiloliter (kL)	1,000 L
1 gram (g)	standard unit	1 liter (L)	standard unit
1 milligram (mg)	0.001 g	1 milliliter (mL)	0.001 L
1 microgram (µg)	0.000001 g	1 mL	1 cm^3

Converting from One Temperature Scale to Another

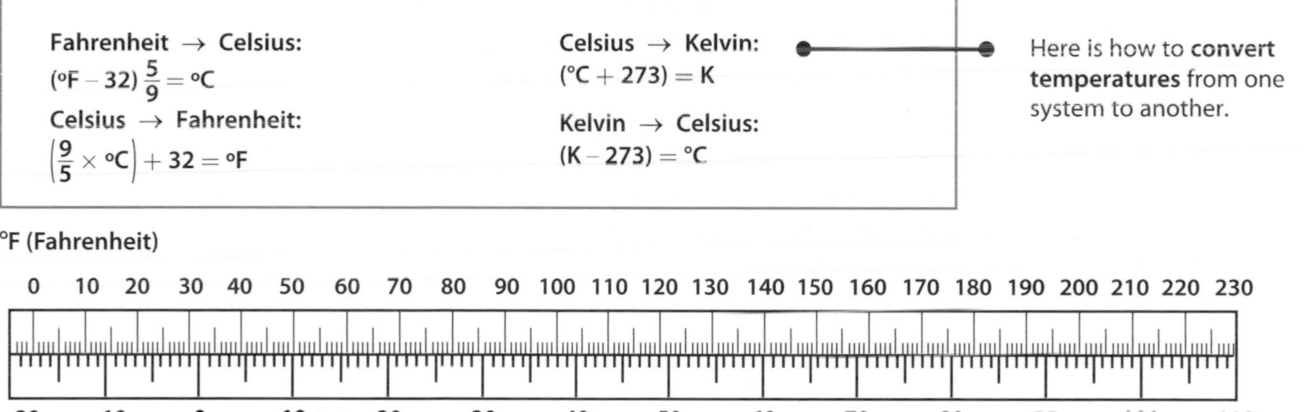

°C (Celsius)

freezing point of water

boiling point of water

Most people in the United States **express temperature** in degrees Fahrenheit. Scientists in the United States and throughout the world, however, express temperature in degrees Celsius. The Celsius scale is based on the freezing point and boiling point of water.

Topographic maps show the shape of earth's surface.

These maps use contour lines to indicate the height above sea level, or elevation, of areas of land. Every point on a given contour line has the same elevation. Topographic maps also use symbols to show the locations of features such as buildings, streams, and vegetation.

The area known as the **Berkshires** stretches through eastern New York and western Massachusetts. The landscape in this region has been shaped by glaciers, running water, and other geologic forces.

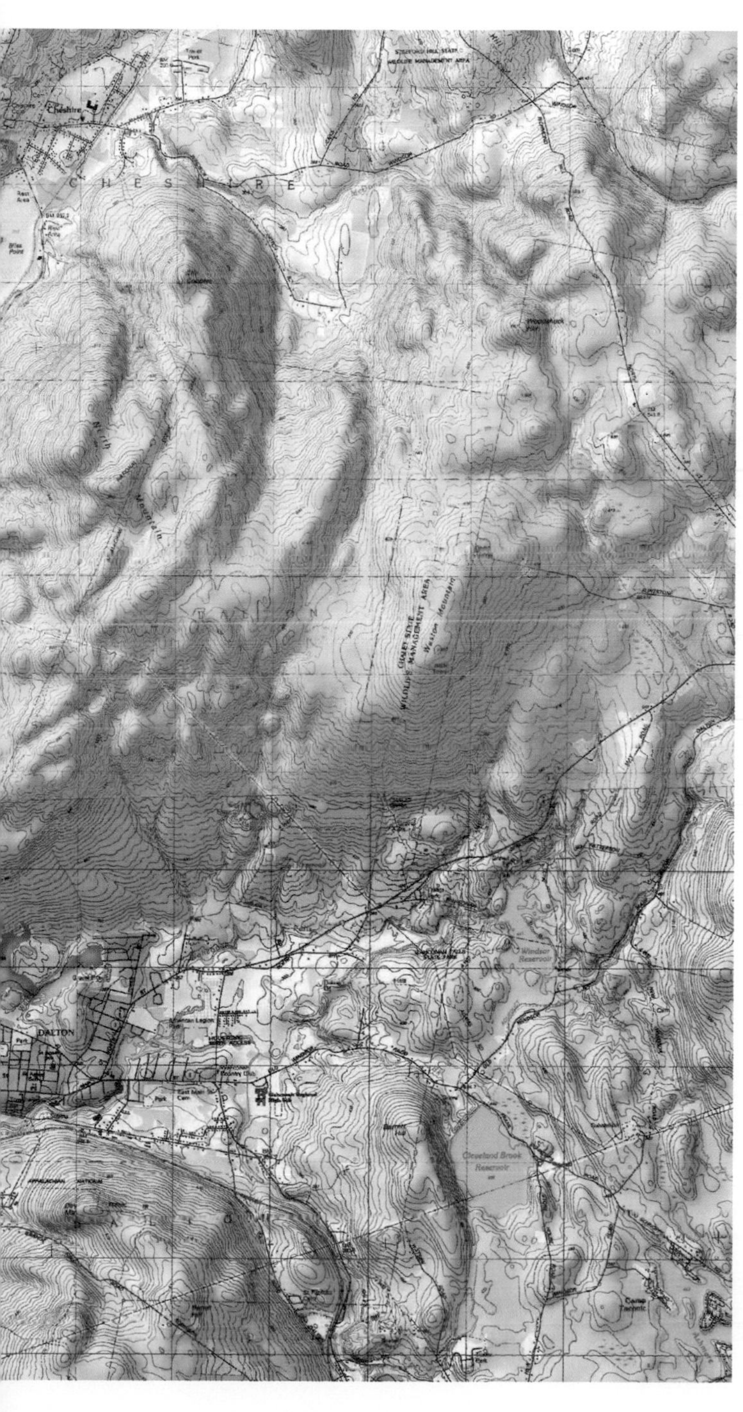

Key for a Typical Topographic Map	
contour line	
index contour line	100
depression contour line	
national boundary	
state boundary	
county, parish, municipal boundary	
township, precinct, town boundary	
incorporated city, village, or town boundary	
national or state reservation boundary	
buildings	
church	
synagogue	
tunnel	
bridge	BM 8025
benchmark	
divided highway	
road	
trail	
railroad	

Geologic maps show rock formations at or near earth's surface.

A *rock formation* is a layer or body of rock with similar age and characteristics. Most geologic maps use color and shading to indicate different rock formations. Symbols show the locations of faults, folds, and other geologic features. Some geologic maps also indicate roads, political boundaries, and major cities.

Geologic Map of Pennsylvania

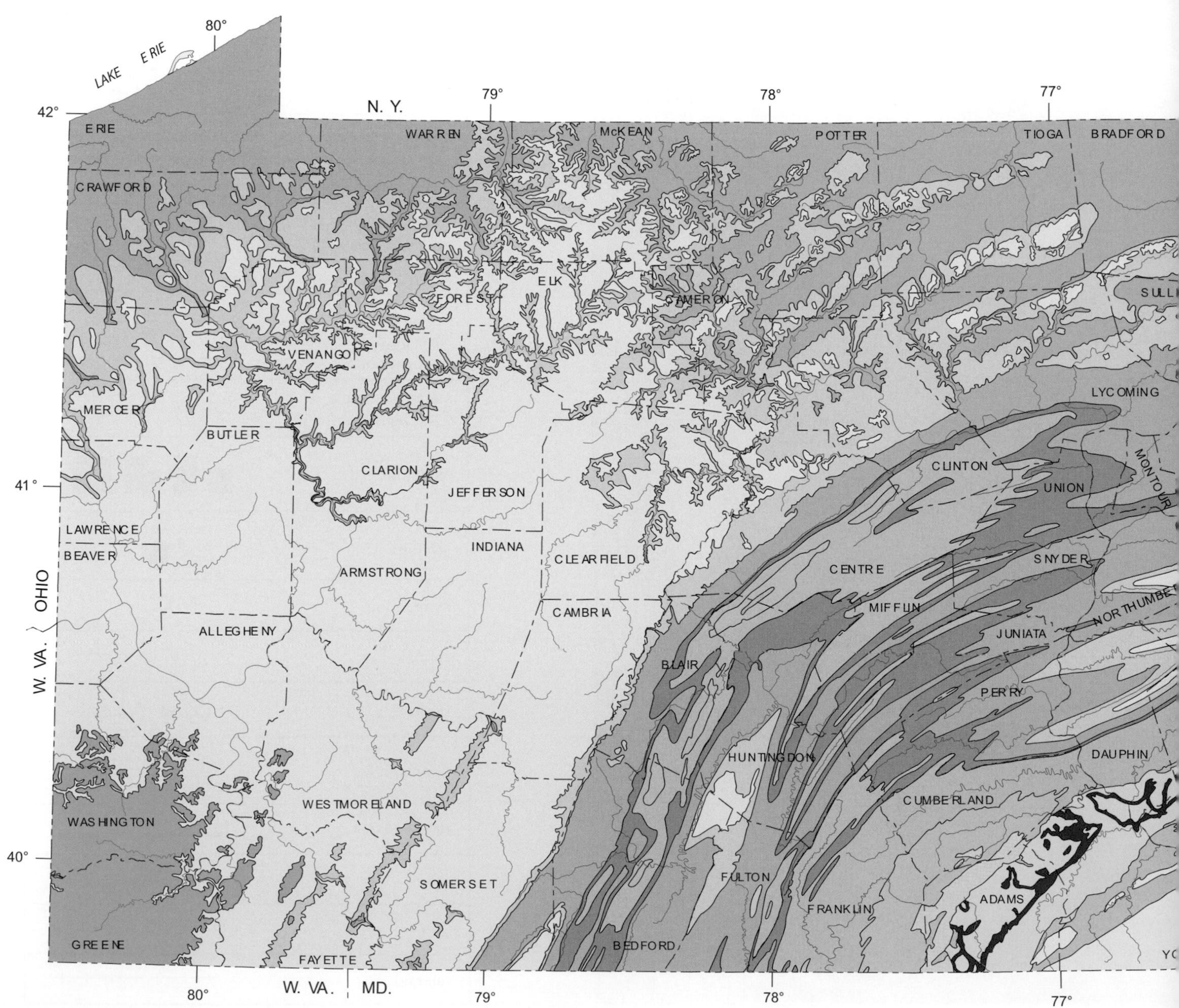

Much of the bedrock of Pennsylvania is sedimentary. The eastern parts of the state, however, also contain many metamorphic and igneous rock formations. Stream erosion has exposed many of those rock formations and, in several locations, has created gorges with cascading waterfalls.

This waterfall on Bear Run is the setting for **Fallingwater**, a house designed by noted architect Frank Lloyd Wright.

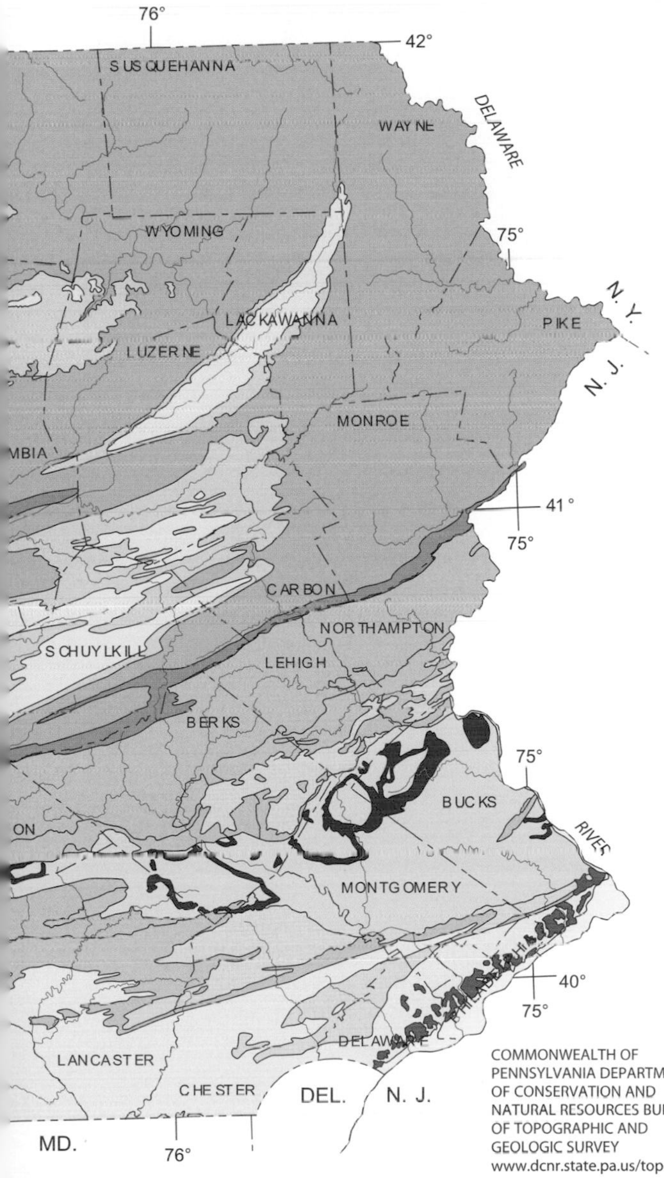

Key	
QUATERNARY sand, gravel, and silt	DEVONIAN red sandstone, gray and black shale, limestone, and chert
TERTIARY sand, gravel, silt, and clay	SILURIAN red and gray sandstone, conglomerate, shale, and limestone
JURASSIC AND TRIASSIC red sandstone, shale, and conglomerate (green), intruded by diabase (red)	ORDOVICIAN shale, limestone, dolomite, and sandstone
PERMIAN cyclic sequences of shale, sandstone, limestone, and coal	CAMBRIAN limestone, dolomite, sandstone, shale, quartzite, and phyllite
PENNSYLVANIAN cyclic sequences of sandstone, red and gray shale, conglomerate, clay, coal, and limestone	LOWER PALEOZOIC schist, quartzite, gneiss, serpentine, slate, and marbl
MISSISSIPPIAN red and gray sandstone, shale, and limestone	PRECAMBRIAN gneiss, granite, anorthosite, metadiabase, metabasalt, metarhyolite, and marble

COMMONWEALTH OF PENNSYLVANIA DEPARTMENT OF CONSERVATION AND NATURAL RESOURCES BUREAU OF TOPOGRAPHIC AND GEOLOGIC SURVEY
www.dcnr.state.pa.us/topogeo

A seismograph is a tool scientists use to measure and record ground motion.

Scientists use seismograms, which are the records produced by seismographs, to study earthquakes and earth's interior.

Seismographs record ground movement as it happens. **Analog seismographs** such as this one record movements on slowly revolving rolls of paper. Digital seismographs record measurements as electronic data on a computer.

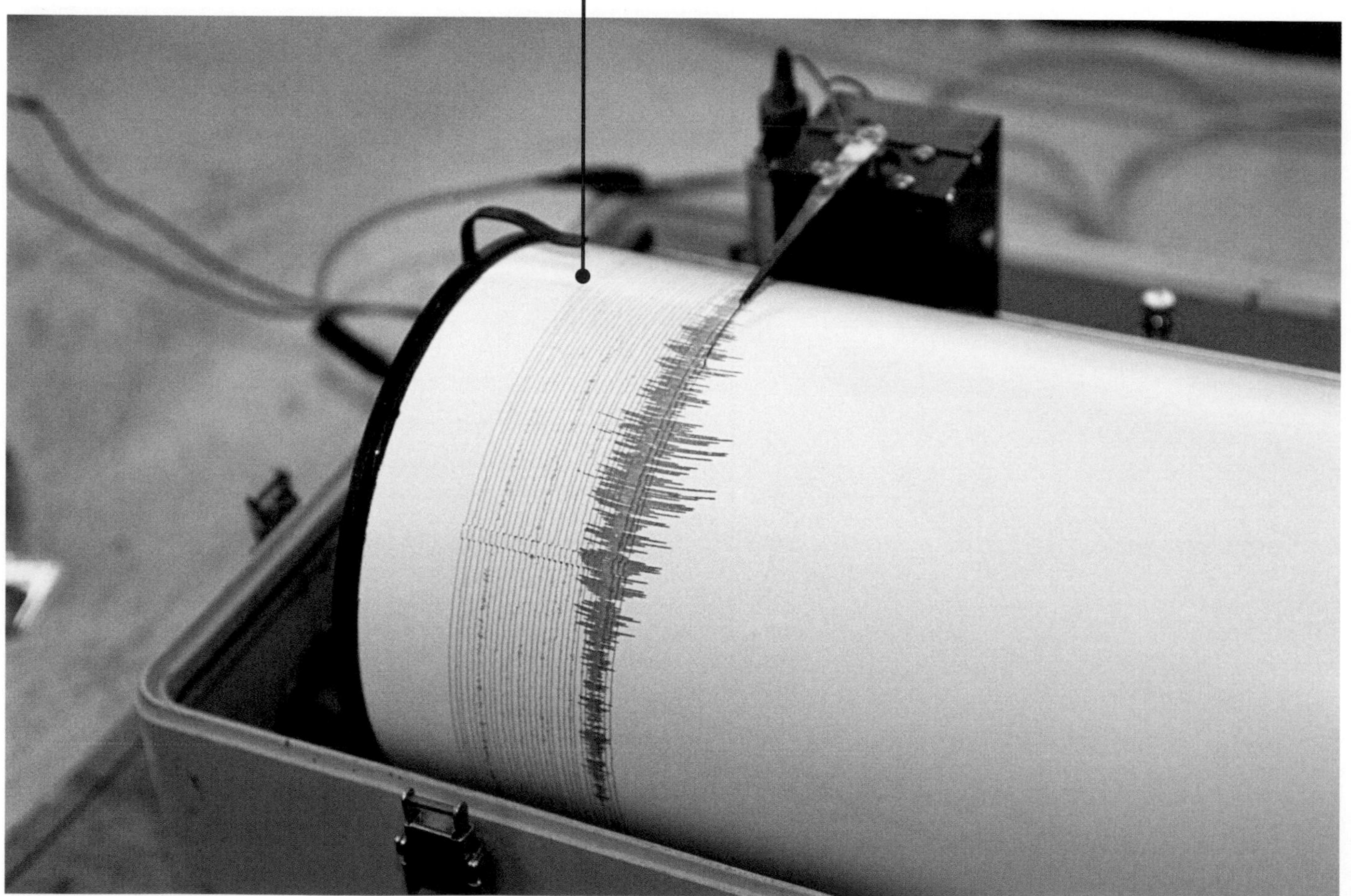

Both analog and digital seismographs receive information from **underground sensors,** which detect ground motion. Many underground sensors are specially designed to respond only to certain types of motion so that they do not detect vibrations from nonearthquake sources, such as passing trains or moving cars.

The **main shock** of a large earthquake may be recorded all the way on the other side of the earth.

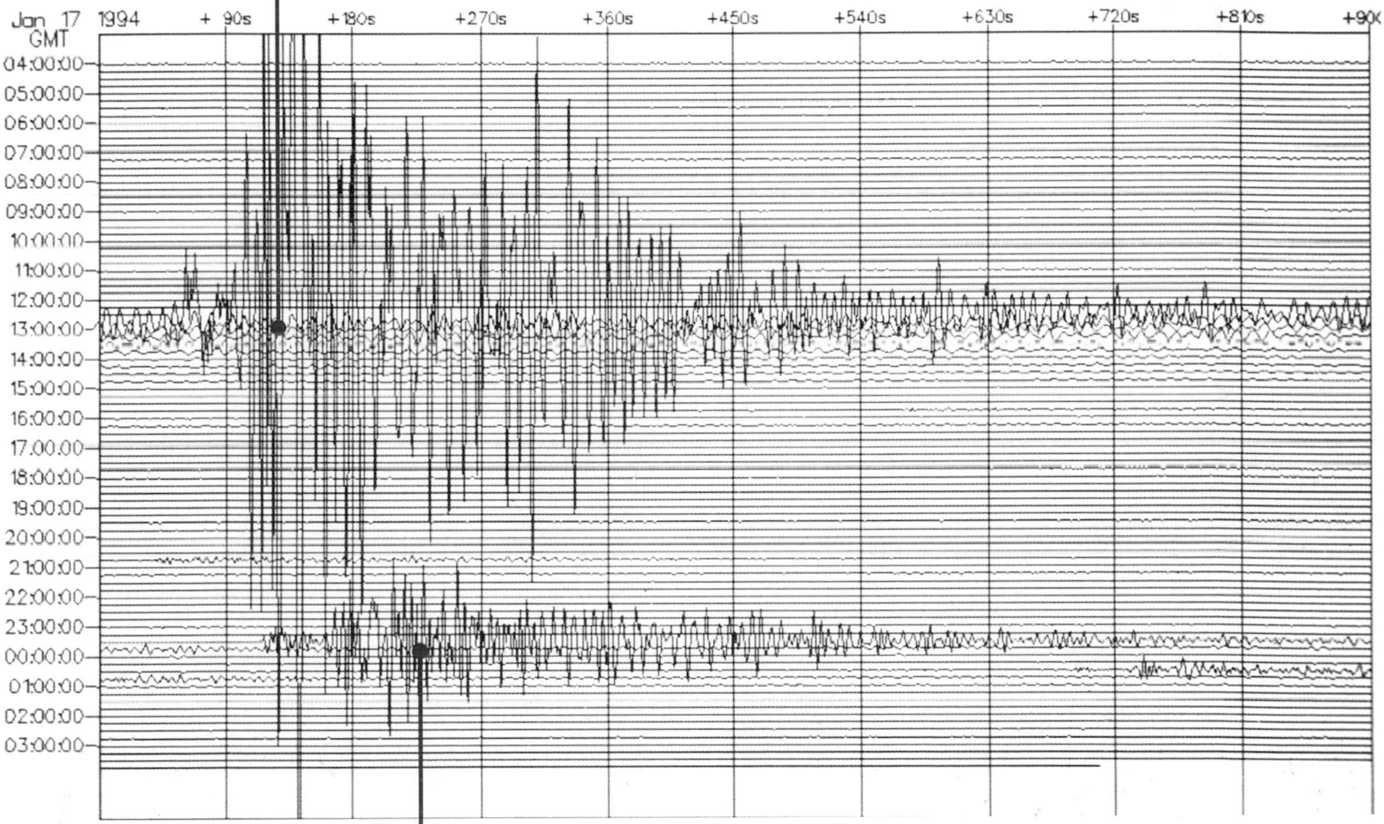

Aftershocks are earthquakes that occur after the main earthquake. They may be as strong as the main quake, so seismographs can often detect them.

There are many ways to do science and not just one scientific method.

For example, many scientists use experiments to test their hypotheses. Not all scientists, however, carry out experiments. Some make observations and use the information they gather to answer questions. The diagram below shows an example of how some scientists carried out an investigation.

Make Observations → **Form a Hypothesis** → **Collect Data** →

In 1977, scientists in the deep submergence vehicle *Alvin* observed giant tube worms living near the deep sea vents along the Galápagos rift.

"I wonder where these tube worms get their energy? Plants and phytoplankton can't survive in this environment because there is not enough sunlight. I hypothesize that the tube worms get their energy from another source."

The scientists used *Alvin's* mechanical arm to slide a probe into some of the tube worms. The probe measured the temperature, acidity, and concentrations of hydrogen and hydrogen sulfide inside each tube worm.

Scientific Theories

In everyday conversation, people often use the word *theory* to mean a guess. In science, however, the word has a very different meaning. A scientific theory represents hypotheses that have been tested repeatedly and confirmed by different scientists. A theory is a general statement backed up by a large body of facts.

Analyze Data

The scientists analyzed the data they collected on temperature, acidity, and hydrogen and hydrogen sulfide concentrations. The scientists searched for patterns in the data to help them discover how the tube worms get their energy.

Draw Conclusions

"Our results indicate that the bacteria inside the tube worms use hydrogen sulfide bubbling out of the vent to produce their food. These bacteria are the base of the food chain at the deep-sea vents."

Alter the Hypothesis/Form a New Hypothesis

Scientists' fascination with hydrothermal vents and the organisms that live near them continues to grow. Each year they design and conduct new experiments to explore this strange environment.

Geologic events such as earthquakes and volcanoes, and factors such as water, wind, and ice, help shape earth's surface.

We can see the results of these events when we look at landforms such as mountains, valleys, and bodies of water. Scientists study land formations to gain a better understanding of geologic processes.

A broad expanse of prairie called the **Great Plains** extends into Canada east of the Rocky Mountains.

Mount Saint Helens is an active volcano in Washington State that began forming about 36,000 years ago, during the Pleistocene epoch. Like the other volcanoes in the Cascades range, it is a result of subduction of the Juan de Fuca plate beneath the North American plate.

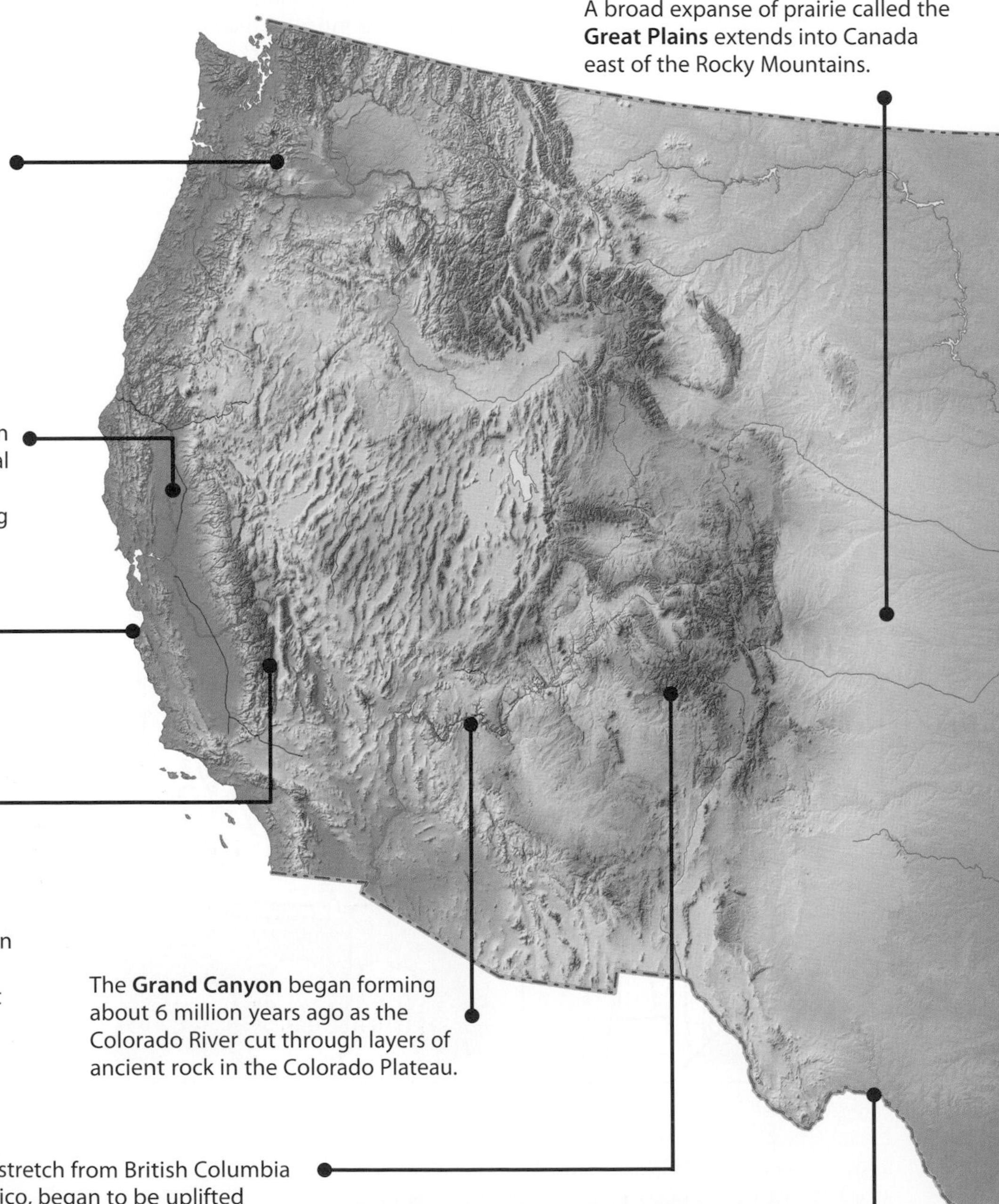

The **Great Central Valley,** nearly 650 km long, is a low-lying region in central California. It is a depositional basin—a place where sediment that is eroded from the surrounding mountains is deposited.

Approximately 800 miles long, the **San Andreas fault zone** is located where the Pacific and North American tectonic plates move past each other.

The **Sierra Nevada**, in eastern California, is a mountain range composed of huge igneous rock bodies that formed deep underground during a tectonic collision about 150 million years ago. Erosion slowly stripped off the overlying rock layers, exposing the rocks that now make up the mountains.

The **Grand Canyon** began forming about 6 million years ago as the Colorado River cut through layers of ancient rock in the Colorado Plateau.

The **Rocky Mountains,** which stretch from British Columbia to the Rio Grande in New Mexico, began to be uplifted about 100 million years ago, during the late Jurassic period.

The **Rio Grande** flows about 3,000 km from Colorado to Mexico. The river follows a rift valley that formed about 30 million years ago.

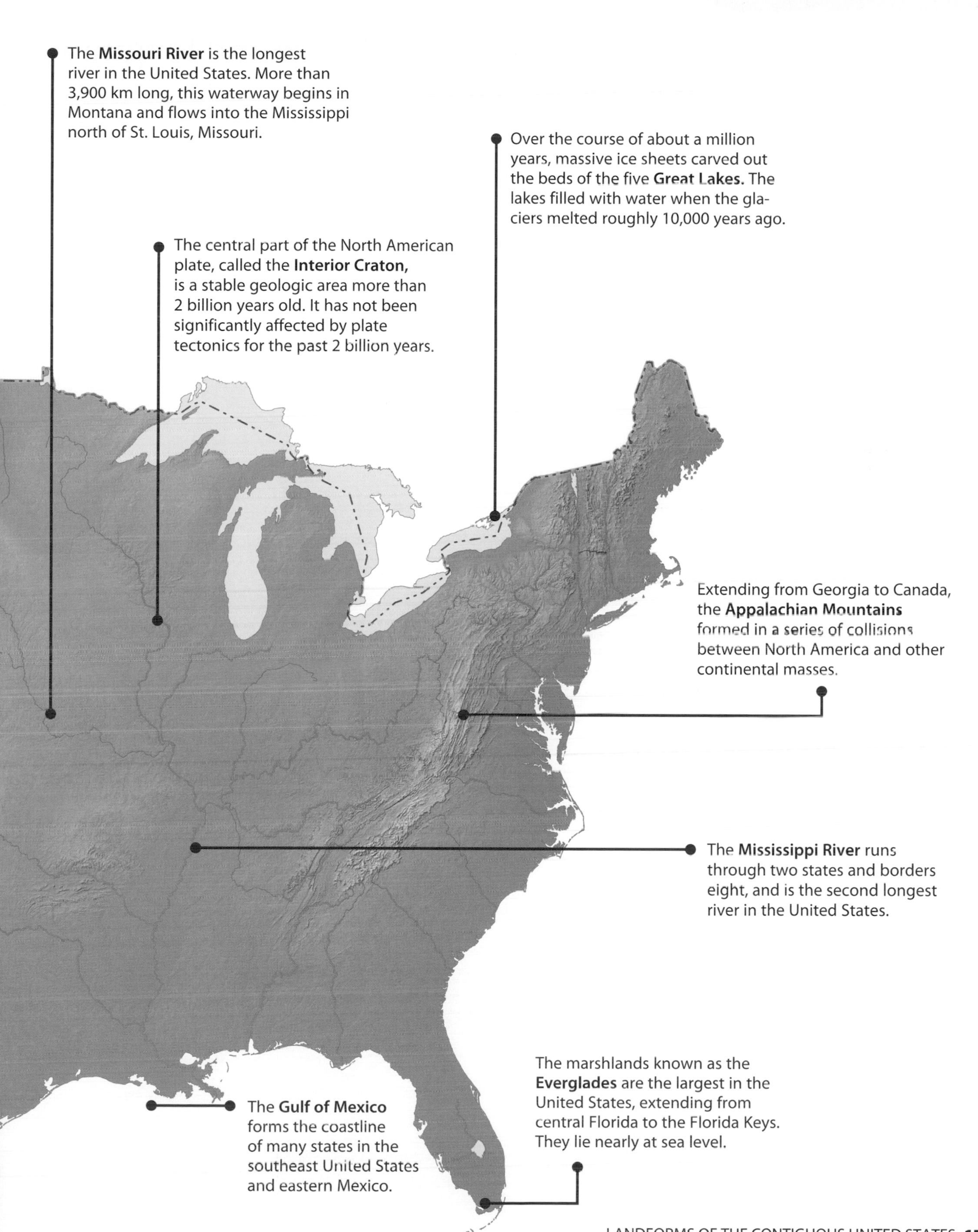

The **Missouri River** is the longest river in the United States. More than 3,900 km long, this waterway begins in Montana and flows into the Mississippi north of St. Louis, Missouri.
Over the course of about a million years, massive ice sheets carved out the beds of the five **Great Lakes.** The lakes filled with water when the glaciers melted roughly 10,000 years ago.
The central part of the North American plate, called the **Interior Craton,** is a stable geologic area more than 2 billion years old. It has not been significantly affected by plate tectonics for the past 2 billion years.
Extending from Georgia to Canada, the **Appalachian Mountains** formed in a series of collisions between North America and other continental masses.
The **Mississippi River** runs through two states and borders eight, and is the second longest river in the United States.
The marshlands known as the **Everglades** are the largest in the United States, extending from central Florida to the Florida Keys. They lie nearly at sea level.
The **Gulf of Mexico** forms the coastline of many states in the southeast United States and eastern Mexico.

Imaginary lines of latitude and longitude encircle the globe.

We use them to describe the exact locations of places on earth.

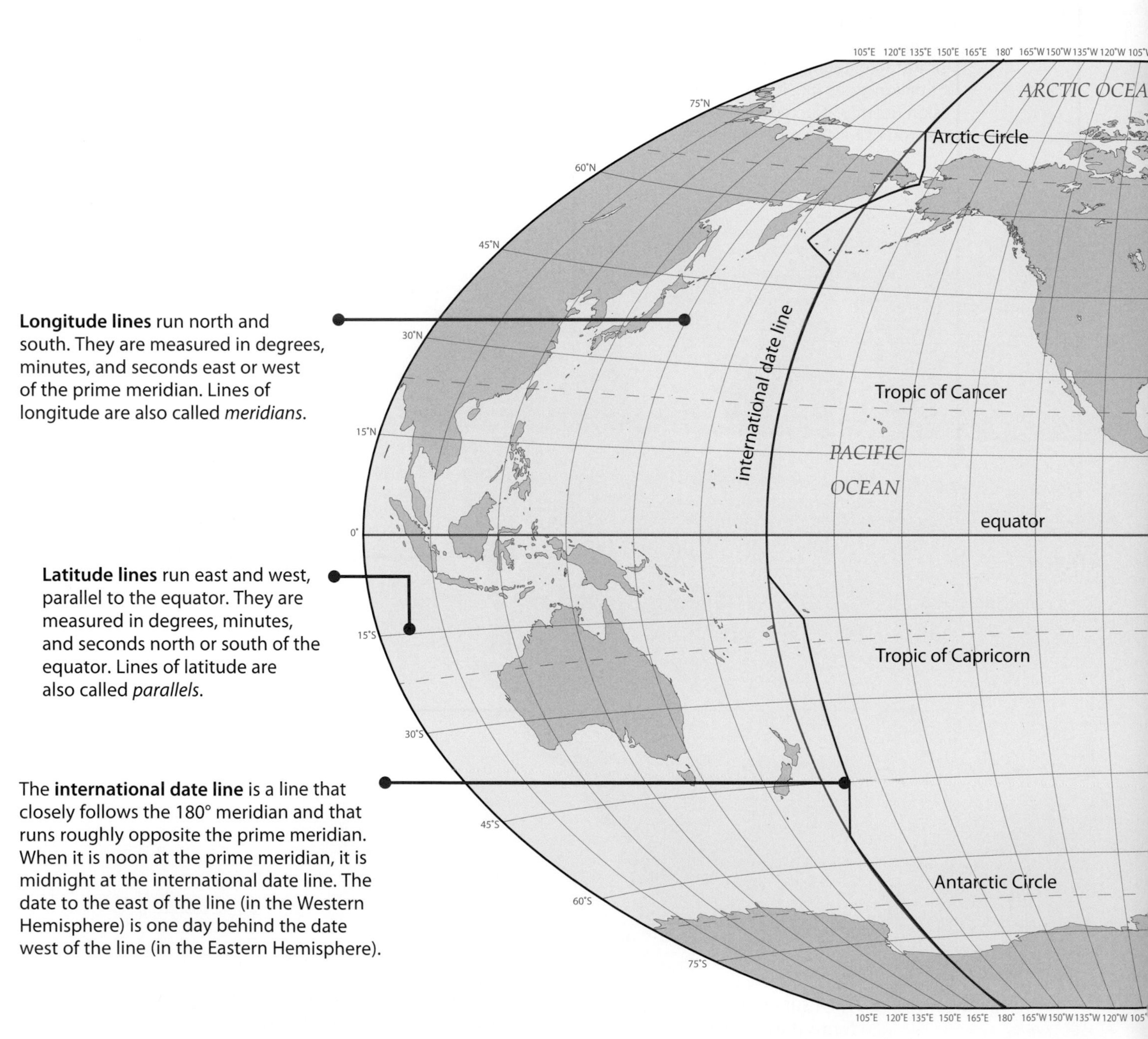

Longitude lines run north and south. They are measured in degrees, minutes, and seconds east or west of the prime meridian. Lines of longitude are also called *meridians*.

Latitude lines run east and west, parallel to the equator. They are measured in degrees, minutes, and seconds north or south of the equator. Lines of latitude are also called *parallels*.

The **international date line** is a line that closely follows the 180° meridian and that runs roughly opposite the prime meridian. When it is noon at the prime meridian, it is midnight at the international date line. The date to the east of the line (in the Western Hemisphere) is one day behind the date west of the line (in the Eastern Hemisphere).

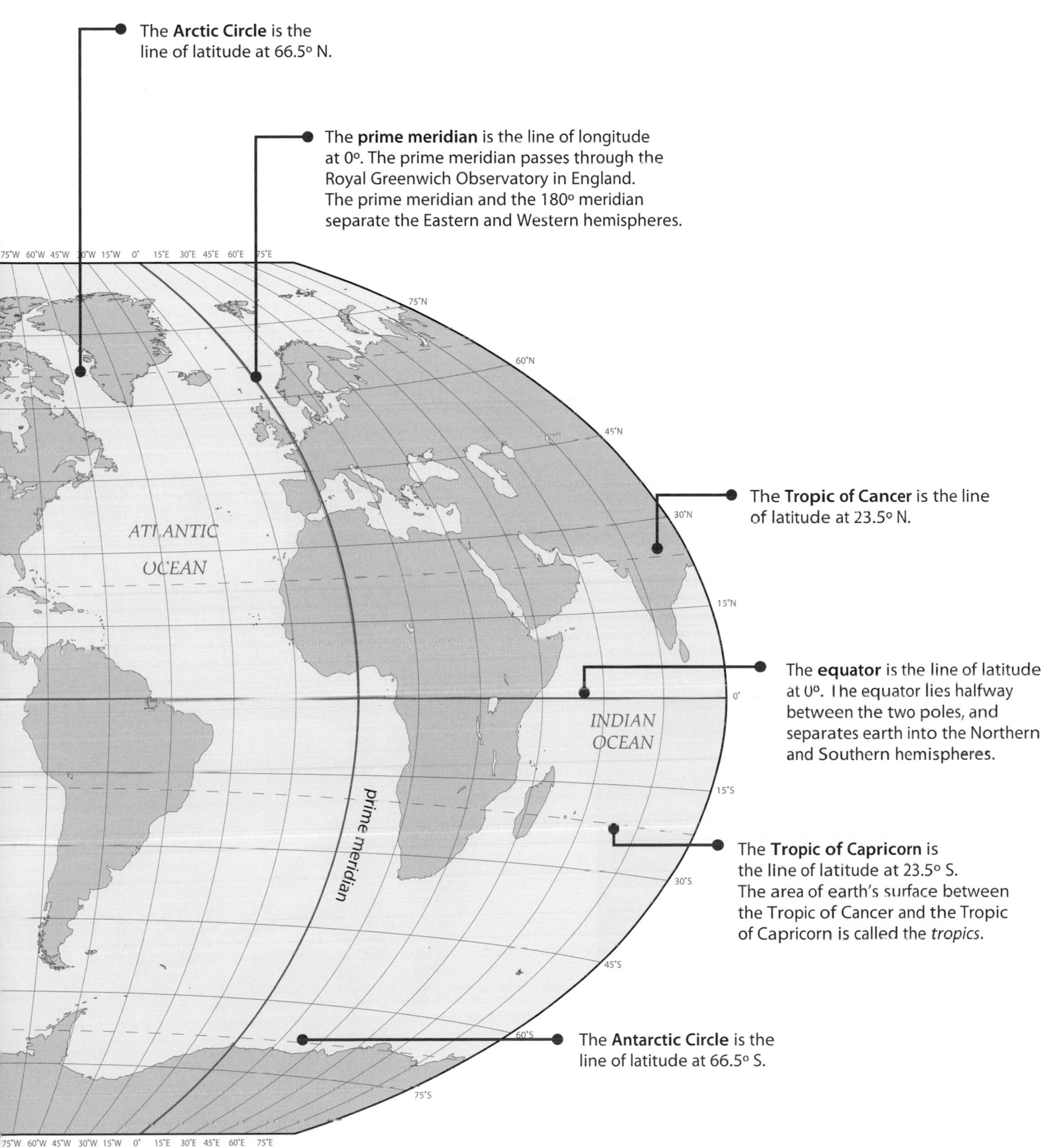
The **Arctic Circle** is the line of latitude at 66.5° N.
The **prime meridian** is the line of longitude at 0°. The prime meridian passes through the Royal Greenwich Observatory in England. The prime meridian and the 180° meridian separate the Eastern and Western hemispheres.
The **Tropic of Cancer** is the line of latitude at 23.5° N.
The **equator** is the line of latitude at 0°. The equator lies halfway between the two poles, and separates earth into the Northern and Southern hemispheres.
The **Tropic of Capricorn** is the line of latitude at 23.5° S. The area of earth's surface between the Tropic of Cancer and the Tropic of Capricorn is called the *tropics*.
The **Antarctic Circle** is the line of latitude at 66.5° S.
ATLANTIC OCEAN
INDIAN OCEAN
prime meridian
75°N
60°N
45°N
30°N
15°N
0°
15°S
30°S
45°S
60°S
75°S
75°W 60°W 45°W 30°W 15°W 0° 15°E 30°E 45°E 60°E 75°E

The earth system is made up of many components, such as air, rock, water, ice, and living things.

Earth scientists have classified the different parts of the earth system into five groups, or spheres: the geosphere, the hydrosphere, the biosphere, the atmosphere, and the cryosphere.

Geosphere

The **geosphere** is all the rock on and just below earth's surface.

Hydrosphere

The **hydrosphere** is all the liquid water on earth's surface.

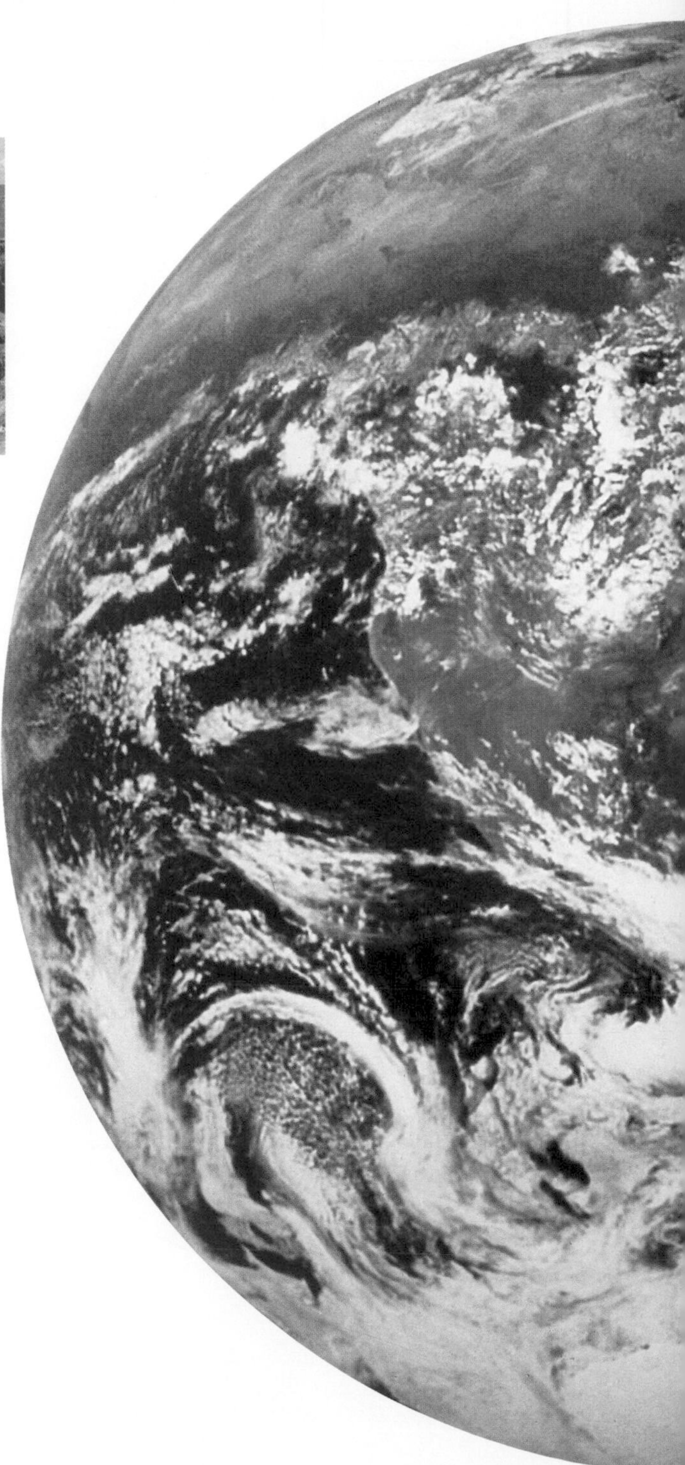

Biosphere

The **biosphere** is all the living organisms on earth.

Atmosphere

The **atmosphere** is the layers of gases surrounding earth that are subject to earth's gravitational force.

Cryosphere

The **cryosphere** is all the frozen water on the earth's surface.

Ten elements make up most of the matter in the earth system.

Scientists have identified 92 elements that occur in nature. In the periodic table below, the elements highlighted in red are some of the most abundant on earth.

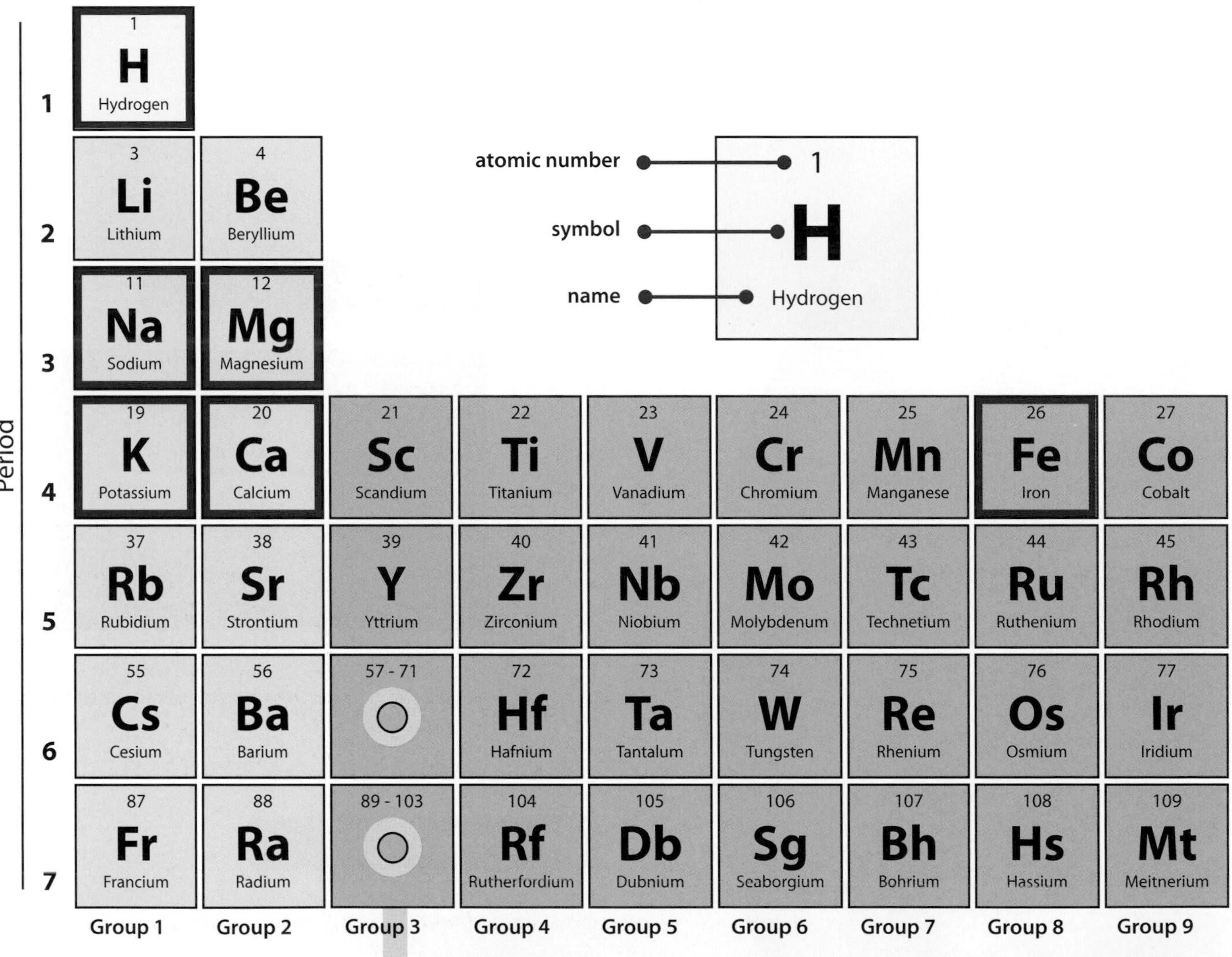

- hydrogen
- semiconductors (also known as metalloids)

Metals

- alkali metals
- alkaline-earth metals
- transition metals
- other metals

Nonmetals

- halogens
- noble gases
- other nonmetal

Group 10	Group 11	Group 12	Group 13	Group 14	Group 15	Group 16	Group 17	Group 18
								2 **He** Helium
			5 **B** Boron	6 **C** Carbon	7 **N** Nitrogen	8 **O** Oxygen	9 **F** Fluorine	10 **Ne** Neon
			13 **Al** Aluminum	14 **Si** Silicon	15 **P** Phosphorus	16 **S** Sulfur	17 **Cl** Chlorine	18 **Ar** Argon
28 **Ni** Nickel	29 **Cu** Copper	30 **Zn** Zinc	31 **Ga** Gallium	32 **Ge** Germanium	33 **As** Arsenic	34 **Se** Selenium	35 **Br** Bromine	36 **Kr** Krypton
46 **Pd** Palladium	47 **Ag** Silver	48 **Cd** Cadmium	49 **In** Indium	50 **Sn** Tin	51 **Sb** Antimony	52 **Te** Tellurium	53 **I** Iodine	54 **Xe** Xenon
78 **Pt** Platinum	79 **Au** Gold	80 **Hg** Mercury	81 **Tl** Thallium	82 **Pb** Lead	83 **Bi** Bismuth	84 **Po** Polonium	85 **At** Astatine	86 **Rn** Radon
110 **Ds** Darmstadtium								

63 **Eu** Europium	64 **Gd** Gadolinium	65 **Tb** Terbium	66 **Dy** Dysprosium	67 **Ho** Holmium	68 **Er** Erbium	69 **Tm** Thulium	70 **Yb** Ytterbium	71 **Lu** Lutetium
95 **Am** Americium	96 **Cm** Curium	97 **Bk** Berkelium	98 **Cf** Californium	99 **Es** Einsteinium	100 **Fm** Fermium	101 **Md** Mendelevium	102 **No** Nobelium	103 **Lr** Lawrencium

Atoms are the basic building blocks of all matter.

An atom is the smallest piece of an element that has all the chemical properties of the element. The images on these pages show the atomic structures of some of the most common elements on earth.

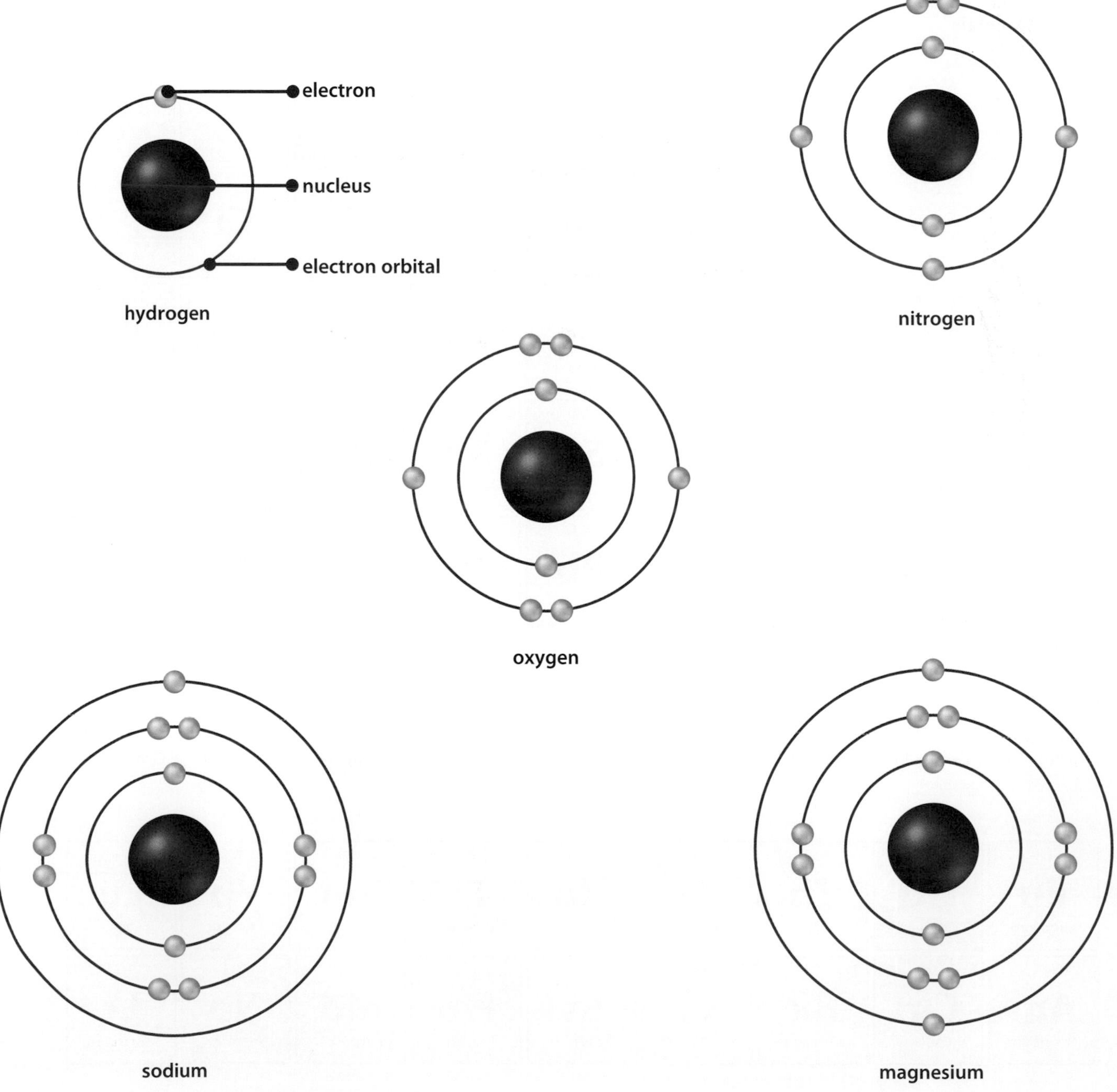

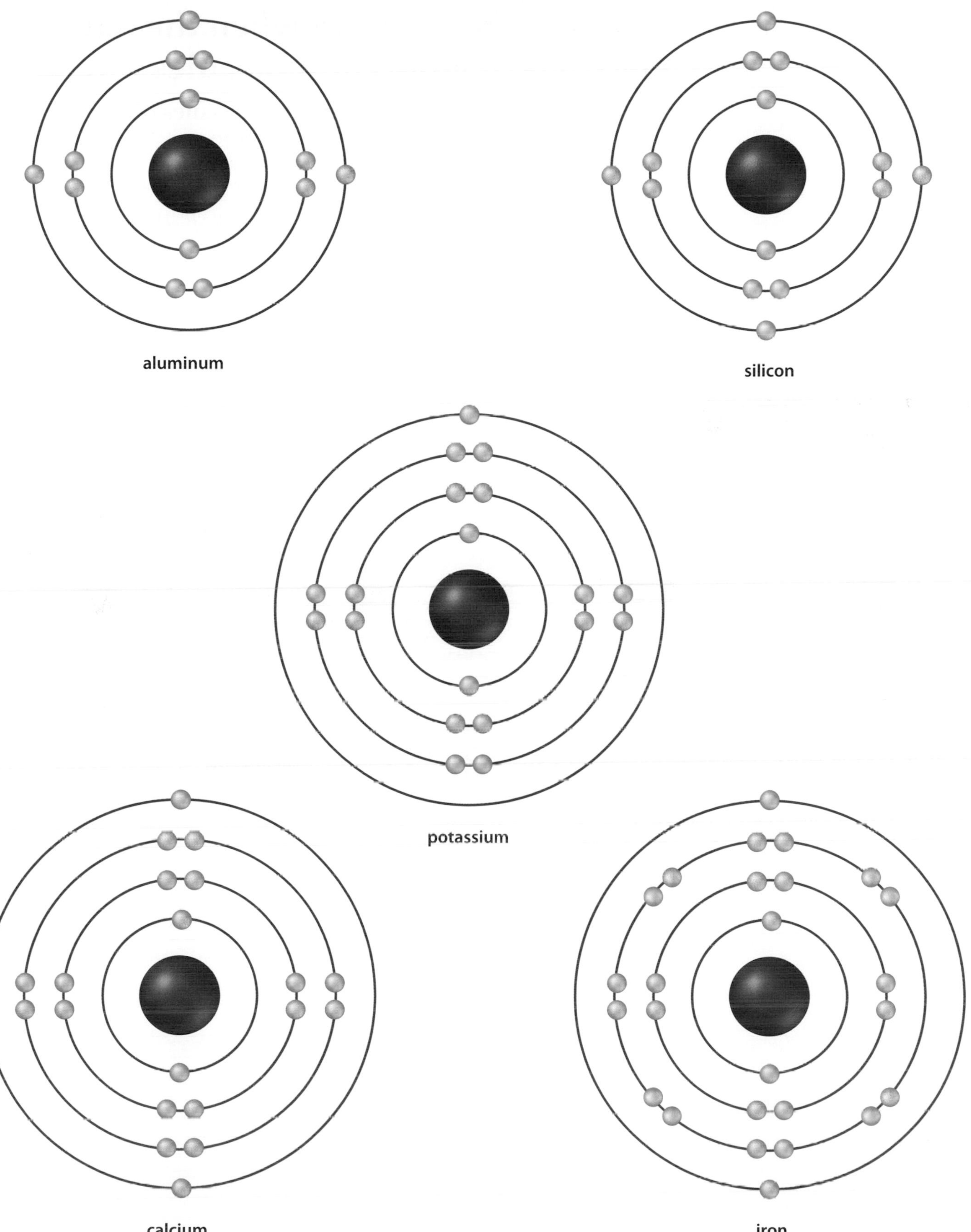
aluminum
silicon
potassium
calcium
iron

Chemical bonds form when the outer electrons of atoms or molecules interact.

The types of bonds between atoms or molecules of a substance affect the properties of the substance. There are three main types of chemical bonds: ionic, covalent, and hydrogen. Each type of bond forms because of a different kind of interaction between the outer electrons, which are called *valence electrons*.

Ionic Bond

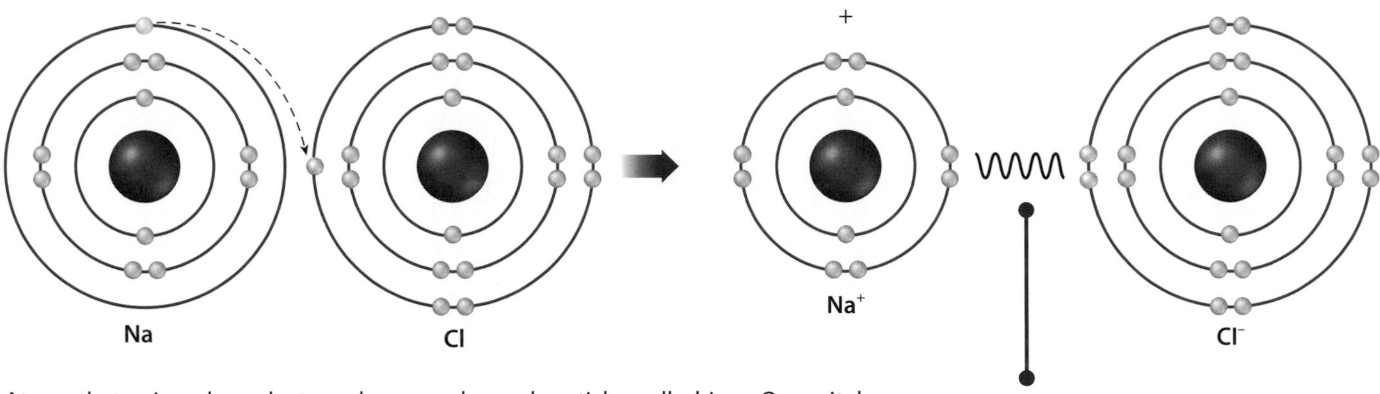

Atoms that gain or lose electrons become charged particles called *ions*. Oppositely charged ions attract each other to form ionic bonds. The electrons lost by one atom are gained by another. For example, an atom of sodium (Na) can easily lose one electron to become the positively charged ion Na^+. Chlorine (Cl) can gain that electron and become the negatively charged chloride ion (Cl^-). The sodium and chloride ions attract each other and form an ionic bond. The resulting compound, sodium chloride (NaCl), is the familiar substance table salt.

Covalent Bond

covalent bond

O O O_2

Covalent bonds are the strongest kind of chemical bond. They form when two atoms share valence electrons. For example, two atoms of oxygen (O) share valence electrons to form the bond in an oxygen molecule. The shared electrons orbit both oxygen atoms in the molecule. Oxygen gas (O_2) is a major component of the air that we breathe.

Hydrogen Bond

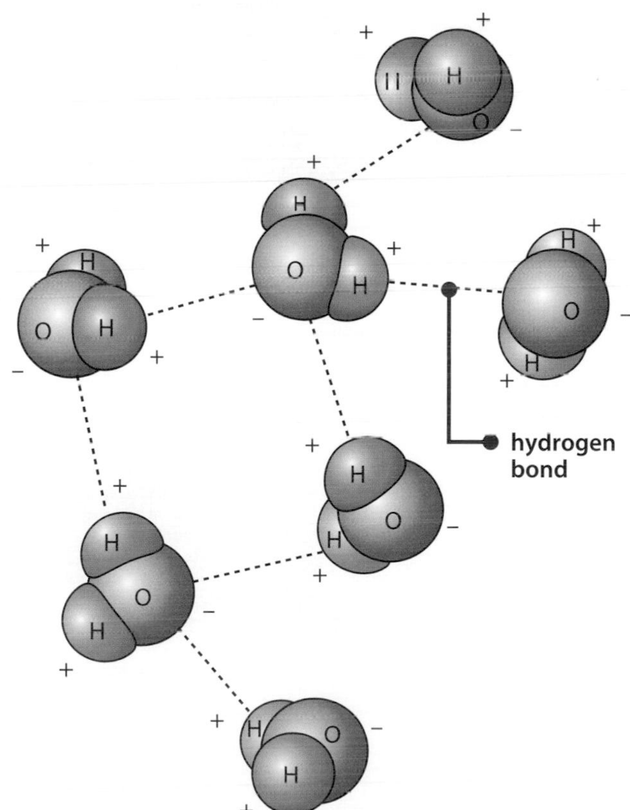

Molecules of water are made of two hydrogen (H) atoms covalently bound to one oxygen (O) atom. The oxygen atom attracts electrons more strongly than the hydrogen atoms, so the oxygen atom has a slight negative charge. Each hydrogen atom has a slight positive charge. The negatively charged oxygen in one water molecule attracts the positively charged hydrogen atoms in another water molecule. This attraction is known as a hydrogen bond. Hydrogen bonds can form between hydrogen atoms and atoms of oxygen, nitrogen, or fluorine on the same molecule or a different molecule. Hydrogen bonds give water many of its unique properties.

A mineral is a solid, naturally occurring substance with a specific chemical composition and crystalline structure.

More than 2,000 known minerals can be found on earth. However, only a few minerals make up most of the rocks in earth's crust. Scientists put minerals into groups based on chemical composition.

Carbonates

Carbonates are minerals that contain the carbonate ion $(CO_3)^{2-}$. Carbonate minerals, such as calcite and dolomite, are most common in sedimentary rocks.

dolomite $(Ca,Mg)CO_3$

calcite $CaCO_3$

Halides

Halides are minerals that contain ions of the halogen elements (such as fluorine, chlorine, bromine, or iodine) bound to metal ions. Many halide minerals crystallize from shallow seas as the water evaporates.

halite $NaCl$

fluorite CaF_2

Native Elements

Native elements are minerals that contain atoms of only one element. Most native elements are metals, such as gold, copper, and silver. However, some native elements are nonmetals, such as carbon (in the form of diamond and graphite) and sulfur.

gold Au

copper Cu

Oxides

Oxides are minerals that contain ions bound to oxygen. Oxide minerals are important economic sources of some metals, such as iron and aluminum.

corundum Al_2O_3

hematite Fe_2O_3

Sulfates

Sulfates are minerals that contain the sulfate ion $(SO_4)^{2-}$. Like halide minerals, many sulfate minerals form when seawater evaporates.

gypsum $CaSO_4 \cdot 2H_2O$

anhydrite $CaSO_4$

Sulfides

Sulfides are minerals that contain compounds of metals and sulfur ions. Like oxide minerals, sulfide minerals can be important ores of useful metals.

galena PbS

pyrite FeS_2

Silicates

Silicates are minerals that contain compounds of silicon and oxygen. Silicate minerals make up most of the rocks on earth.

olivine $(Mg,Fe)_2SiO_4$

plagioclase $CaAl_2Si_2O_8$

Silicon and oxygen are the most abundant elements in earth's crust.

Silicates are minerals that contain silicon-oxygen compounds bound to various other elements. They are some of the most common minerals on earth.

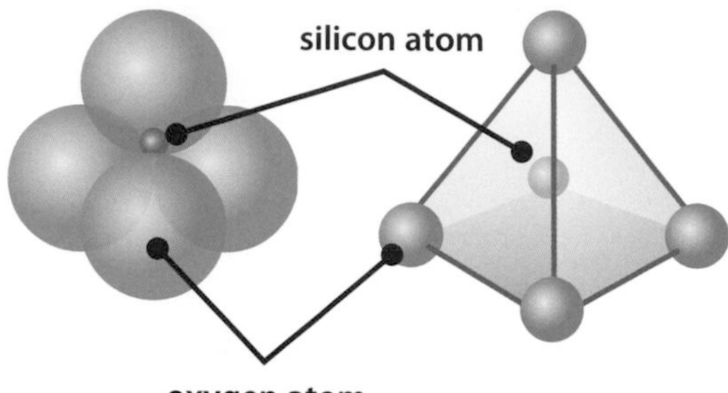

The fundamental unit of silicate minerals is the **silicon-oxygen tetrahedron.** Silicon-oxygen tetrahedrons combine in different ways to form different kinds of silicate minerals.

Single-Tetrahedron Silicates

Some silicate minerals, such as olivine, contain silicon-oxygen tetrahedrons that are not bound to each other. Those minerals form the group of **single-tetrahedron silicate minerals.**

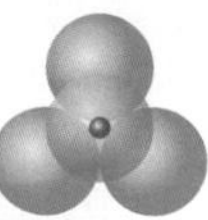

single tetrahedron

olivine

Single-Chain Silicates

Single-chain silicates, such as pyroxene, form when silicon-oxygen tetrahedrons join to form long chains. In single-chain silicates, each tetrahedron is bound to two other tetrahedrons.

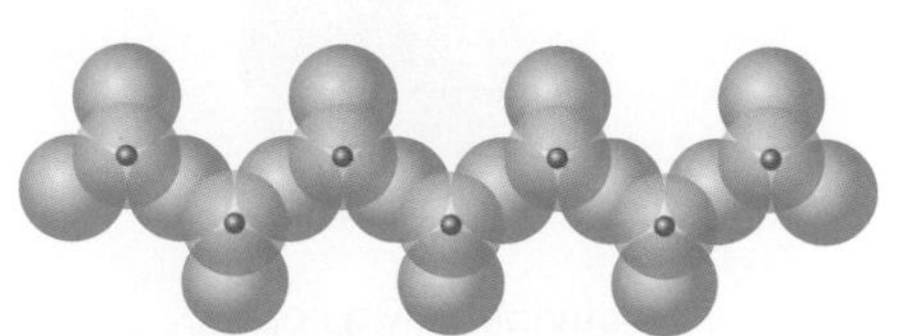

single chain of tetrahedrons

pyroxene

Double-Chain Silicates

Double-chain silicates, such as amphibole, contain silicon-oxygen tetrahedrons bound in pairs of chains. In double-chain silicates, each tetrahedron is bound to two or three other tetrahedrons.

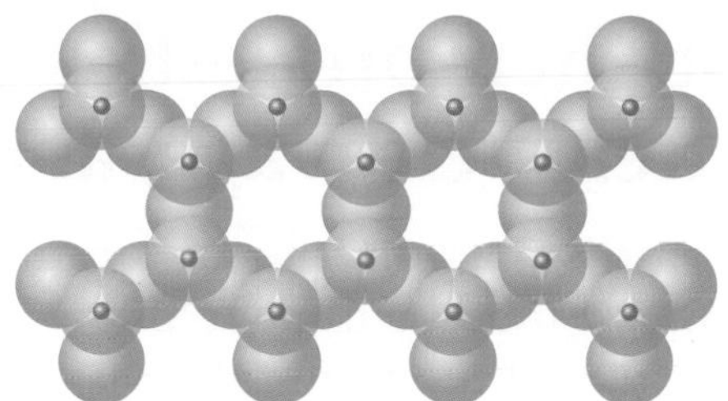

double chain of tetrahedrons

amphibole

Sheet Silicates

Sheet silicates, which include micas, contain sheets of silicon-oxygen tetrahedrons bound together by positive ions. In sheet silicates, each tetrahedron is bound to three other tetrahedrons.

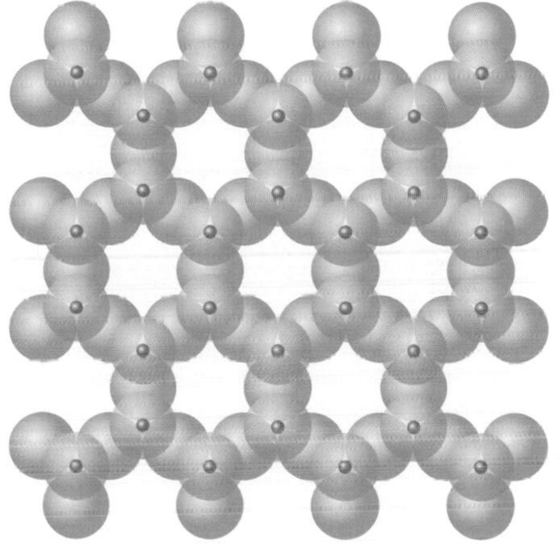

sheet of tetrahedrons

muscovite (mica)

Network Silicates

Network silicates, such as quartz, contain silicon-oxygen tetrahedrons bound together to form a three-dimensional framework. In network silicates, each tetrahedron is bound to four other tetrahedrons.

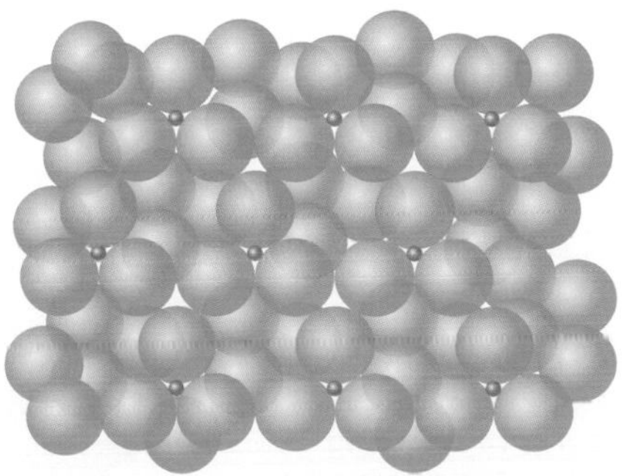

network of tetrahedrons

quartz

Scientists identify minerals based on their properties of hardness, luster, streak, cleavage, and fracture.

Color is not very useful for identifying minerals because the same mineral can have many different colors. For example, crystals of the mineral quartz may be clear, white, pink, purple, gray, green, or even black.

Hardness

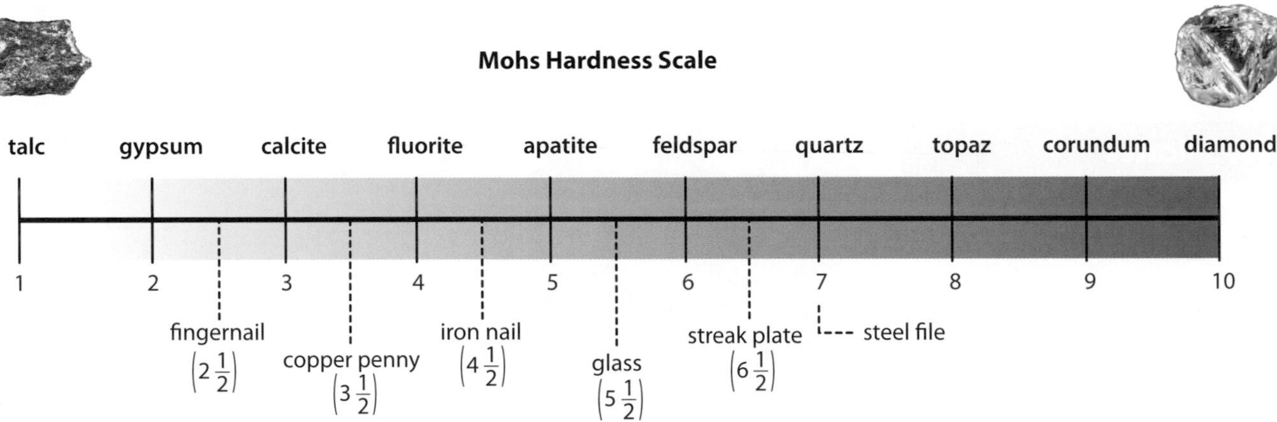

The hardness of a mineral describes how resistant the mineral is to being scratched. Geologists use the Mohs hardness scale to describe the hardness of minerals. Minerals with high hardness ratings will scratch minerals with lower hardness ratings. For example, a piece of quartz will scratch a piece of calcite, but a piece of calcite will not scratch a piece of quartz.

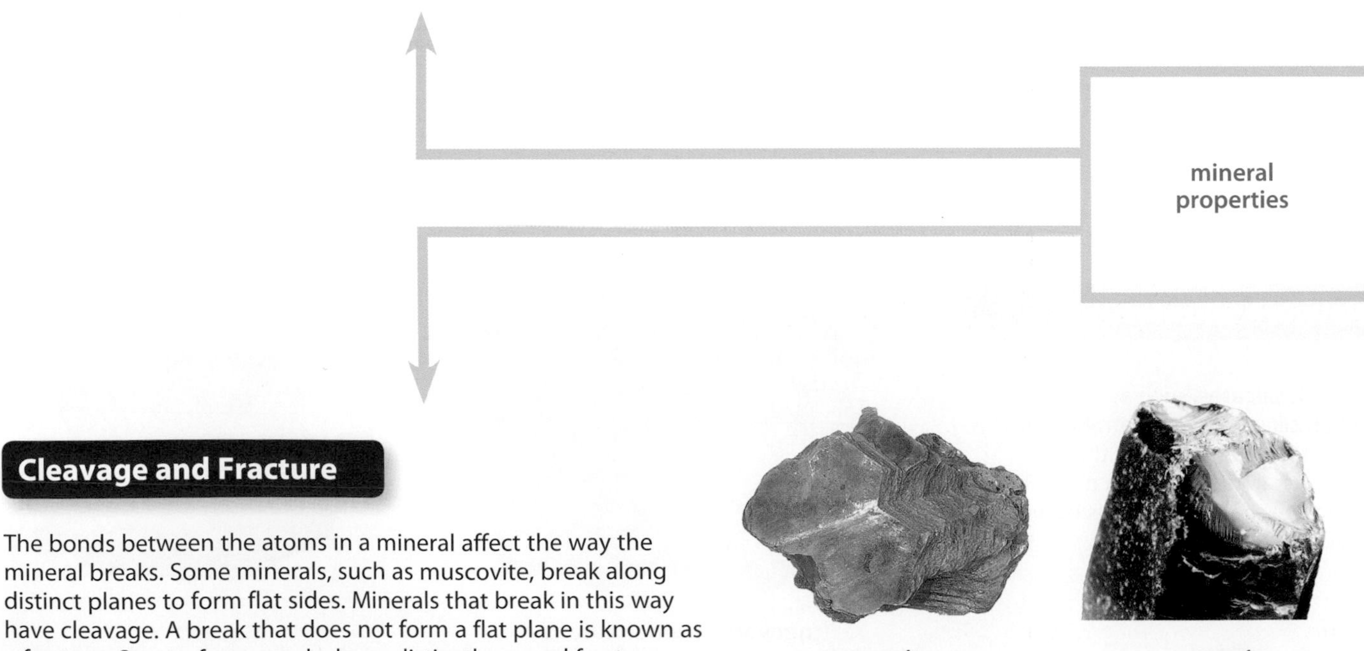

Cleavage and Fracture

The bonds between the atoms in a mineral affect the way the mineral breaks. Some minerals, such as muscovite, break along distinct planes to form flat sides. Minerals that break in this way have cleavage. A break that does not form a flat plane is known as a fracture. Quartz, for example, has a distinctly curved fracture.

muscovite

quartz

Luster

The luster of a mineral is a description of the way the mineral reflects light. Galena, for example, reflects light the way shiny metals do. These minerals are said to have a metallic luster. Other examples of lusters include vitreous (glassy), waxy, resinous (like dull plastic), pearly, submetallic (dull, but reflective), and earthy (dull and nonreflective).

galena (metallic luster)

quartz (vitreous luster)

talc (waxy luster)

Streak

Although color is not that useful for identifying a mineral, the color of the mineral's streak may be very helpful. Scientists determine the streak color of a mineral by rubbing the mineral on a streak plate. The streak color may be very different from the color of the mineral itself and is the same for all samples of the mineral. For example, the mineral hematite may be black, gray, or orange in color, but its streak is always dark red.

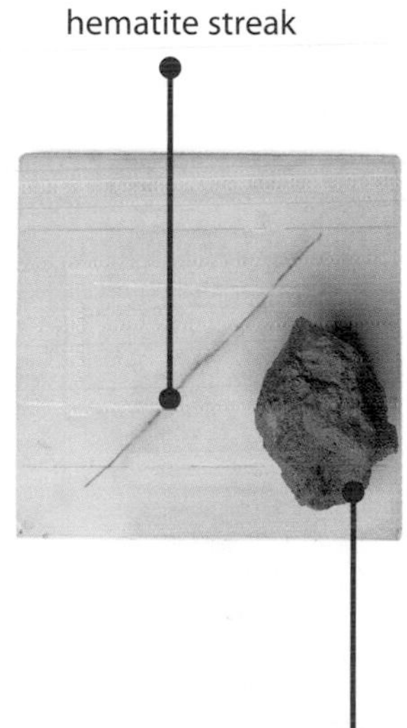

Unusual Properties

Some properties that can be very helpful in identifying certain minerals are not shared by all minerals. For example, calcite and other carbonate minerals will fizz and bubble when they are exposed to acid. Some other unusual mineral properties include radioactivity, magnetism, and fluorescence (glowing under ultraviolet light).

Igneous rocks form when molten rock cools and solidifies or crystallizes.

By studying igneous rocks, geologists can learn about the composition of earth's mantle, the processes of plate tectonics, and the history of earth. The two main types of igneous rocks are extrusive and intrusive.

Extrusive Igneous Rocks

Extrusive igneous rocks form on earth's surface as lava cools rapidly. The rapid cooling typically prevents large crystals from forming in the rock. Bubbles of gas may form in the lava, producing holes in some extrusive igneous rocks.

Pumice is an extrusive igneous rock that contains many holes that form when gases become trapped in lava as it hardens. Pumice is common at many explosive volcanoes, such as Mount Pinatubo in the Philippines.

Basalt is a common extrusive igneous rock. Most of the oceanic lithosphere and many volcanic islands, such as Iceland and Hawaii, are made of basalt.

Extrusive igneous rocks form at volcanoes.

Obsidian is shiny volcanic glass. It cools too quickly for crystals to form. Obsidian can form at volcanoes along convergent boundaries.

Intrusive Igneous Rocks

Intrusive igneous rocks form below earth's surface. Magma below the surface cools very slowly, giving large crystals time to grow. Therefore, most intrusive igneous rocks have large crystals and a coarse-grained texture.

Pegmatite is a very coarse-grained intrusive igneous rock. Many pegmatites form as a result of the interaction between water and magma. Many gemstones form in pegmatites.

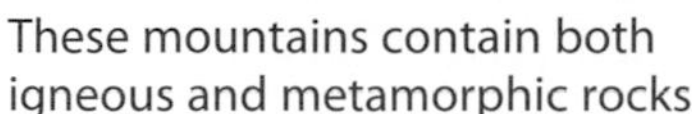

These mountains contain both igneous and metamorphic rocks.

Gabbro is a dark, coarse-grained intrusive igneous rock that is very similar in composition to basalt. It is common in some mountain ranges.

Granite is an intrusive igneous rock that is common in many mountain ranges. Granite commonly contains large crystals because the magma that forms it cools slowly below the surface.

Sedimentary rocks are formed by crystallization or by solidification of sediment.

Geologists classify sedimentary rocks in three different groups—clastic, chemical, and organic or biological—based on what makes up the rocks.

Clastic Sedimentary Rocks

Clastic sedimentary rocks are formed from pieces of other rocks or from minerals. Clastic sedimentary rocks, such as much of the rock in the Grand Canyon, are the most common type of sedimentary rock.

Sandstone is a clastic sedimentary rock. Sandstone consists of sand-sized particles of rocks or minerals, such as quartz and feldspar, cementing together.

Shale is a fine-grained clastic sedimentary rock. Clay and mud particles harden and cement together to form shale.

Conglomerate is a clastic sedimentary rock largely made up of rounded, gravel-size fragments. These large particles are cemented together by finer-grained sediment.

Biological Sedimentary Rocks

Organic, or biological, sedimentary rocks form from the remains of living things. The rock created from past generations of coral is an example of organic sedimentary rock.

Coquina is an organic sedimentary rock with a chemical composition similar to limestone. It forms when shell fragments are compacted and cemented together.

Chemical Sedimentary Rocks

Chemical sedimentary rocks form when minerals crystallize from water solutions.

Evaporite is a chemical sedimentary rock. It forms when minerals such as halite and gypsum crystallize out of warm, shallow seas.

Metamorphic rocks form when the chemical composition of a rock changes due to heat and pressure.

Many metamorphic rocks form in mountain ranges, where temperature and pressure are high. The two main groups of metamorphic rock are foliated and nonfoliated.

Foliated Metamorphic Rocks

Foliated metamorphic rocks contain visible layers or sheets of minerals. The layers may be different colors, depending on the minerals that make them up.

Gneiss is a strongly foliated metamorphic rock that commonly contains alternating bands of light and dark minerals. The bands in gneiss may be folded or bent because of the high temperatures and pressures under which gneiss forms.

Slate is a fine-grained, foliated metamorphic rock that forms under lower temperature and pressure. The foliations in slate allow it to break into thin, flat sheets. For this reason, slate is used for tiling in walkways and roofs.

Schist is a moderately foliated metamorphic rock that can form when slate is put under increased heat and pressure. Schist generally contains bands of mica minerals, which make the rocks look shiny or sparkly.

Nonfoliated Metamorphic Rocks

Nonfoliated metamorphic rocks do not contain visible layers or sheets of minerals. Many nonfoliated metamorphic rocks form when minerals in the original rock recrystallize.

Marble is a nonfoliated metamorphic rock that is formed by the recrystallization of small calcite crystals in limestone.

The rock cycle includes all the processes that cause rock to form and break down.

These processes can affect any kind of rock. Therefore, rocks can follow many different paths through the rock cycle.

Igneous Rock

heat and pressure

melting and cooling

melting and cooling

Although many **igneous rocks** form when sedimentary or metamorphic rock melts and cools, igneous rock can also remelt and cool to form new igneous rock.

Sedimentary Rock

When **sedimentary rocks** break down, the remains can eventually form new sedimentary rocks.

erosion and deposition

erosion and deposition

melting and cooling

Metamorphic Rock

heat and pressure

erosion and deposition

heat and pressure

When a **metamorphic rock** is heated or exposed to a lot of pressure, its minerals can recrystallize, forming a new metamorphic rock. The heat and pressure exerted on these rocks can cause them to fold.

Rocks can break or bend when they are exposed to stress or pressure.

The place where a rock body breaks is called a *fault*. A bend in a rock that is caused by stress or pressure is called a *fold*.

Folds

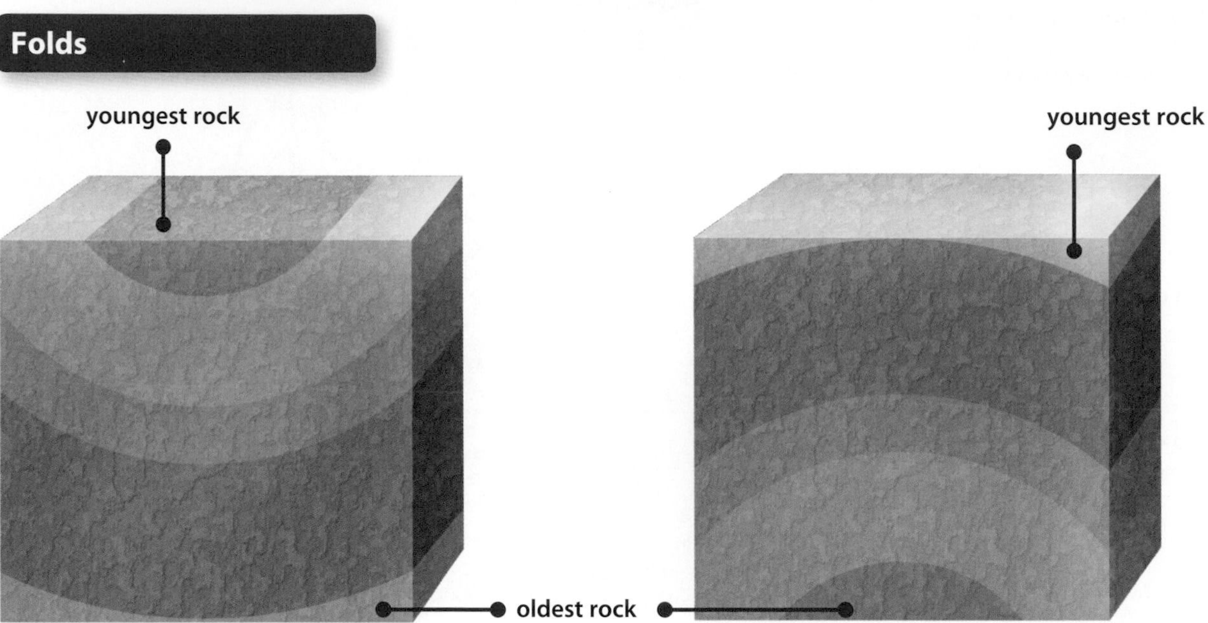

A **syncline** is a fold in which the youngest rock layers are located in the interior of the fold. Most synclines have a shape like this: ∪.

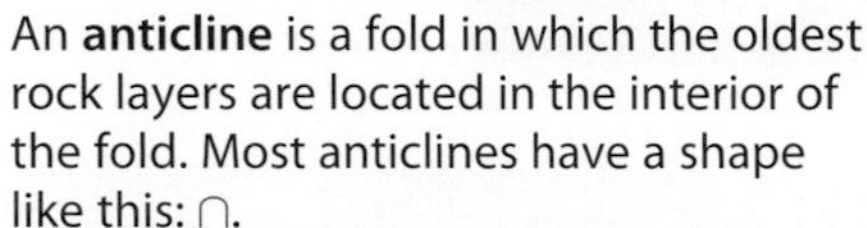

An **anticline** is a fold in which the oldest rock layers are located in the interior of the fold. Most anticlines have a shape like this: ∩.

A **monocline** is a fold that looks like a step. The rock layers on one side of the fold are lower than the layers on the other side.

A **recumbent fold** is a syncline or anticline that is turned on its side. If the youngest rock layers are in the interior of the fold, the fold is a recumbent syncline. If the oldest rock layers are in the interior of the fold, the fold is a recumbent anticline.

Faults

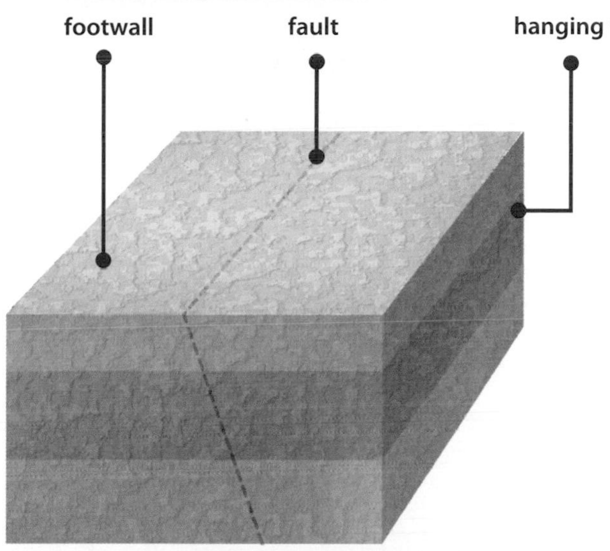

Faults are planes that divide two bodies of rock. The rock bodies, or **fault blocks,** can move along the fault. Many faults form at an angle to the surface. This forms two fault blocks, one above the fault and one below the fault. The block above the fault is called the hanging wall. The block below the fault is called the *footwall*.

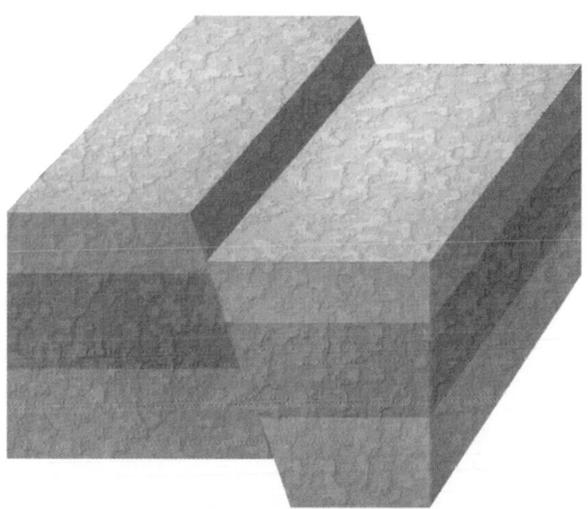

A **normal fault** is one in which the hanging wall moves downward relative to the footwall. Most normal faults form in places where rock is pulled apart, such as at divergent plate boundaries.

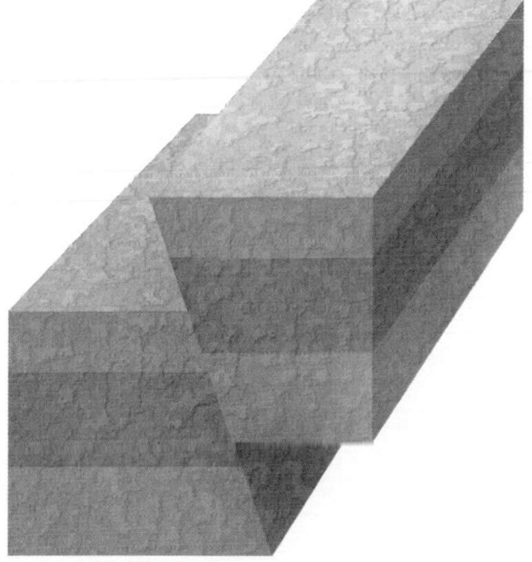

A **reverse fault** is one in which the hanging wall moves upward relative to the footwall. Most reverse faults form in places where rock is pushed together, such as at convergent plate boundaries.

A **strike-slip fault** forms where two bodies of rock move horizontally in opposite directions. Most strike-slip faults form perpendicular to earth's surface.

Soils are mixtures of rock pieces, air, water, organic matter, and living things.

There are many different types of soil, each characterized by specific amounts of these components. The climate, bedrock, and organisms that live in a region affect the type of soil that forms there.

Alfisols are nutrient-dense soils that form mainly in humid tropical, subtropical, and midlatitude regions. Many alfisols have a horizon, or layer, of clay below their surfaces. They are common in the northeastern United States, much of northern Europe, and parts of sub-Saharan Africa.

Gelisols form in cold regions. The top meter or two of soil in a gelisol can thaw during some parts of the year, but all gelisols contain a layer of permafrost—soil that is frozen year-round.

Aridisols form in arid regions. Few organisms can survive in these dry conditions, so aridisols typically contain little organic matter. Aridisols can be very fertile soils if they are properly irrigated. However, if irrigation is used poorly, the large amounts of sodium in the soil can leach out to form salty crusts.

Andisols are soils that form mainly from volcanic ash. Because volcanic ash contains many different minerals and trace nutrients, andisols tend to be very fertile soils if they receive enough rainfall to allow plants to grow.

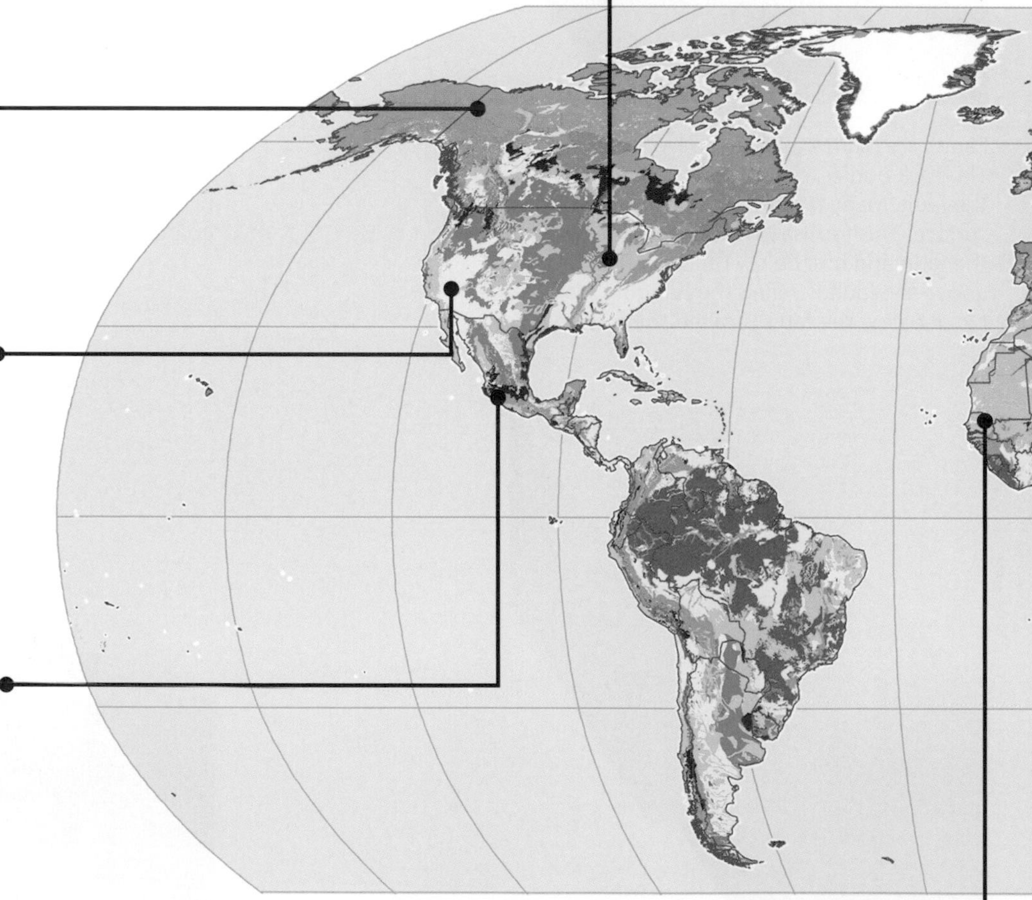

Entisols are very young soils that have a composition very similar to the bedrock beneath them. Many entisols form where rocks break up, but are carried away quickly before they have time to develop into mature soils. Entisols are also common in desert areas where the main material is quartz sand, because quartz does not easily weather into soil.

Spodosols form in coniferous forests in cool, humid climates. The leaves of many conifers are very acidic. When these leaves drop to the ground and decay, the acids move into the soil. Spodosols are acidic.

Histosols are rich in organic matter from dead organisms. Decomposers in the soil use oxygen to break down dead organisms, leaving little oxygen in the soil. Oxygen content can get so low that the decomposers cannot break down organic matter, and the remains of the organisms may be preserved—sometimes for thousands of years.

Mollisols are very fertile soils that form in grasslands in humid continental regions. They contain a great deal of organic matter. Therefore, their upper horizons may appear dark brown or black. Mollisols are fertile soils for agriculture.

Soil Types	
alfisols	oxisols
andisols	spodosols
aridisols	ultisols
entisols	vertisols
gelisols	rocky land
histosols	shifting sand
inceptisols	ice/glacier
mollisols	

Inceptisols are young soils just beginning to form. They are found in many different climates but are particularly common in floodplains and deltas.

Ultisols contain significant amounts of clay. Many have a surface horizon that is rich in organic matter. Leaching in ultisols causes them to lose many soil nutrients. However, certain fertilizers can make them suitable for growing crops.

Vertisols contain extremely high concentrations of certain kinds of clay minerals that expand when they are wet. When these clays dry out, they contract, causing cracks. Vertisols are most common in regions that have distinct wet and dry seasons. They may be nutrient rich, but they are difficult to cultivate because of their high clay content.

Oxisols form in warm, humid areas that receive a great deal of rainfall. This climate can support a great deal of diversity, so many competing organisms and heavy rainwater leach most nutrients from the soil. Oxisols are generally nutrient poor. Most oxisols are a bright, deep red color.

Earth's surface is constantly being built up and worn down by different geologic processes.

Weathering and erosion are two of the most important processes that wear down earth's surface. Weathering happens when rocks break down in place. Erosion happens when pieces of rock are transported over earth's surface.

Ways That Earth's Surface Can Change

Two main kinds of weathering are chemical weathering and mechanical weathering. **Chemical weathering** occurs when the chemical composition of a rock changes. Most chemical weathering is caused by running water, which can dissolve certain minerals in the rock and wash them away.

Mass wasting occurs when gravity pulls large amounts of rock and soil downhill. Heavy rainfall, earthquakes, and volcanic eruptions can trigger mass-wasting events on unstable slopes.

The Sierra Nevada have been shaped by weathering and erosion.

Mechanical weathering occurs when a rock is broken into smaller pieces by physical processes. For example, water can seep into cracks in rocks. When the water freezes, it expands. This puts pressure on the cracks, causing the rock to split apart.

Glaciers are huge bodies of moving ice. As glaciers slowly move over earth's surface, they can pick up rock pieces and carry them over long distances. Glaciers covered much of North America about 12,000 years ago and left behind boulders the size of houses.

Running water is the most important agent, or cause, of erosion on earth. Water in rivers, streams, and oceans can carry sediment from one place to another. For example, the cliffs in this photograph have been eroded by ocean waves.

Wind is an important agent of erosion in certain areas, carrying small particles of rock over long distances. For example, wind can carry dust from African deserts to North, Central, and South America. If the wind is strong enough, the sediment particles can act like sandpaper and wear away rocks and other objects. Rocks shaped by windblown sediment are called *ventifacts*.

Scientists use magnitude and intensity to describe the severity of an earthquake.

The magnitude of an earthquake is related to the amount of energy that is released at the focus. An earthquake's intensity is based on the effects of the earthquake observed in different areas.

Seismographs are used to measure how much the ground moves during an earthquake. Scientists use this information to determine the earthquake's magnitude.

Scientists collect information about the **amount of damage** that occurs in different areas during an earthquake. They use this information to determine the intensity of the earthquake.

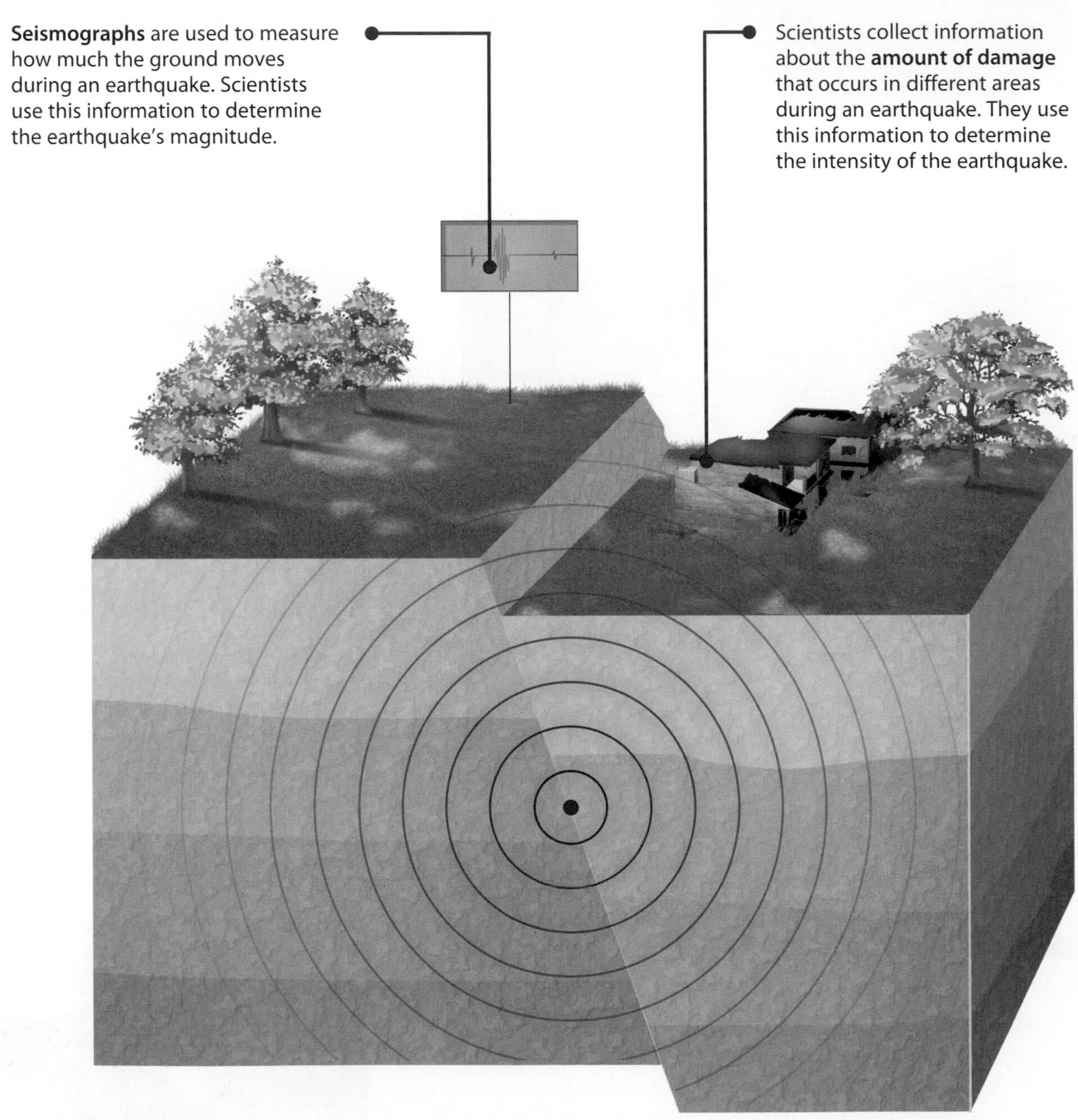

ABBREVIATED RICHTER SCALE FOR EARTHQUAKE MAGNITUDE

DESCRIPTION	MAGNITUDE	AMOUNT OF TNT EQUIVALENT TO THE ENERGY RELEASED BY THE EARTHQUAKE	FREQUENCY OF OCCURRENCE
Micro	1.9 or less	less than 13 pounds	~8,000 daily
Very minor	2.0–2.9	13 pounds to 390 pounds	~1,000 daily
Minor	3.0–3.9	390 pounds to 1,000 pounds	49,000 yearly
Light	4.0–4.9	6 tons (equal to a small atomic bomb) to 200 tons	6,200 yearly
Moderate	5.0–5.9	200 tons to 6,200 tons	800 yearly
Strong	6.0–6.9	6,200 tons to 199,000 tons	120 yearly
Major	7.0–7.9	199,000 tons to 6 million tons	18 yearly
Great	8.0–8.9	6 million tons to 200 million tons	1 yearly
Rare great	9.0 or greater	more than 200 million tons	1 per 20 years

ABBREVIATED DESCRIPTION OF THE TWELVE LEVELS OF THE MODIFIED MERCALLI INTENSITY SCALE

INTENSITY	DESCRIPTION OF DAMAGE
I	felt by very few
II	felt by a few people, mostly in the top floors of buildings
III	felt by many people indoors
IV	felt by most people indoors and some people outdoors; dishes disturbed
V	felt by most people indoors and outdoors; some windows broken
VI	felt by all; some furniture moves; slight damage to buildings
VII	considerable damage in poorly constructed buildings
VIII	considerable damage in ordinary buildings; substantial damage in poorly constructed buildings
IX	considerable damage even in specially designed buildings
X	substantial damage to most buildings
XI	bridges destroyed; few objects remain standing
XII	total damage; objects thrown into the air

Earthquakes occur when earth's plates move.

Most earthquakes are so weak that we cannot feel them. However, strong earthquakes can be very destructive. The earthquakes listed here were some of the most powerful during the twentieth and early twenty-first centuries. The accompanying number gives the magnitude of the earthquake on the Richter scale.

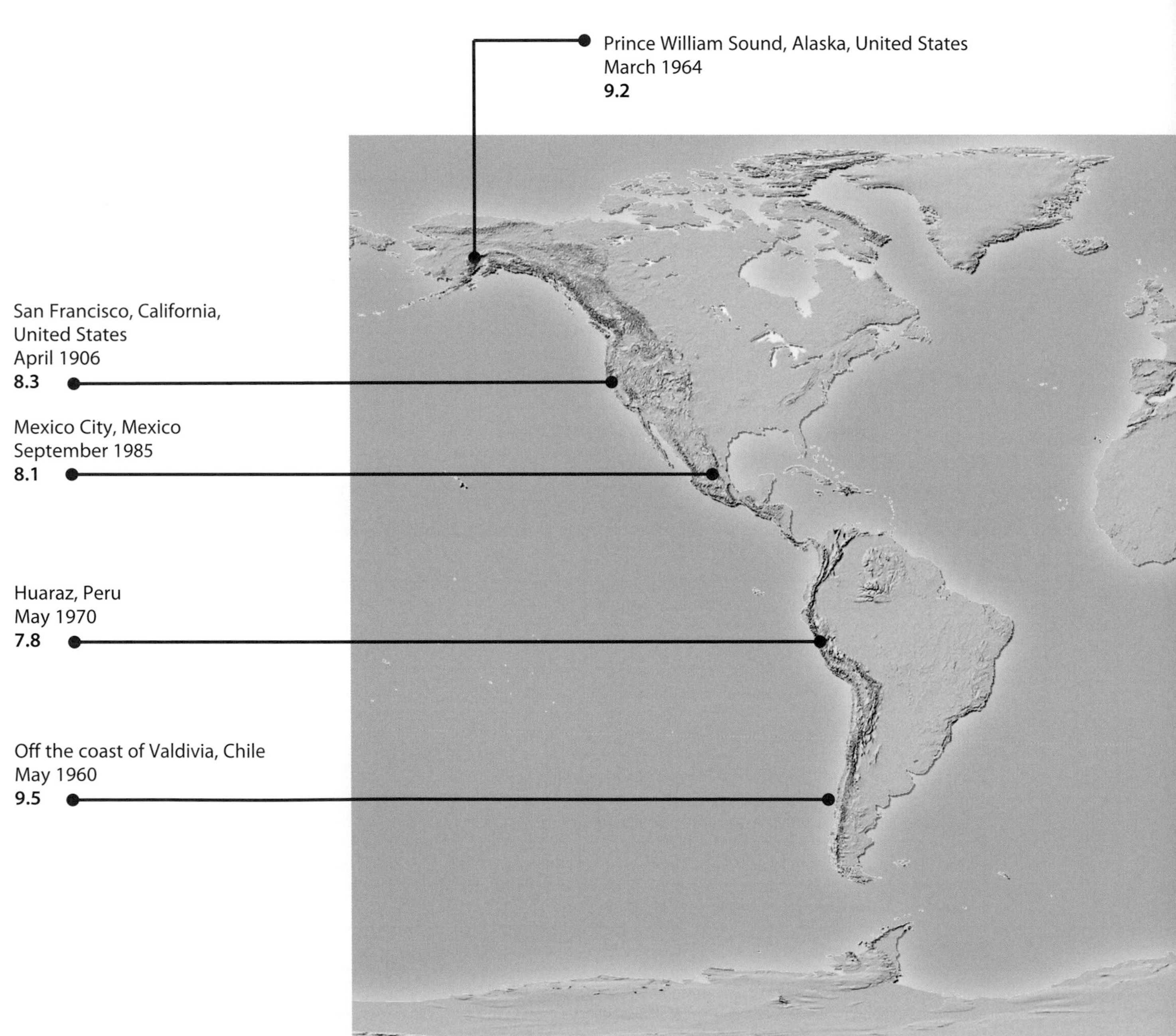

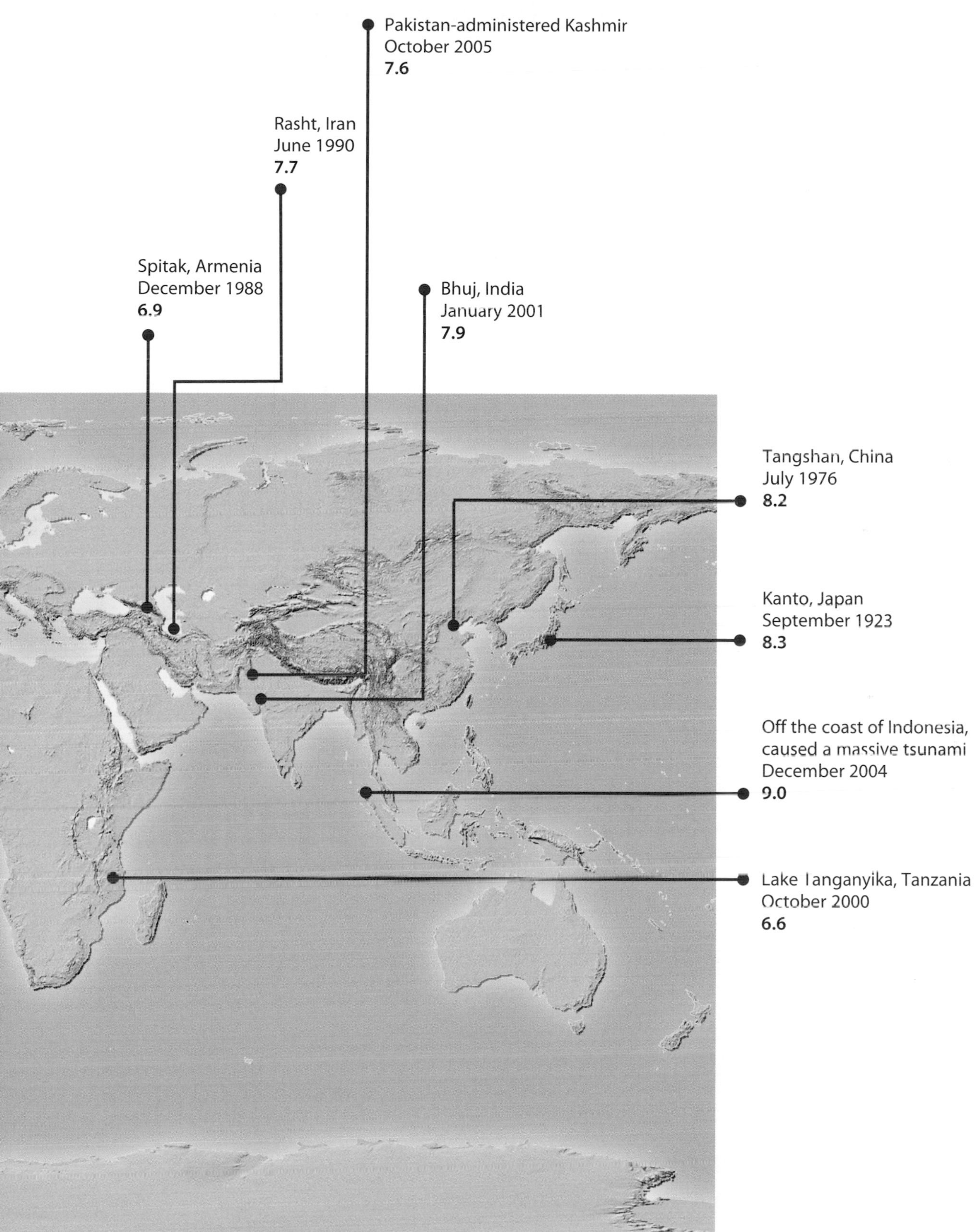
Pakistan-administered Kashmir
October 2005
7.6
Rasht, Iran
June 1990
7.7
Spitak, Armenia
December 1988
6.9
Bhuj, India
January 2001
7.9
Tangshan, China
July 1976
8.2
Kanto, Japan
September 1923
8.3
Off the coast of Indonesia, caused a massive tsunami
December 2004
9.0
Lake Tanganyika, Tanzania
October 2000
6.6

Earthquakes happen when earth's lithosphere breaks because of stress.

Earthquakes release huge amounts of energy, which travel through the earth in the form of waves called *seismic waves*. The three main types of seismic waves are P waves, S waves, and surface waves.

This **seismogram** is a recording of how the ground at a particular place moved as different types of seismic waves passed through it. P waves arrive first because they travel the most quickly. Surface waves arrive last because they travel the most slowly.

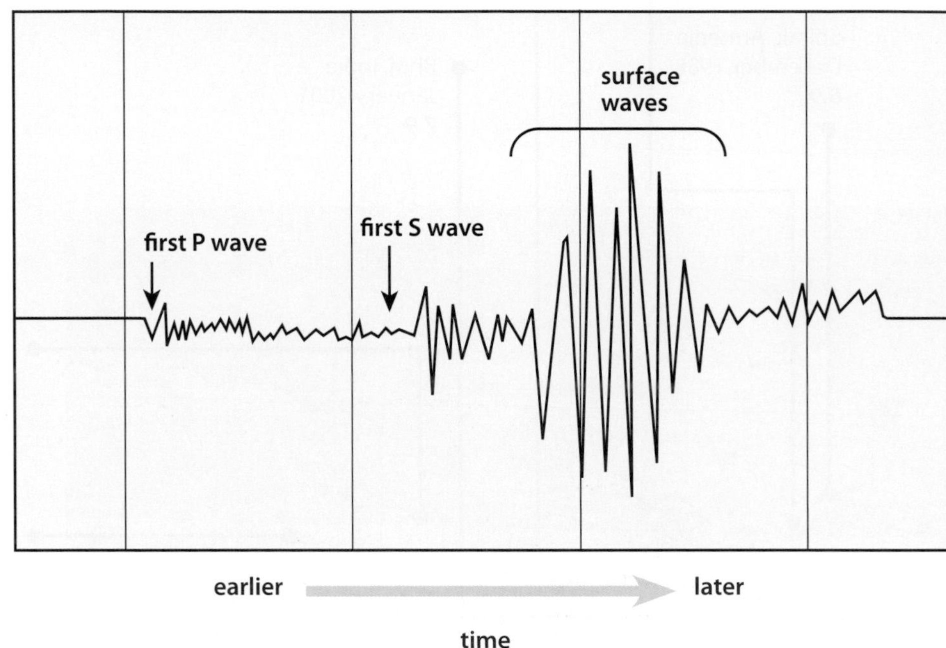

P waves, also called *pressure or primary waves*, are seismic waves that move particles back and forth, causing compression and extension. Particles in the rocks vibrate parallel to the direction that the waves are traveling.

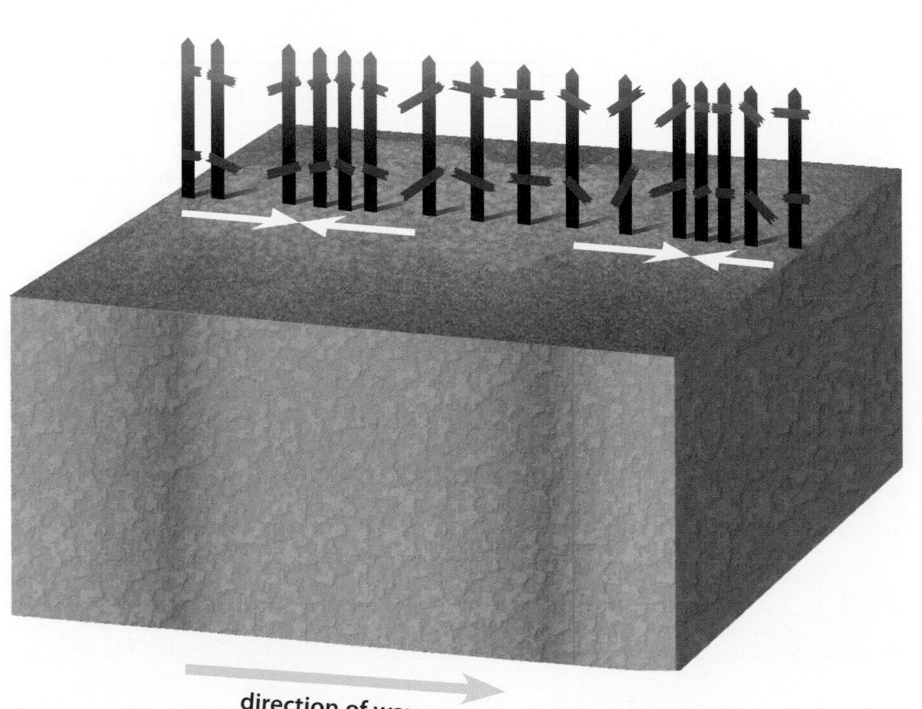

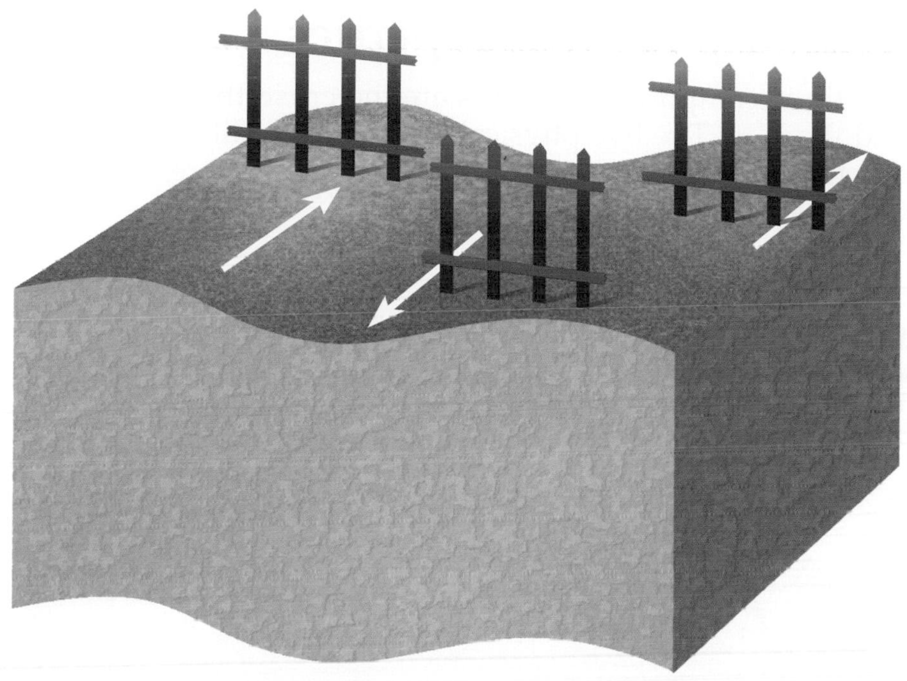

S waves, also called *shear* or *secondary waves*, are seismic waves that move particles from side to side. The particles vibrate perpendicular to the direction that the waves are traveling.

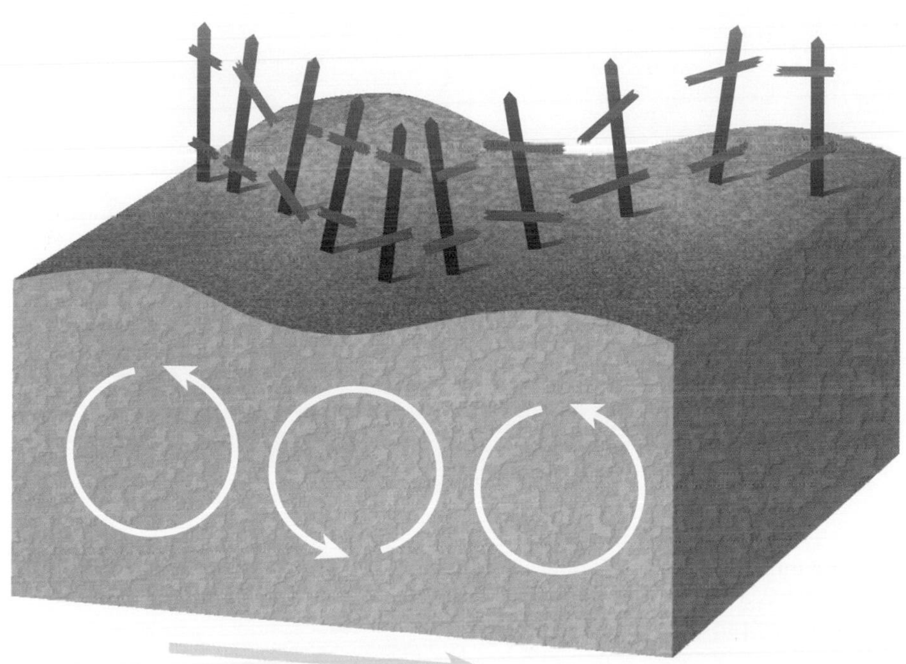

Surface waves are seismic waves that cause particles to move in a circular motion. Surface waves travel very slowly and affect only areas near the epicenter of an earthquake. However, they can cause serious damage.

Scientists define the layers of earth's interior based on differences in their composition and differences in their physical properties.

The three compositional layers are the crust, the mantle, and the core. The five physical layers are the lithosphere, the asthenosphere, the mesosphere, the outer core, and the inner core.

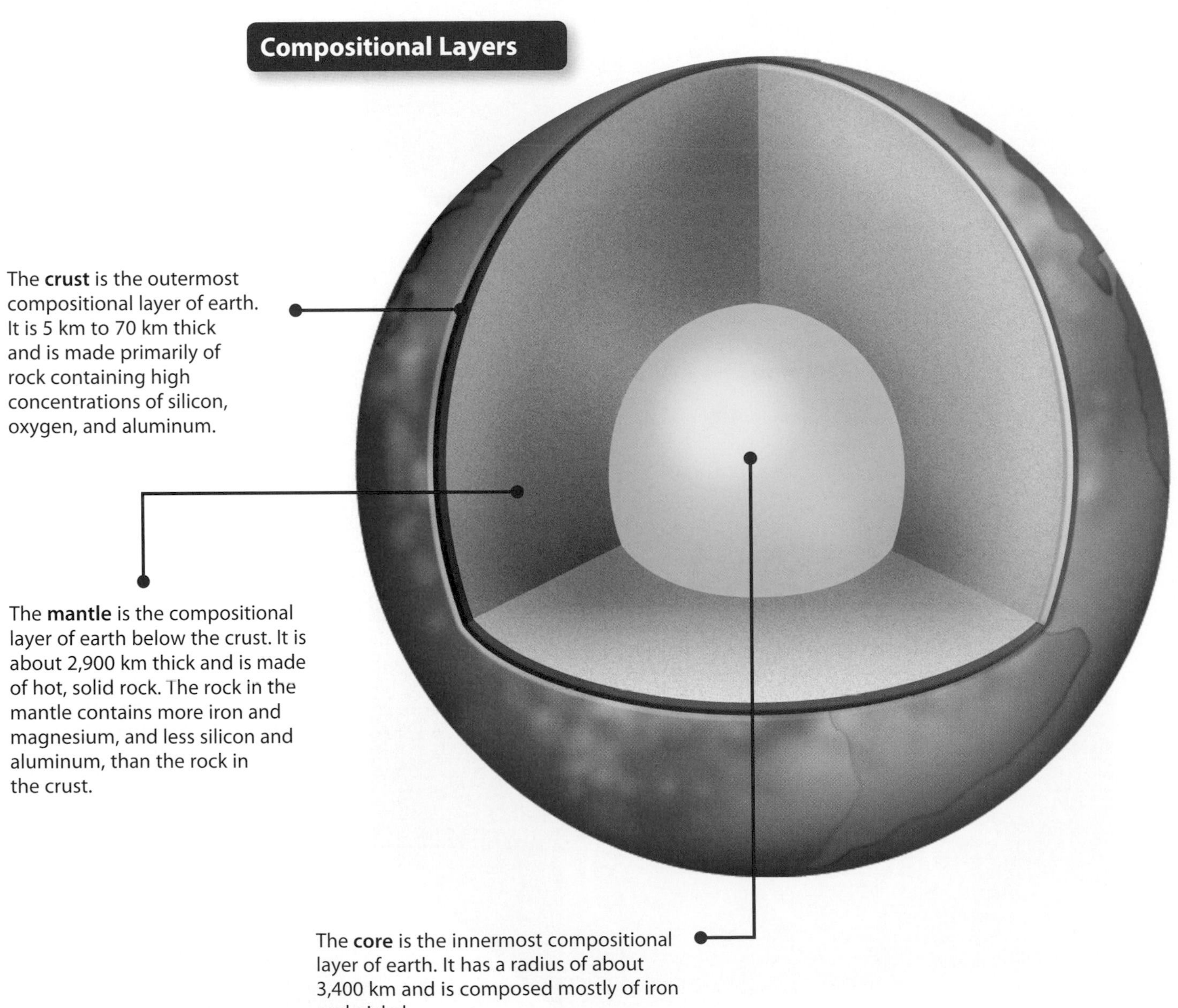

Physical Layers

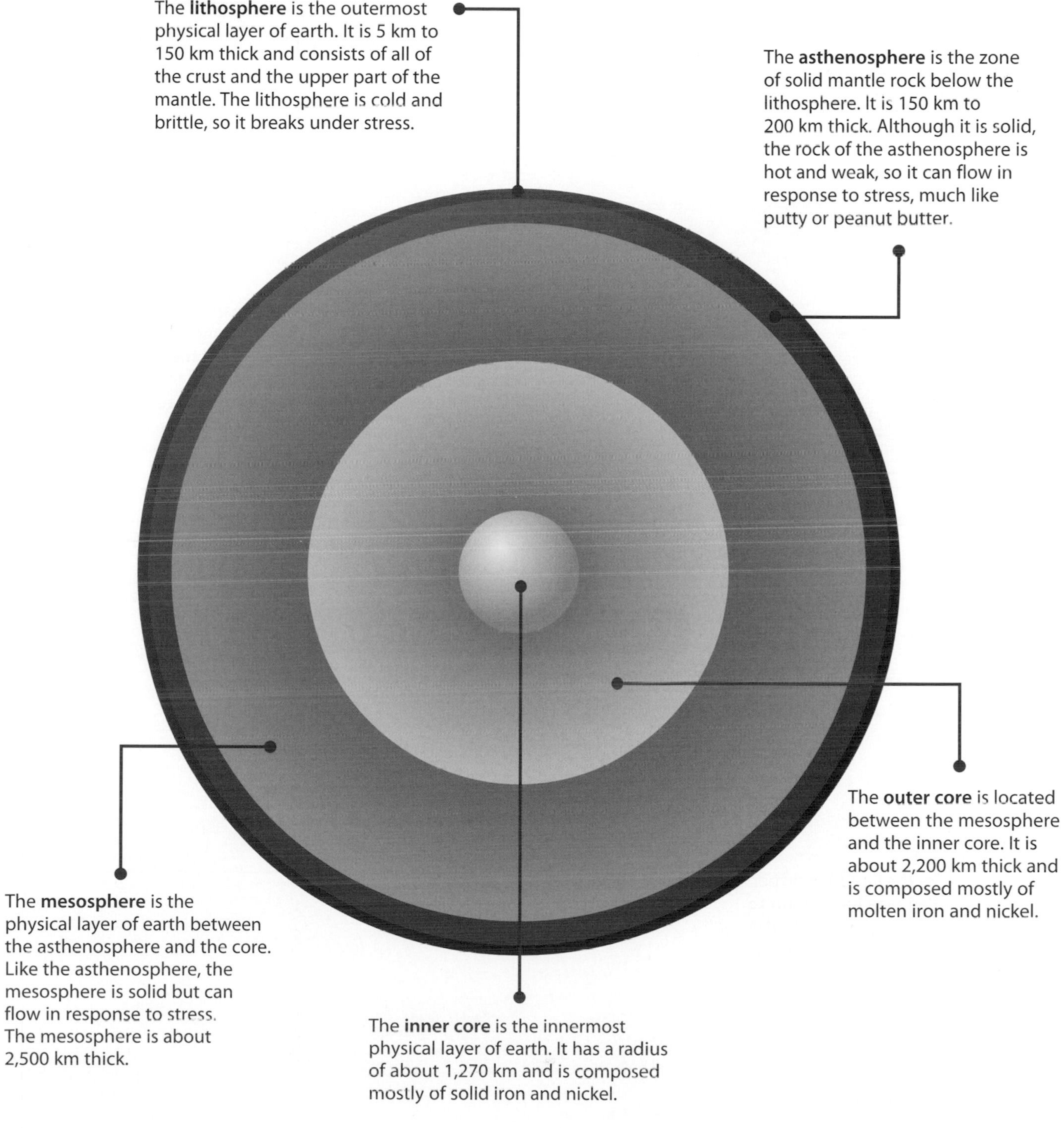

Earth's lithosphere is broken into many tectonic plates that move slowly over the planet's surface.

These movements cause the locations of the continents and oceans to shift over time. Many other surface features, such as mountains and volcanic chains, are also the result of tectonic plate movements.

Early Jurassic Period (200 mya)

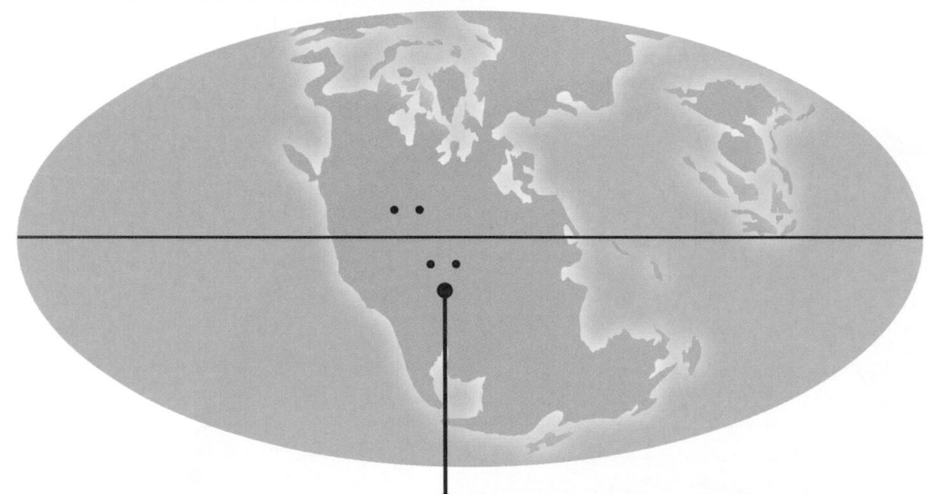

More than 500 million years ago, the movements of the plates **brought all the continents together**. Scientists call this supercontinent *Pangaea*. Pangaea existed until the Jurassic period.

These dots represent locations that were close together on Pangaea. Notice on subsequent maps that these locations have moved apart over time.

Late Jurassic Period (150 mya)

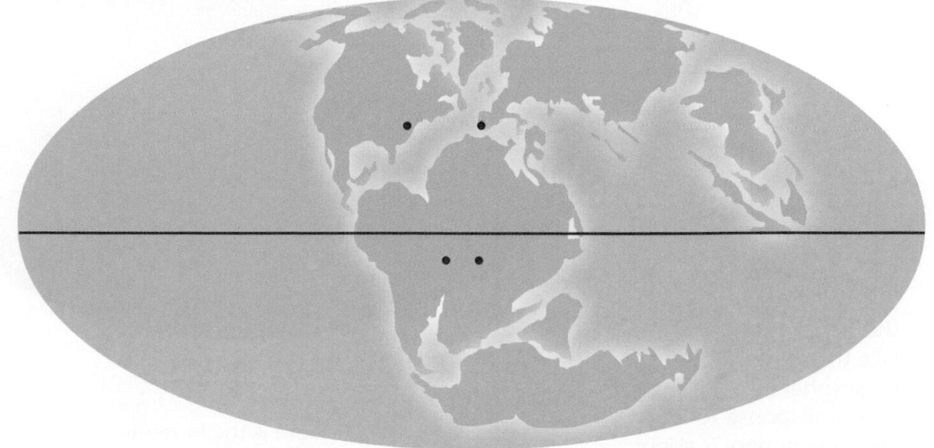

By the late Jurassic period, Pangaea **had begun to break apart** due to the movements of the plates.

mya = millions of years ago

Eocene Epoch (20 mya)

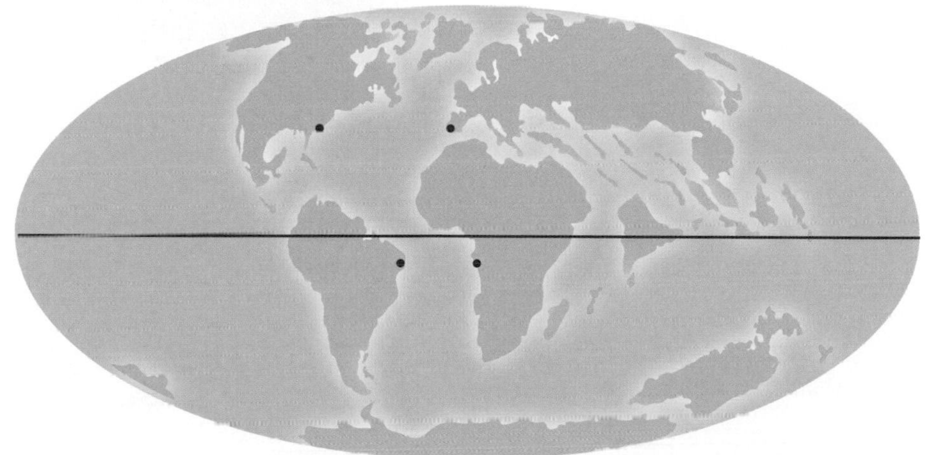

During the Eocene epoch, the Asian and Indian plates **began to collide**.

Present

Earth's tectonic plates **continue to move**. For example, the Atlantic Ocean becomes about 4 cm wider each year because of plate motions.

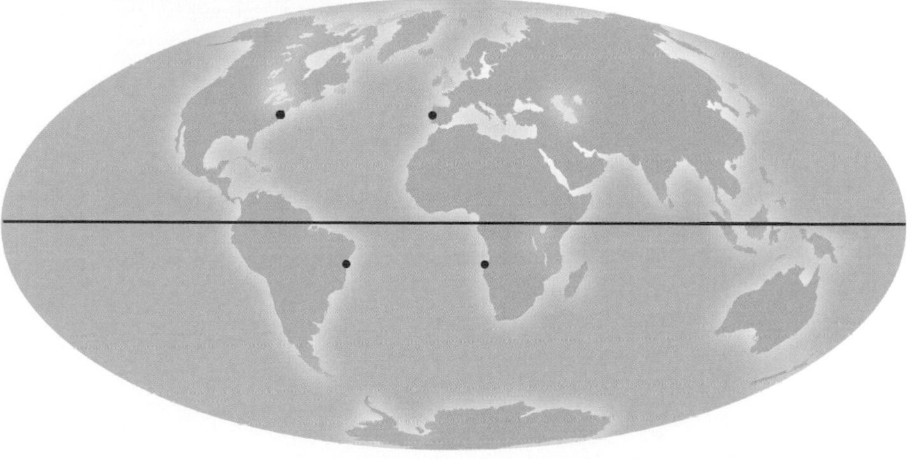

Earth's lithosphere is broken into many large and small tectonic plates.

Some plates are made entirely of oceanic lithosphere. Others consist of both oceanic and continental lithosphere. Although some plates are made primarily of continental lithosphere, all have at least some oceanic lithosphere.

The **floor of the Atlantic Ocean** lies on five different tectonic plates: the North American plate, the South American plate, the Eurasian plate, the African plate, and the Caribbean plate.

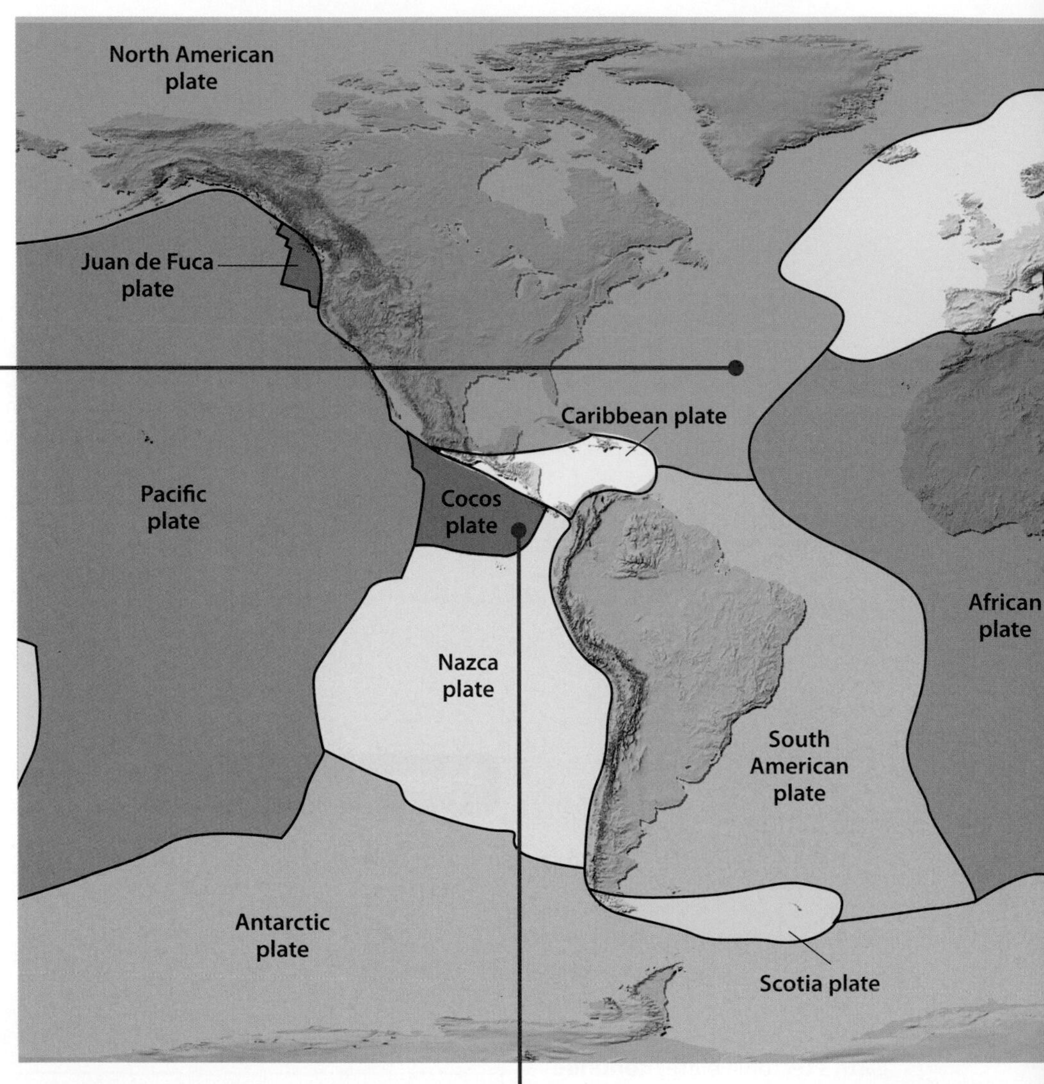

The **Cocos plate** is a small tectonic plate that consists of only oceanic lithosphere.

The **African plate** is a large tectonic plate that consists of both oceanic and continental lithosphere.

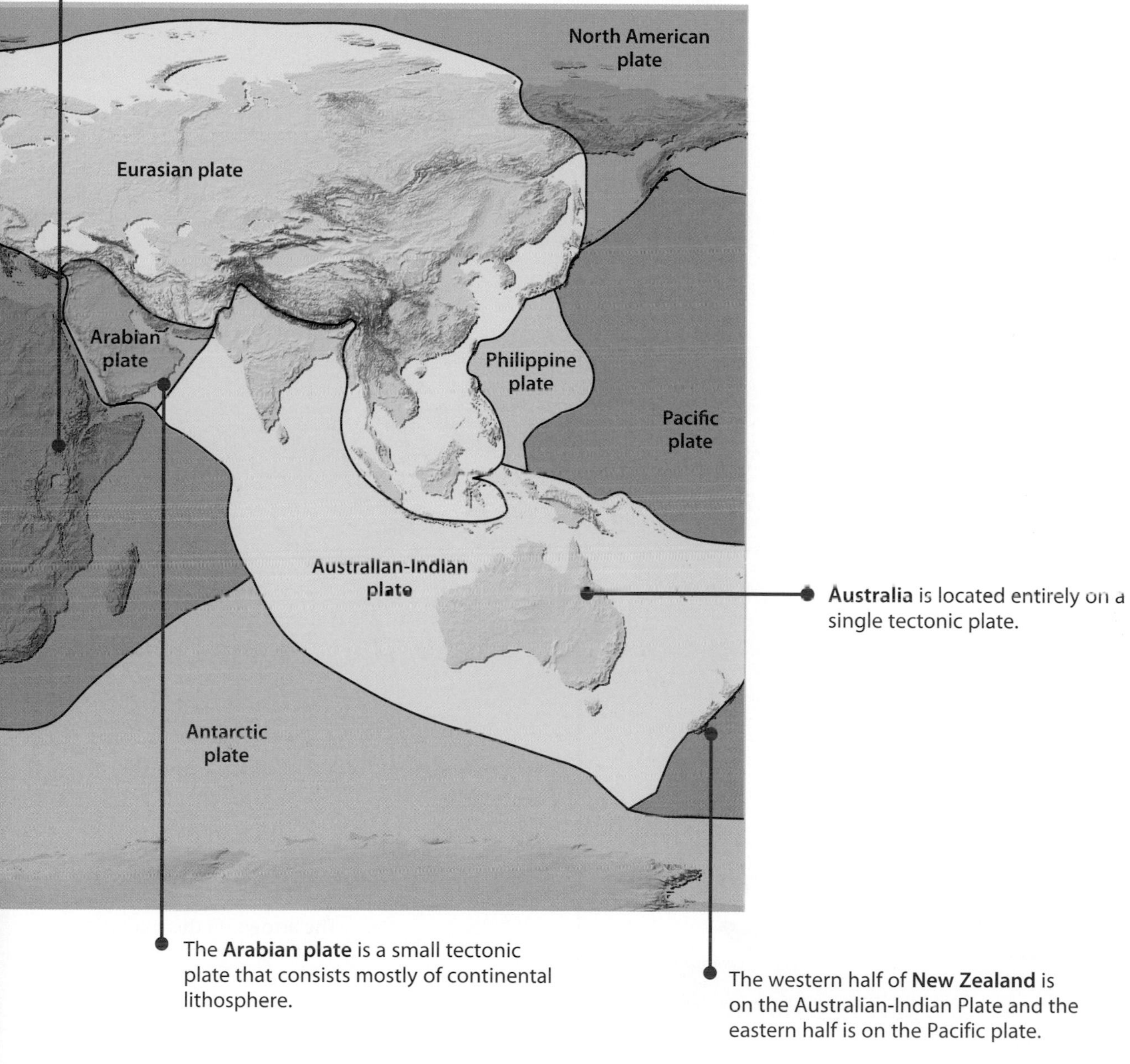

Australia is located entirely on a single tectonic plate.

The **Arabian plate** is a small tectonic plate that consists mostly of continental lithosphere.

The western half of **New Zealand** is on the Australian-Indian Plate and the eastern half is on the Pacific plate.

Tectonic plate boundaries are the sites of significant geologic structures and processes.

The three main types of plate boundaries are convergent, divergent, and transform.

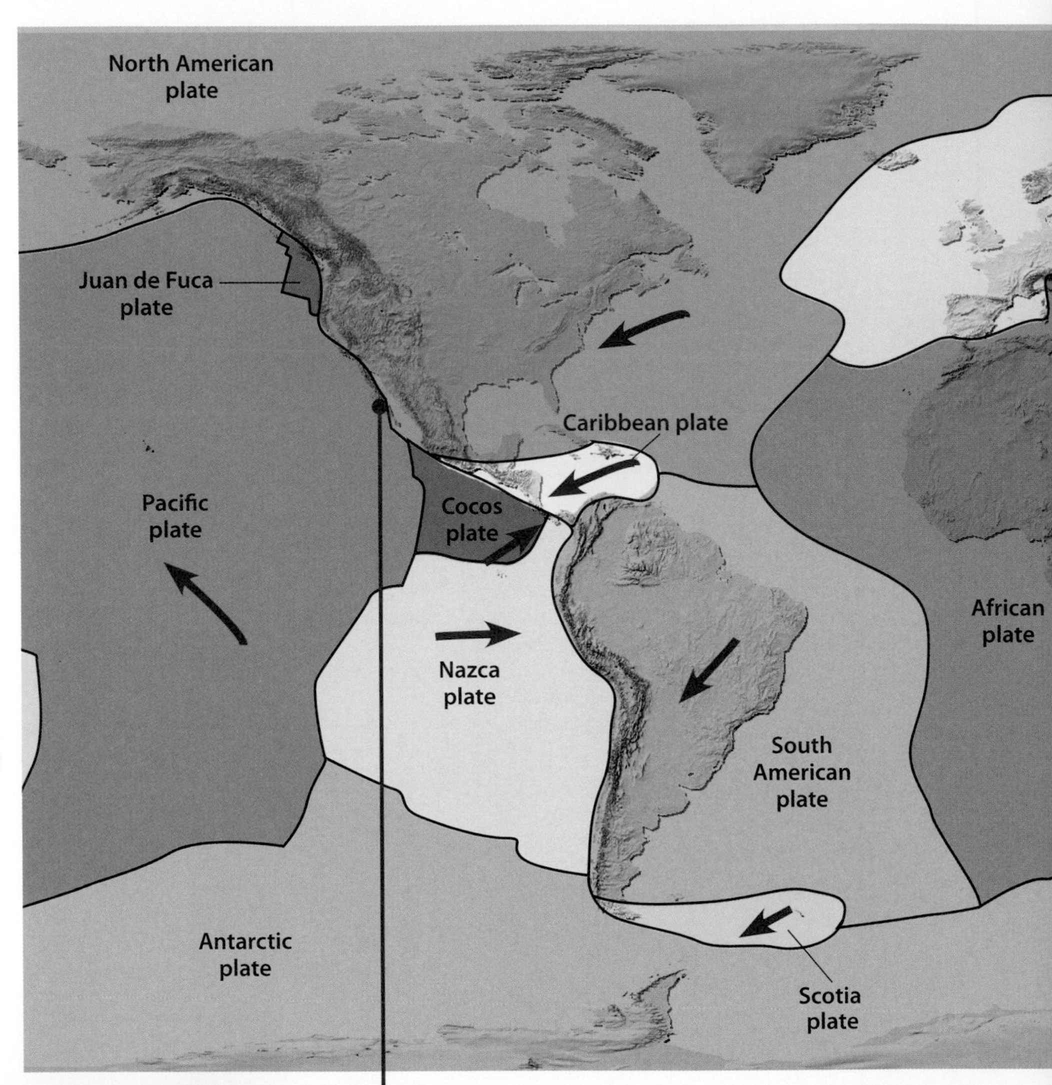

The arrows on this map show the directions in which the plates are moving.

Transform Boundaries

At a **transform plate boundary**, two plates slide past each other horizontally. The San Andreas fault zone is part of a transform boundary between the Pacific and North American plates. Transform boundaries are also common on the ocean floor, where they connect segments of mid-ocean ridges.

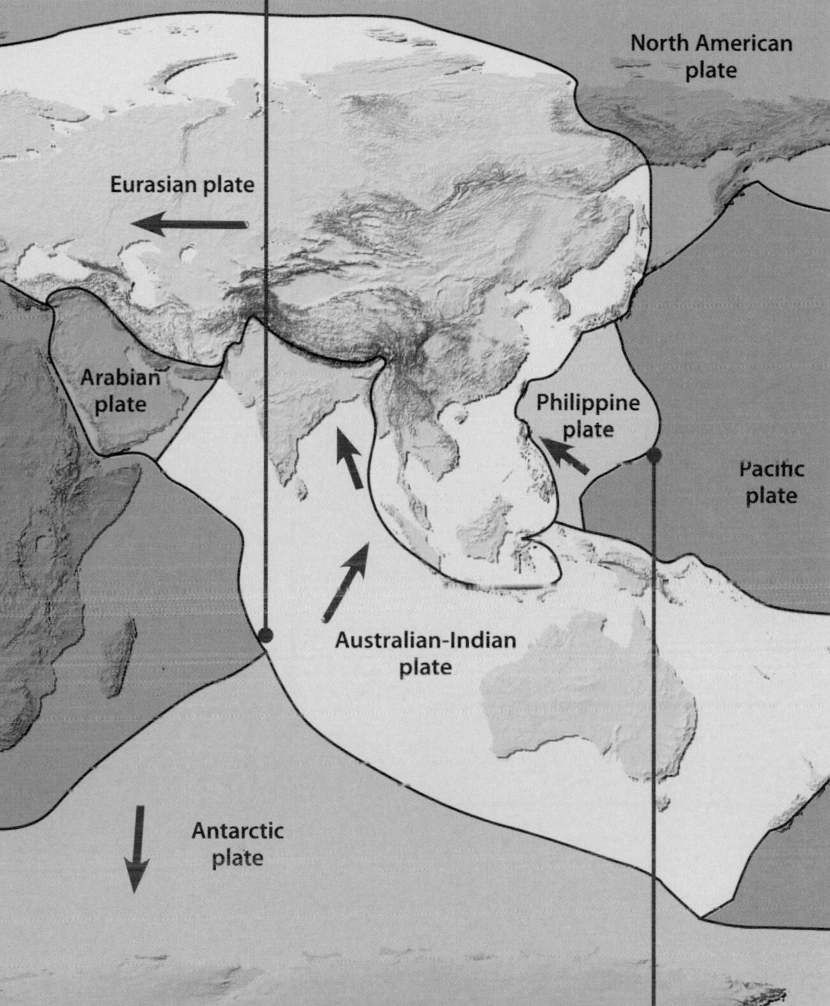

Divergent Boundaries

At a **divergent plate boundary,** two tectonic plates move away from each other. New lithosphere forms at divergent plate boundaries when material rises from beneath earth's surface. Today, most divergent plate boundaries, such as the Southeast Indian Ridge, are located beneath the oceans.

Convergent Boundaries

At a **convergent plate boundary,** two plates move together. The lithosphere can be pulled into the asthenosphere at convergent plate boundaries. The Mariana Arc is a chain of volcanoes that has formed along the convergent boundary between the Pacific plate and the Philippine plate.

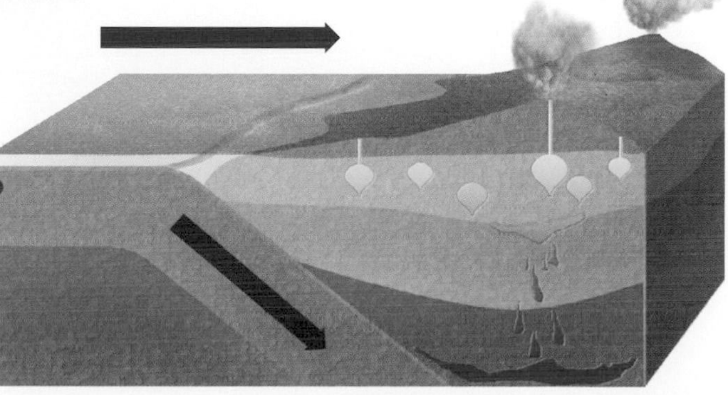

Volcanoes are cracks in earth's crust through which magma, ash, and gases erupt.

Most volcanoes form along tectonic plate boundaries, but some develop far from plate boundaries. A few of the thousands of active volcanoes are shown on the map below.

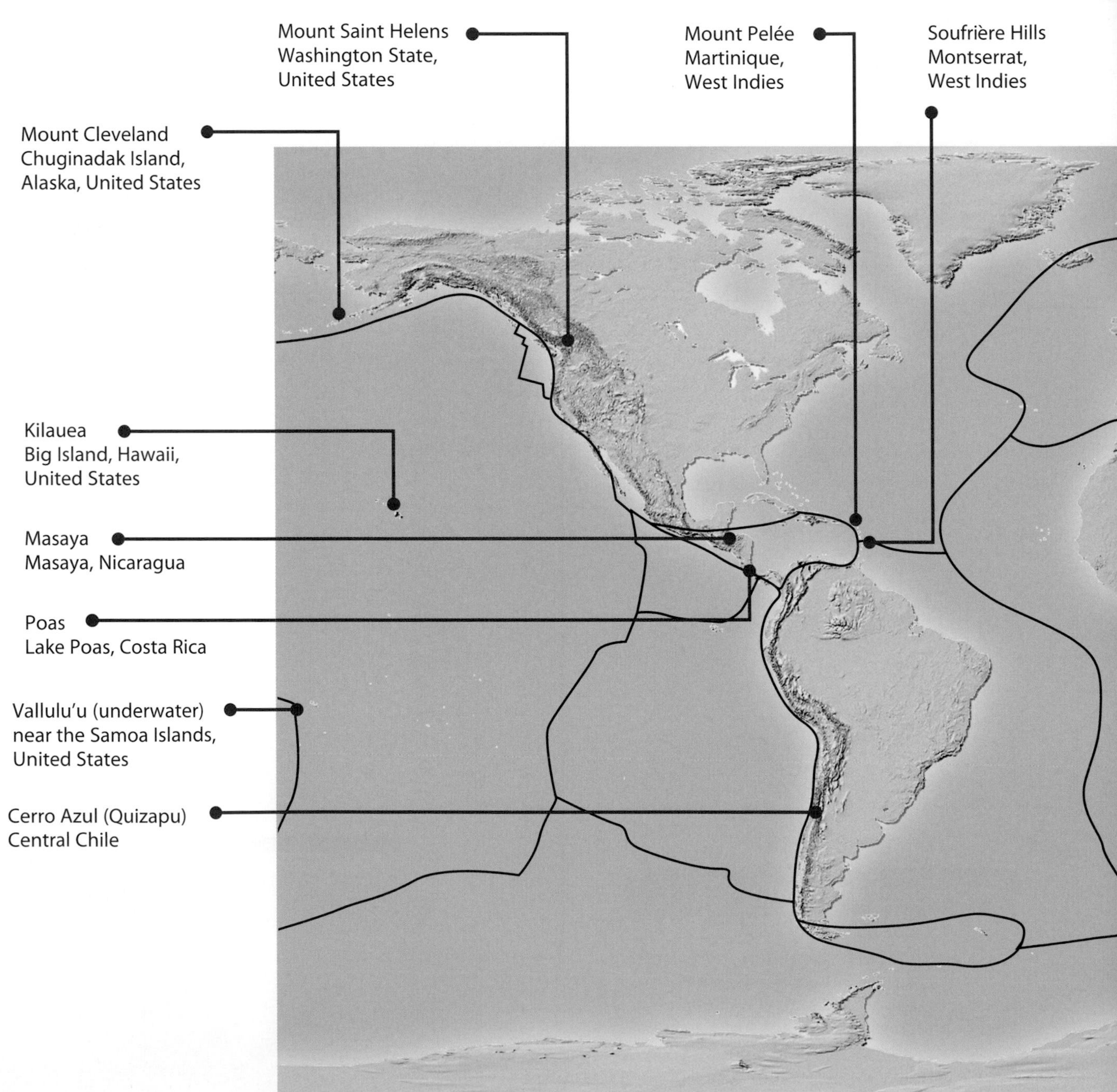

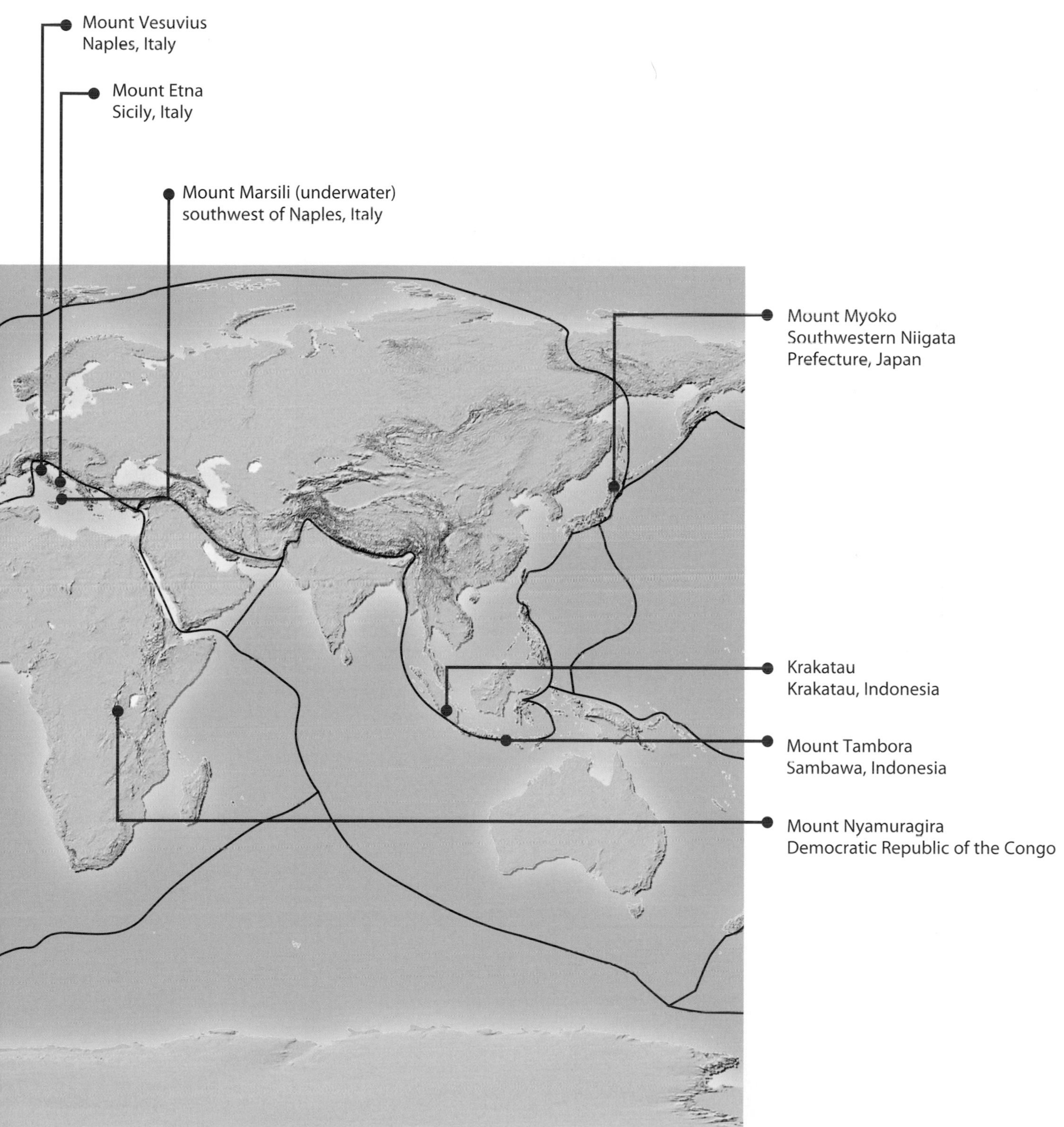
Mount Vesuvius
Naples, Italy
Mount Etna
Sicily, Italy
Mount Marsili (underwater)
southwest of Naples, Italy
Mount Myoko
Southwestern Niigata
Prefecture, Japan
Krakatau
Krakatau, Indonesia
Mount Tambora
Sambawa, Indonesia
Mount Nyamuragira
Democratic Republic of the Congo

Tall mountains form when tectonic plates collide.

For example, the Himalayas are growing as the Indian and Asian plates collide. The plates crumple and fold as they come together. Because the continental lithosphere of each plate is so thick, the collision produces extremely tall mountains.

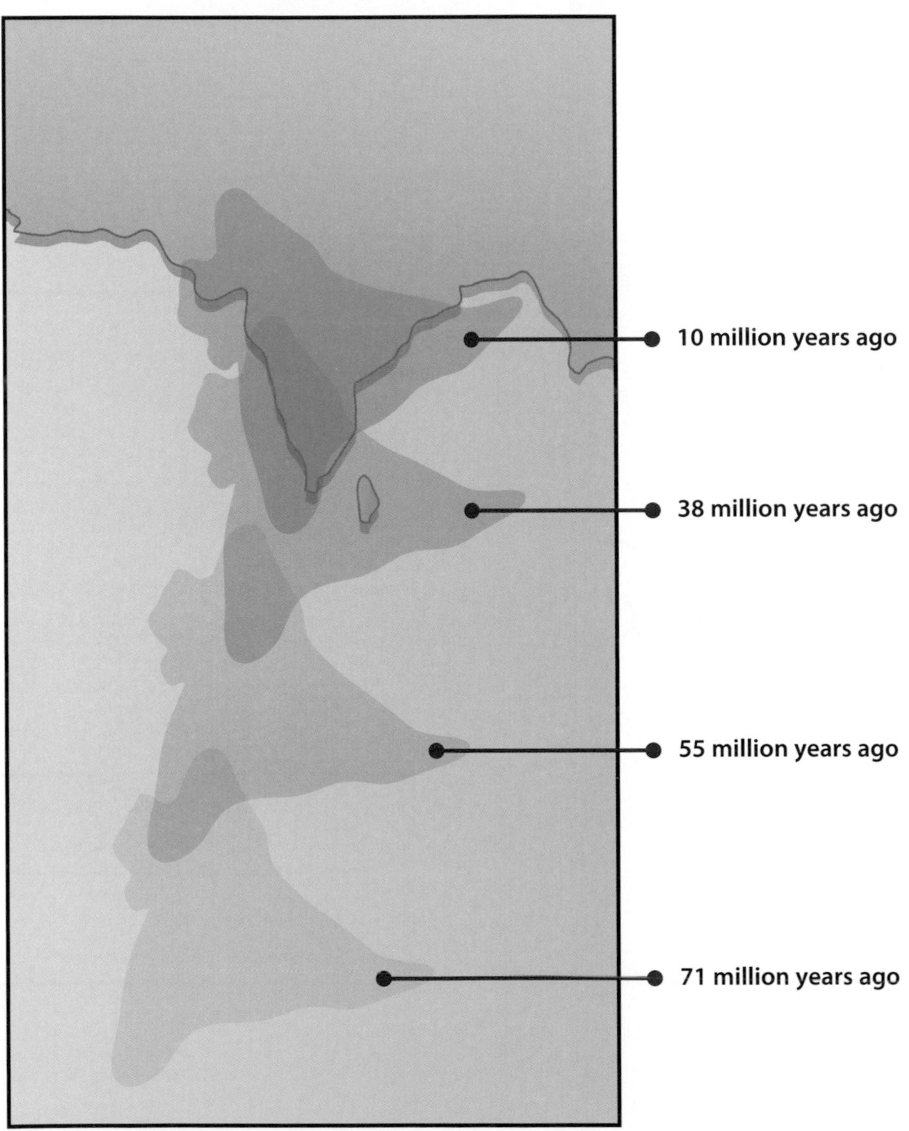

The Indian plate has moved over time and has collided with the Asian plate. This diagram shows the relative locations of the **Indian and Asian plates** over time.

More than 50 million years ago, the northern part of the **Indian plate** (right) consisted of oceanic lithosphere. As the Indian and Asian plates collided, the oceanic lithosphere on the Indian plate subducted beneath the continental lithosphere of the Asian plate, producing a chain of volcanic mountains similar to the modern Andes.

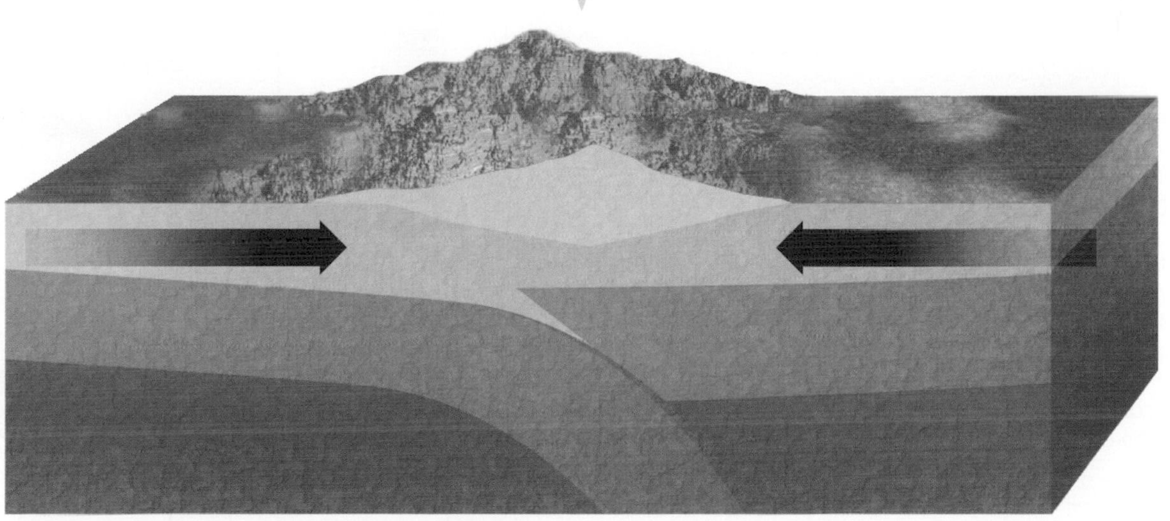

Today, the Indian and Asian plates are still colliding. This collision is causing the **Himalayas** (center) to grow slowly over time. The continental lithosphere of the Himalayas is extremely thick—more than 70 km.

The theory of plate tectonics connects and explains many observations about earth's surface features and history that geologists once thought were unrelated.

The theory states that earth's lithosphere is broken into many pieces called tectonic plates. These plates move slowly over earth's surface.

Before the mid-1900s, geologists thought that the continents and oceans on earth's surface had always looked much as they do today.

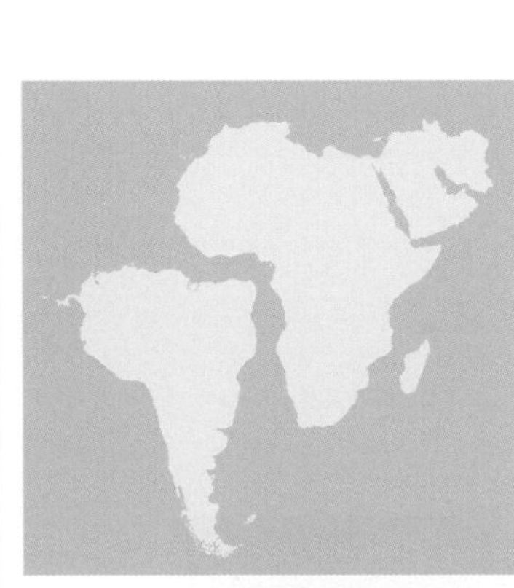

In 1912, **Alfred Wegener** proposed the hypothesis of continental drift. Wegener's idea was that the continents were once connected in one large landmass. He collected a great deal of evidence to support his idea, but he could not explain how the continents could move through the ocean basins.

By the 1920s, the first seismometers were being used to record and study earthquake activity. Scientists soon discovered that earthquakes are most common in several distinct bands in earth's lithosphere.

In 1929, **Motonori Matuyama** discovered patterns of magnetism in rocks. He showed that some rocks on earth today have magnetic fields that are aligned in the opposite direction to earth's current magnetic field.

Brunhes normal

Matuyama reversed

Gauss normal

Gilbert reversed

In the 1930s, **David Griggs** demonstrated that mantle rock could flow like putty when exposed to extremely high temperatures and pressures. This research provided a possible explanation for how the continents move and helped support Wegener's continental drift hypothesis.

EARLY 1900s	1912	1920s	1929	1930s

In 1947, scientists on the research vessel ***Atlantis*** found that ocean floor sediment was much thinner than expected. These findings suggest that the ocean floor is younger than much of the continental land mass.

In 1962, **Harry Hess** proposed that new oceanic lithosphere may be created at mid-ocean ridges, such as the Mid-Atlantic Ridge, and that lithosphere may be consumed into the mantle at oceanic trenches. These ideas explained the origins of many features of the ocean floor and became part of the theory of plate tectonics.

In 1963, **Fred Vine** and **Drummond Matthews** helped support Harry Hess's proposal of seafloor spreading and subduction by studying the patterns of magnetic rock on the seafloor.

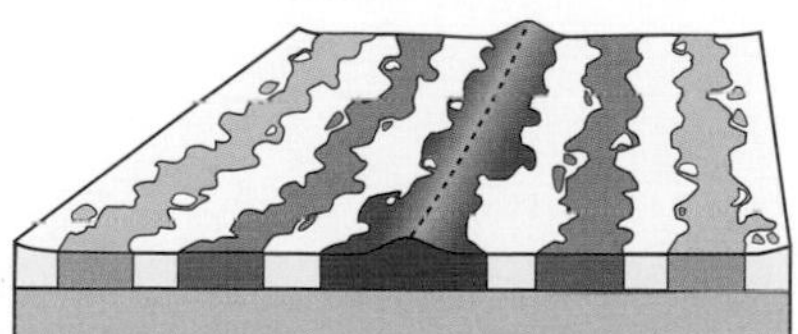

In the late 1960s, **Xavier Le Pichon, W. Jason Morgan,** and several other geologists formally proposed the theory of plate tectonics. They included many pieces of evidence to support their theory, including the magnetism, sediment thicknesses, and topography of the ocean floor.

In the late 1960s, the research vessel ***Glomar Challenger*** drilled core samples deep beneath the ocean floor. The samples show that oceanic lithosphere is youngest near ocean ridges and oldest near ocean trenches. These observations support the idea that oceanic lithosphere is created at ridges and consumed at trenches.

Research on the forces that drive plate motions continues today. As scientists gather more data, their ideas about how the plates move continue to change.

1947 | **1962** | **1963** | **LATE 1960s** | **TODAY**

Volcanoes form at convergent boundaries, where tectonic plates collide.

In 1943, a volcano formed almost overnight near such a boundary in the middle of a cornfield in Paricutín, Mexico.

1. At convergent plate boundaries, **two plates move toward each other**. If one of the plates is made of oceanic lithosphere, it can subduct, or sink below the other plate.

4. When the magma reaches the surface and erupts, a **volcano forms**. Chains of volcanoes are common at convergent margins.

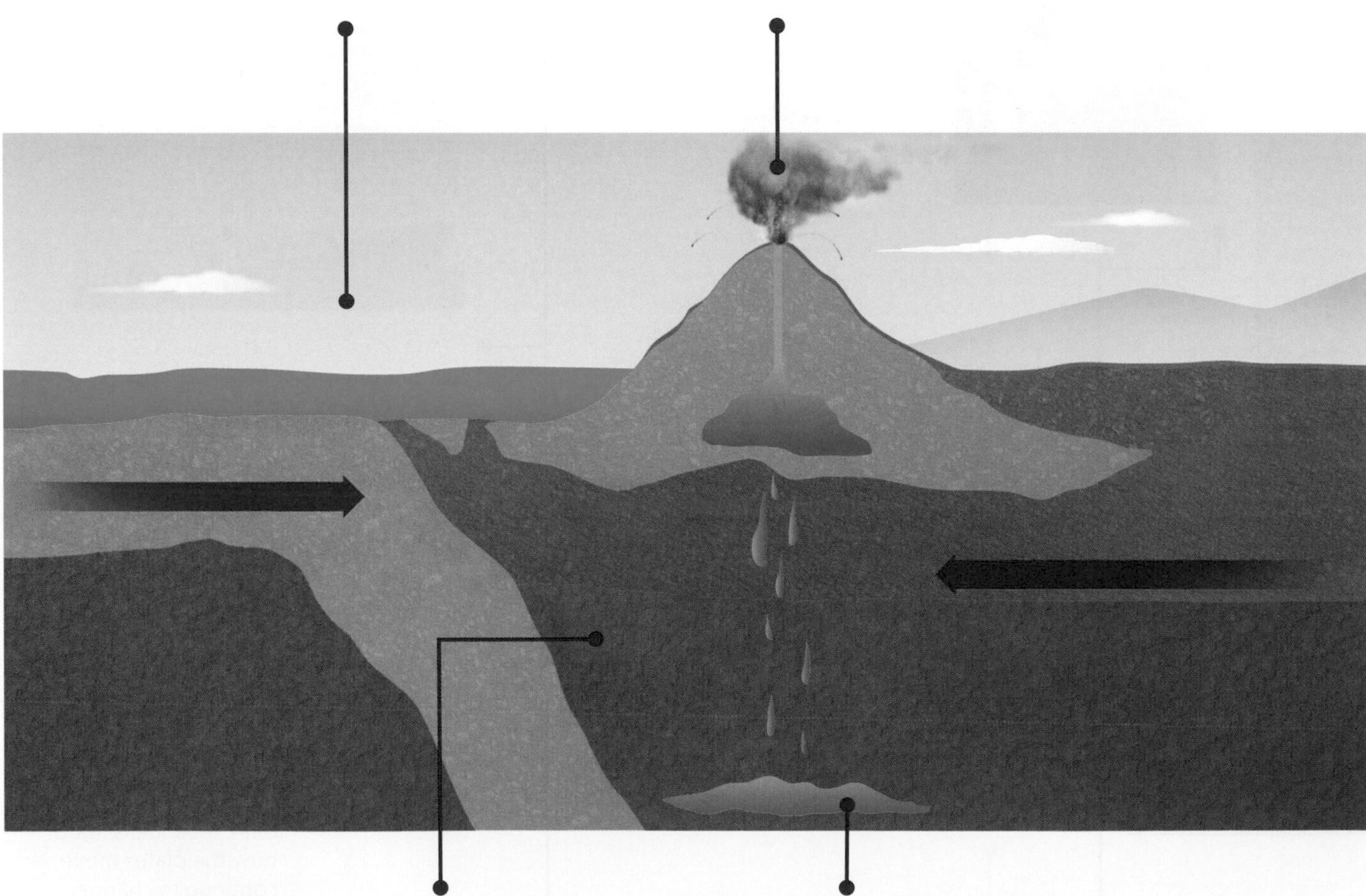

2. As the oceanic lithosphere subducts, the **pressure and temperature increase.** This causes water in the plate to be released into the mantle above.

3. When the water mixes with the mantle rock, the **rock begins to melt**. The magma is less dense than the surrounding rock, so it rises toward the surface.

Paricutín volcano in Mexico formed over a period of only 9 years. During this time, the volcano grew from a small vent in the ground to a large mountain more than 2,500 m tall.

Volcanoes come in different sizes, shapes, and compositions, depending on how they form.

The three main types of volcanoes are shield, composite, and cinder cone. Each type forms in a different way.

Shield volcanoes are made of multiple layers of lava deposited by repeated, nonexplosive eruptions. Shield volcanoes typically have gently sloping sides because the lava is very fluid and spreads out over a large area. Mauna Loa in Hawaii is an example of a shield volcano.

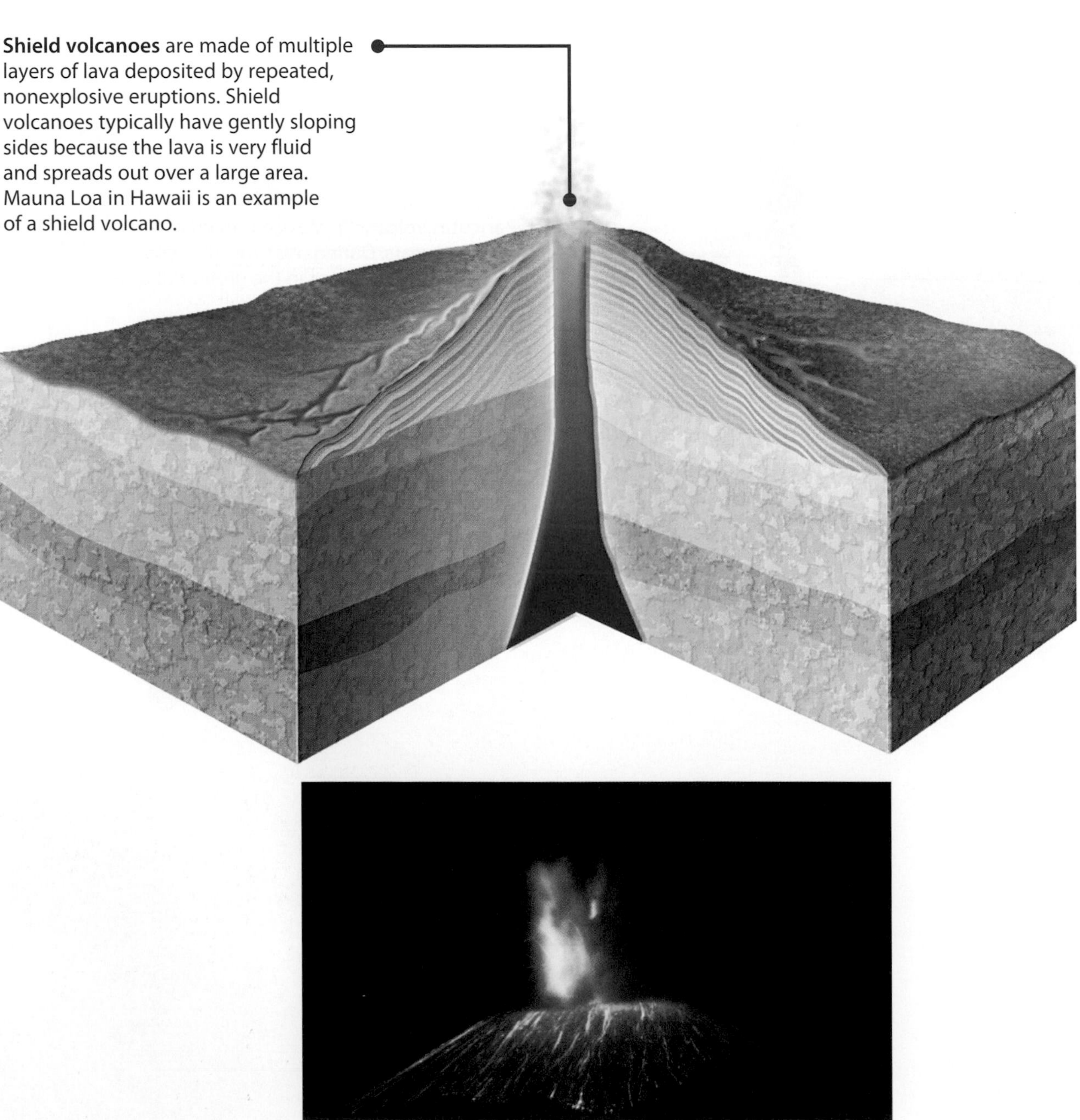

Composite volcanoes, or stratovolcanoes, form from alternating explosive and nonexplosive eruptions. Consequently, they have alternating layers of lava and pyroclastic material (small particles that are not cemented together). Most composite volcanoes have a very broad base, and the sides grow increasingly steeper toward the top of the cone. Mount Saint Helens in Washington State is an example of a composite volcano.

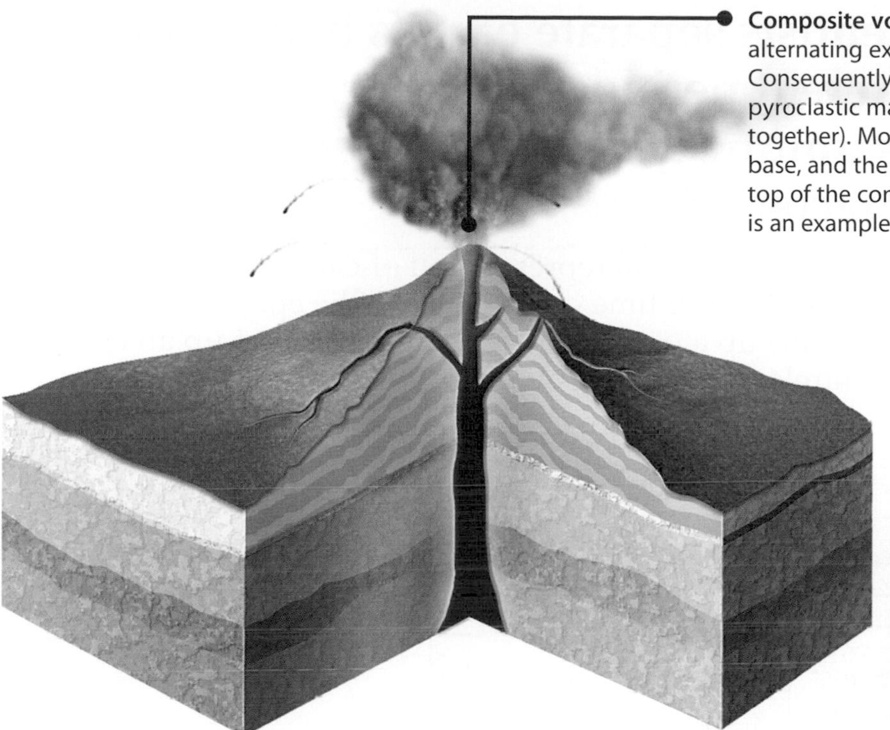

Cinder cone volcanoes are often fairly small and develop as a result of moderately explosive eruptions. Cinder cone volcanoes sometimes form in clusters, and most have a distinctive cone shape. Paricutín in Mexico is an example of a cinder cone volcano.

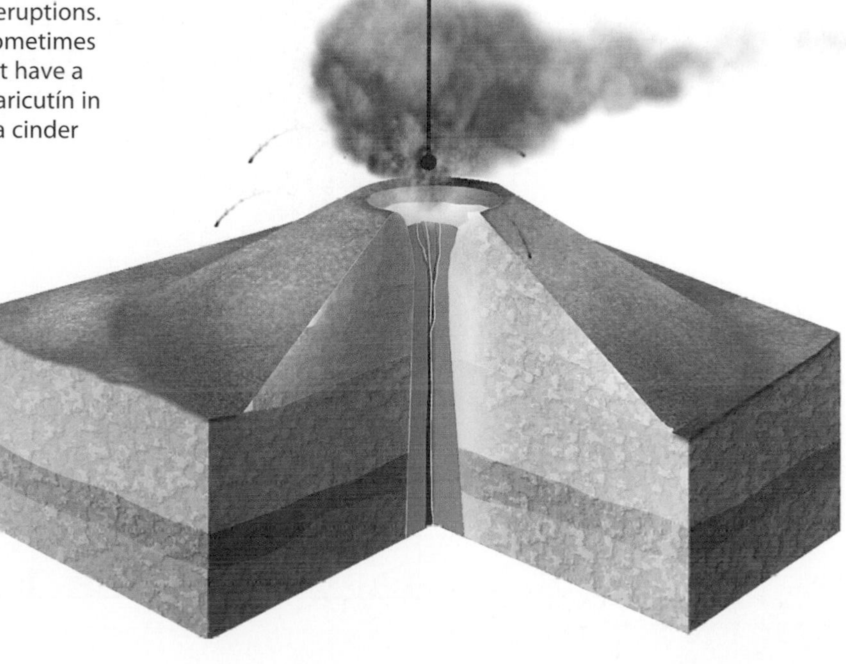

Scientists separate earth's history into different periods of time, each of which has unique geologic and biologic characteristics.

The geologic time scale represents all of earth's history. The smallest unit of time on this scale is the epoch. Multiple epochs make up a period, multiple periods make up an era, and multiple eras make up an eon.

Eon		Era	Millions of years ago
Phanerozoic		Cenozoic	65
		Mezozoic	248
		Paleozoic	540
Precambrian	Proterozoic	Late	900
		Middle	1600
		Early	2500
	Archean	Late	3000
		Middle	3400
		Early	3800
	Hadean		4500

Era	Period	Epoch	Millions of years ago
Cenozoic	Quaternary	Holocene	0.01
		Pleistocene	1.8
	Tertiary	Pliocene	5.3
		Miocene	23.8
		Oligocene	33.7
		Eocene	54.8
		Paleocene	65.0
Mesozoic	Cretaceous		144
	Jurassic		206
	Triassic		248
Paleozoic	Permian		290
	Carboniferous	Pennsylvanian	323
		Mississippian	354
	Devonian		417
	Silurian		443
	Ordovician		490
	Cambrian		540
Precambrian			

There is strong evidence that life on earth has changed over time.

The first organisms appeared on earth about 3.5 billion years ago. They were tiny bacteria-like cells that did not need oxygen to survive. All life on earth evolved from these early organisms.

Era	Period	Epoch	Millions of years ago
Cenozoic	Quaternary	Holocene	0.01
Cenozoic	Quaternary	Pleistocene	1.8
Cenozoic	Tertiary	Pliocene	5.3
Cenozoic	Tertiary	Miocene	23.8
Cenozoic	Tertiary	Oligocene	33.7
Cenozoic	Tertiary	Eocene	54.8
Cenozoic	Tertiary	Paleocene	65.0
Mesozoic	Cretaceous		144
Mesozoic	Jurassic		206
Mesozoic	Triassic		248
Paleozoic	Permian		290
Paleozoic	Carboniferous	Pennsylvanian	323
Paleozoic	Carboniferous	Mississippian	354
Paleozoic	Devonian		417
Paleozoic	Silurian		443
Paleozoic	Ordovician		490
Paleozoic	Cambrian		540
Precambrian			

The oldest known fossils of **hominids**, or humanlike primates, are between 5 million and 8 million years old. The earliest modern humans, *Homo sapiens*, evolved more than 130,000 years ago.

Hyracotherium (Eohippus), an early ancestor of the modern horse, lived 60 million years ago. The Cenozoic era, in which most modern mammals (including horses) evolved, is sometimes called the Age of Mammals.

The end of the Cretaceous period (about 65 million years ago) is defined by the **extinction** of nearly half of all animal and plant species on earth.

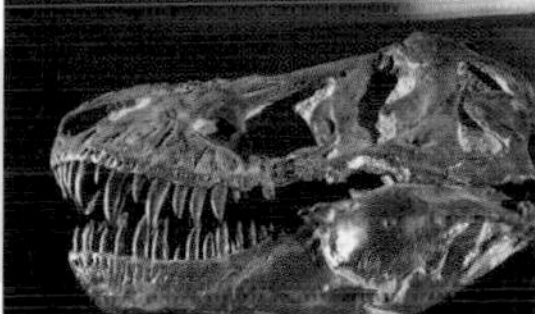

Tyrannosaurus rex lived during the Cretaceous period (between about 144 million and 65 million years ago). All dinosaurs, including *Tyrannosaurus rex*, lived during the Mesozoic era. For this reason, the Mesozoic era is sometimes called the Age of Reptiles.

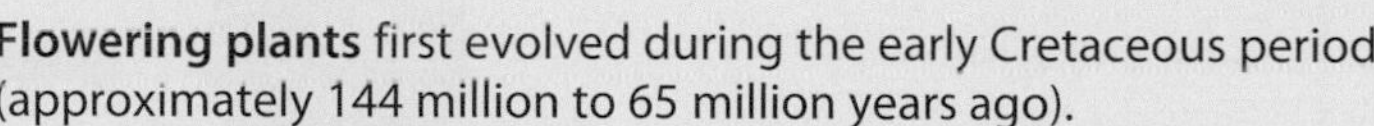

Flowering plants first evolved during the early Cretaceous period (approximately 144 million to 65 million years ago).

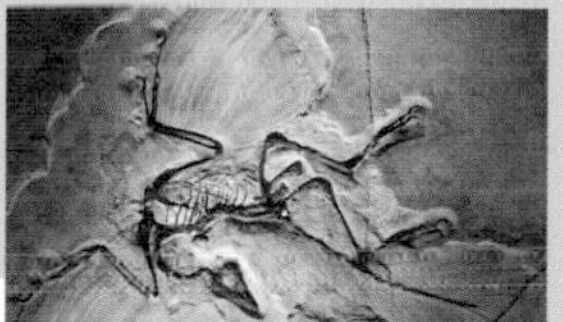

Archaeopteryx lived during the late Jurassic and early Cretaceous periods (about 206 million to 144 million years ago). It had characteristics of both modern birds and of dinosaurs, suggesting that modern birds evolved from dinosaurs.

The largest **mass extinction** event in earth's history occurred at the end of the Permian period (about 250 million years ago). About 90 percent of the marine organisms at the time went extinct.

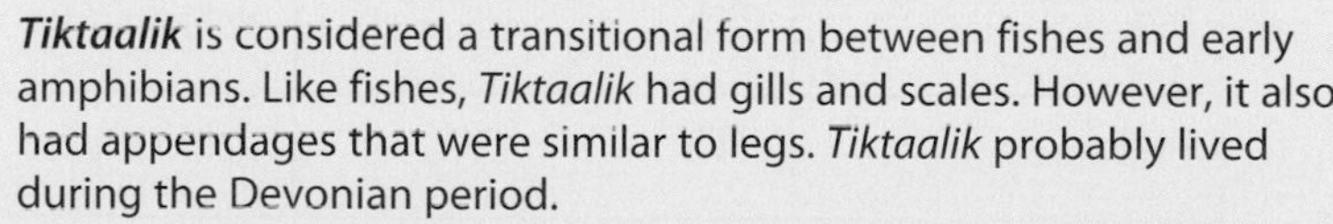

During the Carboniferous period (354 million to 290 million years ago), **huge swamps** filled with many kinds of plant life covered earth. When those plants died, they were buried in sediment and eventually became massive coal deposits.

Tiktaalik is considered a transitional form between fishes and early amphibians. Like fishes, *Tiktaalik* had gills and scales. However, it also had appendages that were similar to legs. *Tiktaalik* probably lived during the Devonian period.

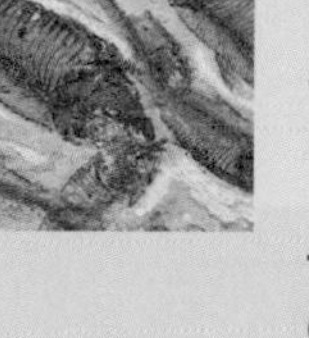

The first **fishes** evolved during the Ordovician and Silurian periods (between 490 million and 417 million years ago). Fishes dominated the oceans during the Devonian period (between 417 million and 354 million years ago), so this period is sometimes called the Age of Fishes.

Trilobites, an extinct class of arthropods, flourished during the Cambrian period (between 540 million and 490 million years ago) and died out at the end of the Permian period (between 290 million and 248 million years ago).

Some isotopes break down and form other isotopes through radioactive decay.

Parent isotopes break down, or decay, to form daughter isotopes. Parent isotopes decay in different ways, including alpha emission, beta emission, and electron capture.

Alpha emission occurs when a parent isotope breaks down into a daughter isotope and an alpha particle. An alpha particle is made of two protons and two neutrons. It is identical to the nucleus of a ^{4}He atom.

Beta emission occurs when a parent isotope breaks down into a daughter isotope and a beta particle. A beta particle is an electron.

During **electron capture**, an electron combines with a proton in the parent isotope to form a daughter isotope.

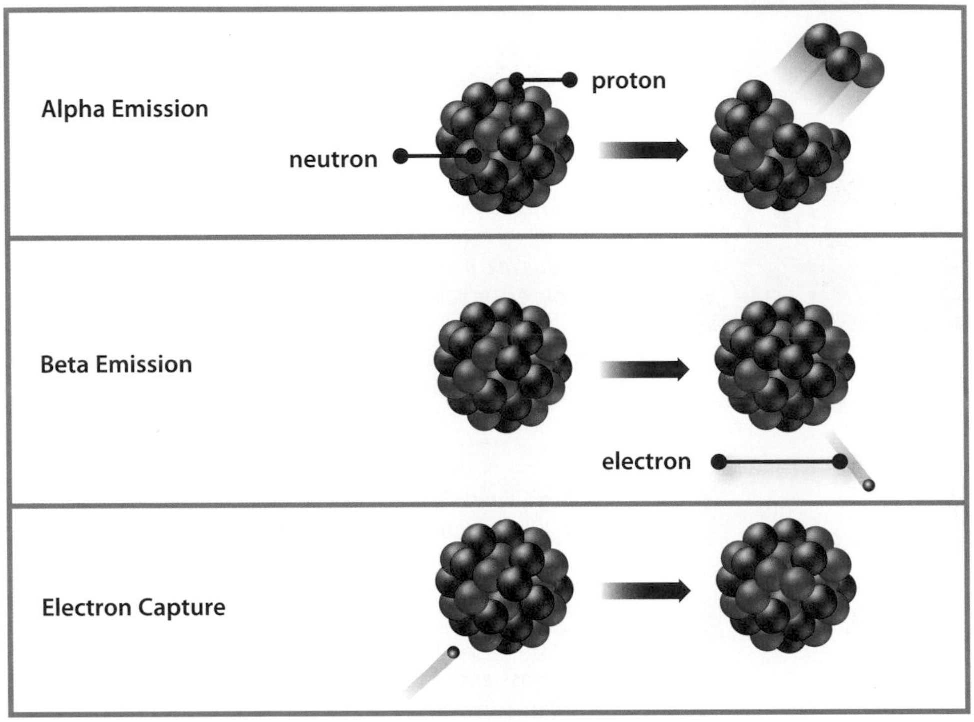

Alpha Emission

proton

neutron

Beta Emission

electron

Electron Capture

There are **many radioactive isotopes**, but only a few are commonly used by scientists to study earth materials. The isotopes in this table are some of the most commonly used isotopes.

Uranium-238 (^{238}U), uranium-235 (^{235}U), and thorium-232 (^{232}Th) all produce several daughter isotopes when they decay. These daughter isotopes are also radioactive and can decay. The lead isotopes listed in the table are the final, stable daughter isotopes that form during the decay of these parent isotopes. Potassium-40 (^{40}K) decays in two different ways. Each type of decay produces a different daughter isotope.

PARENT ISOTOPE	DAUGHTER ISOTOPE	TYPE OF DECAY
^{238}U	^{206}Pb	alpha emission
^{235}U	^{207}Pb	alpha emission
^{232}Th	^{208}Pb	alpha emission
^{147}Sm	^{143}Nd	alpha emission
^{87}Rb	^{87}Sr	beta emission
^{40}K	^{40}Ca	beta emission
^{40}K	^{40}Ar	electron capture
^{14}C	^{14}N	beta emission

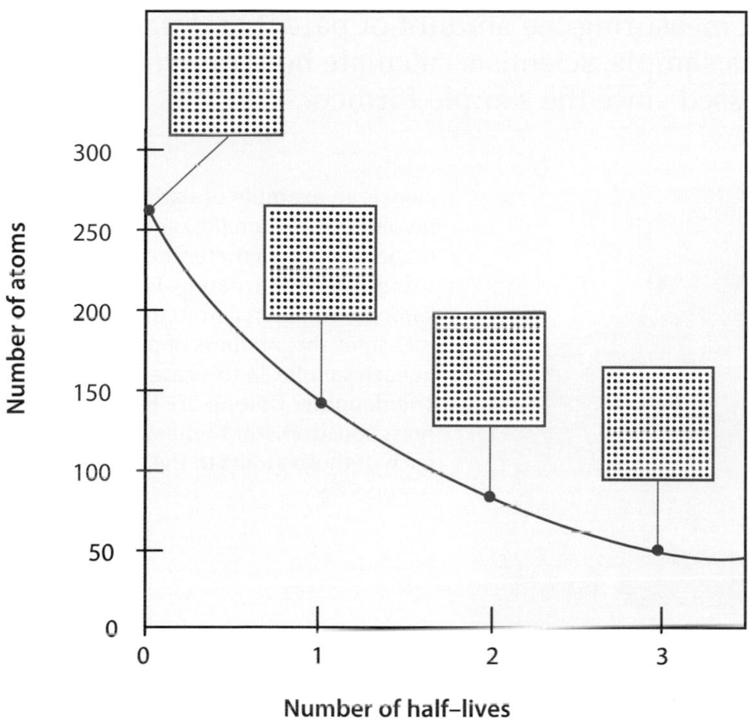

The amount of time it takes for half the atoms of a parent isotope in a sample to decay is the isotope's half–life.

For example, imagine a rock that contains 256 atoms of a parent isotope. The parent isotope has a half–life of 1,000 years. After one half–life (1,000 years), there will be 128 atoms of the parent isotope left in the rock. After two half–lives (2,000 years), there will be 64 atoms of the parent isotope left in the rock.

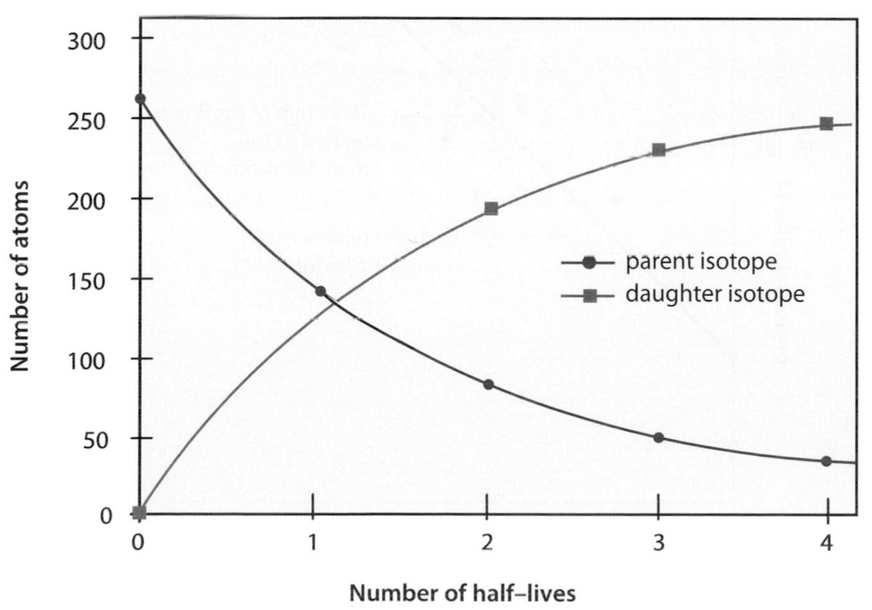

As the number of parent isotope atoms decreases, the number of daughter isotope atoms increases.

Scientists use radiometric dating to learn the ages of earth materials.

Radioactive isotopes decay at a specific known rate. By measuring the amount of parent and daughter isotopes in a sample, scientists calculate how much time has passed since the sample formed.

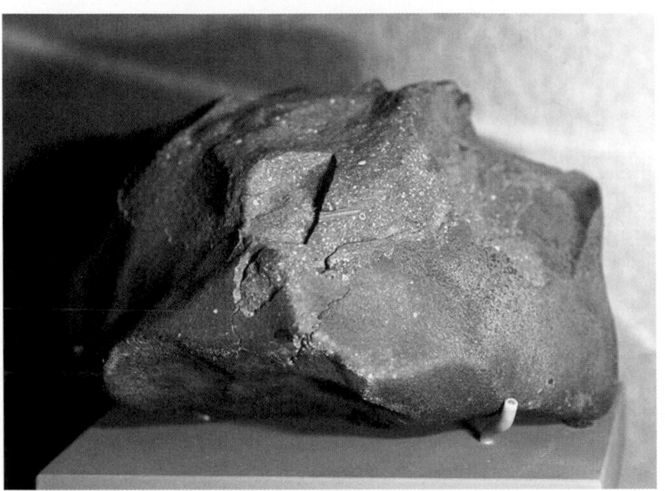

Here is an **example of dating a meteorite** using a system involving rubidium (Rb) and strontium (Sr). Imagine that a scientist finds a meteorite and wishes to learn its age using radiometric dating. First, the scientist takes several samples of material from the meteorite. The scientist measures the amounts of parent and daughter isotopes in each sample. In this case, ^{87}Rb is the parent isotope. The daughter isotope of ^{87}Rb is ^{87}Sr. The scientist uses a mass spectrometer to measure the ratio of the number of each of those atoms to the number of atoms of a stable isotope, ^{86}Sr.

Once the scientist has measured the $\frac{^{87}Rb}{^{87}Sr}$ and $\frac{^{86}Sr}{^{87}Sr}$ ratios in each sample, she plots them on a **graph**. Each sample is represented by one point on the graph. Then, the scientist calculates a straight line that best fits the points.

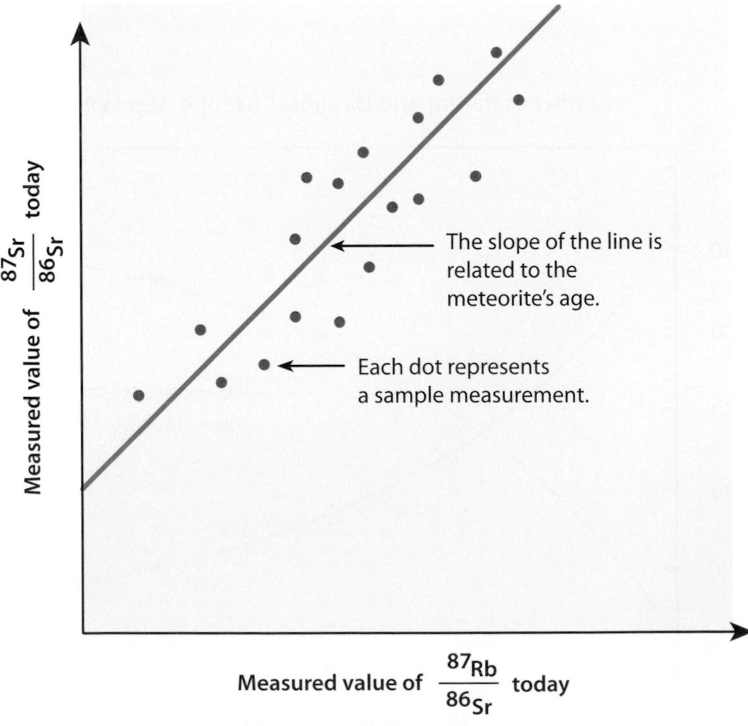

The **slope of the line** is related to the meteorite's age. The steeper the slope, the older the meteorite. If the scientist knows the equation for the line, she can calculate the age of the meteorite.

Scientists use many radioactive isotopes for radiometric dating. The half–life of each isotope determines the ages of rocks it can be used to date. Isotopes with larger half–lives can be used to date older rocks.

To determine which isotope to use to date a sample, a scientist must make an educated guess about how old the sample is. The scientist also determines which isotopes are likely to be present in the sample. Then, the scientist chooses an available isotope that can be used to date material of that age.

PARENT ISOTOPE	HALF–LIFE OF PARENT ISOTOPE	DAUGHTER ISOTOPE	AGE OF SAMPLES THAT CAN BE DATED USING THIS ISOTOPE
^{238}U	4.5 billion years	^{206}Pb	more than 10 million years old
^{235}U	704 million years	^{204}Pb	more than 10 million years old
^{232}Th	14.0 billion years	^{207}Pb	more than 10 million years old
^{87}Rb	48.8 billion years	^{87}Sr	more than 10 million years old
^{40}K	1.25 billion years	^{40}Ar	more than 50,000 years old
^{14}C	5,730 years	^{14}N	less than 70,000 years old

We study aspects of earth's history by examining the layers of rocks on earth's surface.

The Grand Canyon is a colorful, steep-sided gorge in northern Arizona. By studying the rocks in each layer of the Grand Canyon, scientists have been able to determine what the area was like hundreds of millions of years ago, when the rocks formed.

The **Grand Canyon** began to form about 65 million years ago, as the Colorado River cut through the rock in the Colorado Plateau. Some rocks in the Grand Canyon are more than 2 billion years old, but they have been exposed only in the last 65 million years as the river has cut through them.

The **Supai Group** consists of several bodies of rock that were deposited between the early Pennsylvanian period (about 315 million years ago) and the early Permian period (about 285 million years ago). These sediments were probably deposited in a delta environment, where a river ran into the sea.

The **Bright Angel Shale** is approximately 515 million years old. As sea level rose, the area that was a beach during the early Cambrian period became a nearshore marine environment. Fossils of early marine animals that lived in this environment can be found in the rocks of this formation. When the animals died, they were buried and preserved in the mud and silt that formed the shale.

The **Vishnu Schist** is made up of metamorphic rock. The schist formed about 1.7 billion years ago, during an episode of mountain building, when previously deposited marine sediments were exposed to extreme heat and pressure.

The **Zoroaster Granite** formed during a mountain-building episode about 1.7 billion years ago. The huge granite intrusions formed when magma formed and cooled deep underground.

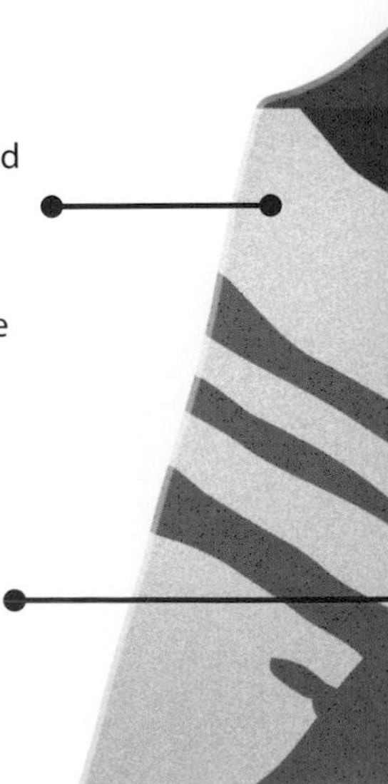

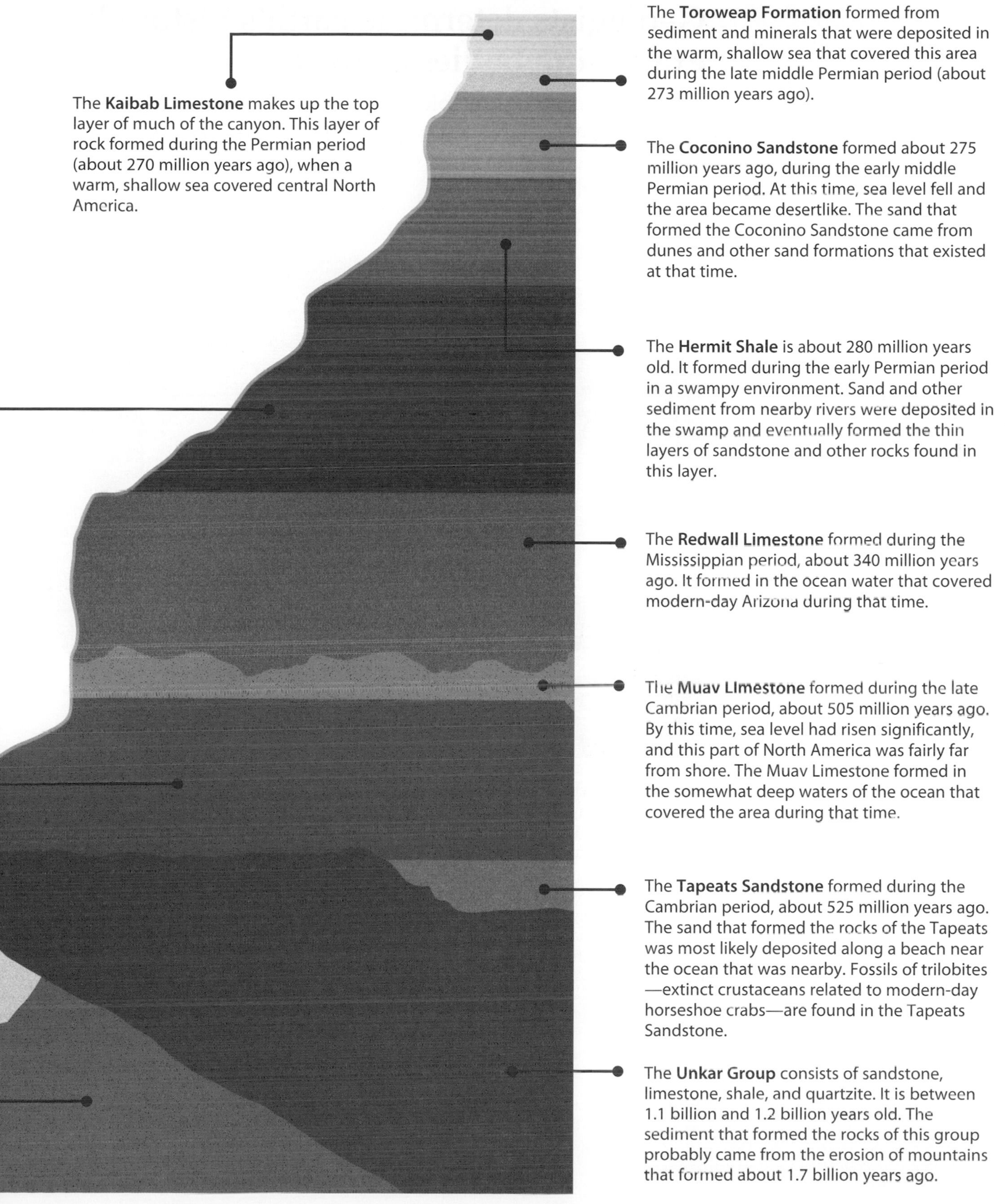

The **Kaibab Limestone** makes up the top layer of much of the canyon. This layer of rock formed during the Permian period (about 270 million years ago), when a warm, shallow sea covered central North America.

The **Toroweap Formation** formed from sediment and minerals that were deposited in the warm, shallow sea that covered this area during the late middle Permian period (about 273 million years ago).

The **Coconino Sandstone** formed about 275 million years ago, during the early middle Permian period. At this time, sea level fell and the area became desertlike. The sand that formed the Coconino Sandstone came from dunes and other sand formations that existed at that time.

The **Hermit Shale** is about 280 million years old. It formed during the early Permian period in a swampy environment. Sand and other sediment from nearby rivers were deposited in the swamp and eventually formed the thin layers of sandstone and other rocks found in this layer.

The **Redwall Limestone** formed during the Mississippian period, about 340 million years ago. It formed in the ocean water that covered modern-day Arizona during that time.

The **Muav Limestone** formed during the late Cambrian period, about 505 million years ago. By this time, sea level had risen significantly, and this part of North America was fairly far from shore. The Muav Limestone formed in the somewhat deep waters of the ocean that covered the area during that time.

The **Tapeats Sandstone** formed during the Cambrian period, about 525 million years ago. The sand that formed the rocks of the Tapeats was most likely deposited along a beach near the ocean that was nearby. Fossils of trilobites —extinct crustaceans related to modern-day horseshoe crabs—are found in the Tapeats Sandstone.

The **Unkar Group** consists of sandstone, limestone, shale, and quartzite. It is between 1.1 billion and 1.2 billion years old. The sediment that formed the rocks of this group probably came from the erosion of mountains that formed about 1.7 billion years ago.

Geologists determine earth's history by examining bodies of rock.

Geologists use the principle of superposition and the law of crosscutting relationships to reconstruct the history of a rock body.

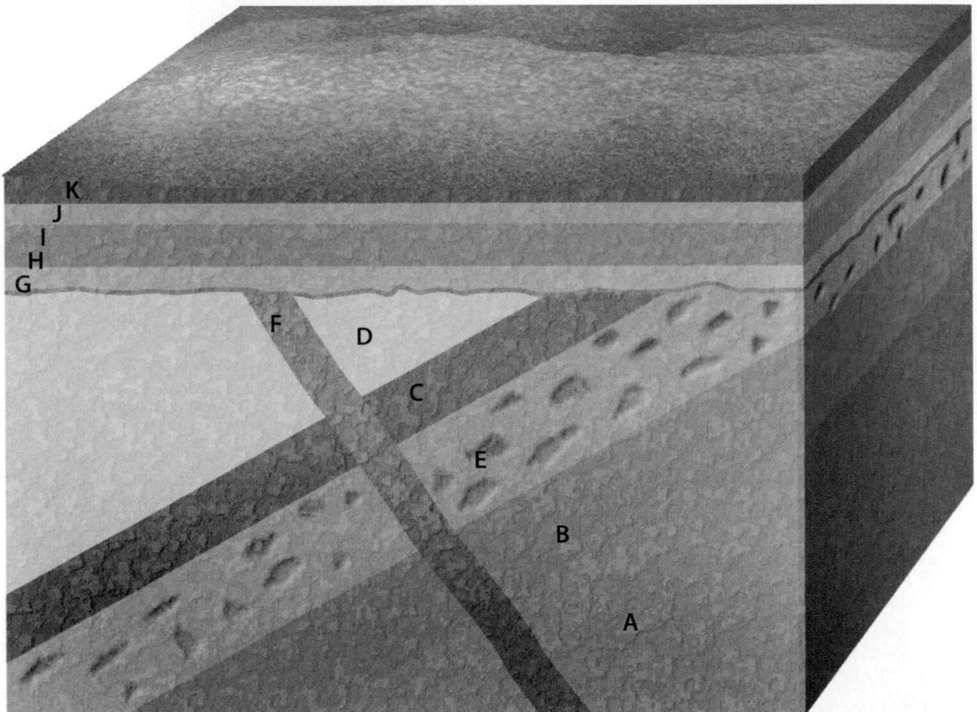

Rock bodies, such as the one shown here, can be very complicated. They may contain many different layers and various features that cut through those layers. By applying the law of crosscutting relationships and the principle of superposition, geologists can often determine how complicated rock bodies formed.

The principle of superposition states that the oldest rock layers in a body of rock are at the bottom, and the youngest layers are at the top. The principle of superposition applies only to rock bodies that are not folded or tilted. The law of crosscutting relationships states that a feature that cuts across rock layers, such as a fault, must be younger than the rock layers it cuts across.

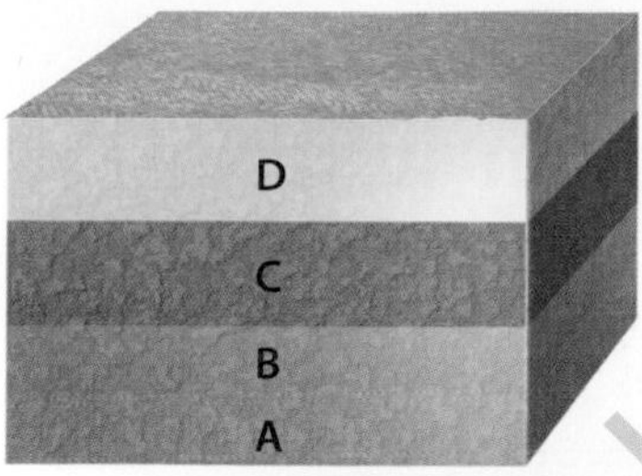

Rock layers A, B, C, and D must be the oldest rocks in the rock body, because all of the other features cut through them or lie above them. Layer A must be the oldest of the four rock layers. Therefore, the rock body must once have looked like this.

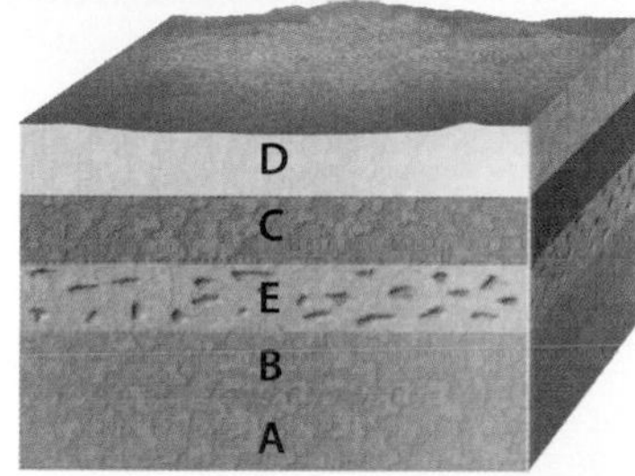

Layer E is a sill, a body of igneous rock that forms when magma flows between two rock layers. Layer E contains pieces of layers B and C. Therefore, layer E must be younger than layers B and C. It must have formed after those other layers.

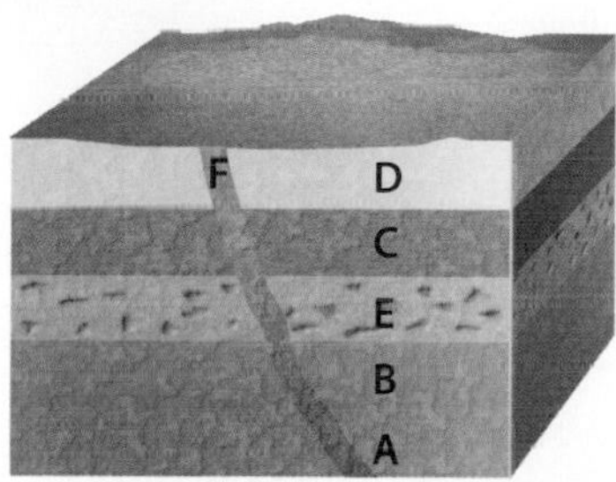

Layer F is a dike, a body of igneous rock that forms when magma cuts through cracks in rock layers. Layer F cuts through layers A, B, C, D, and E, so it must be younger than all those layers.

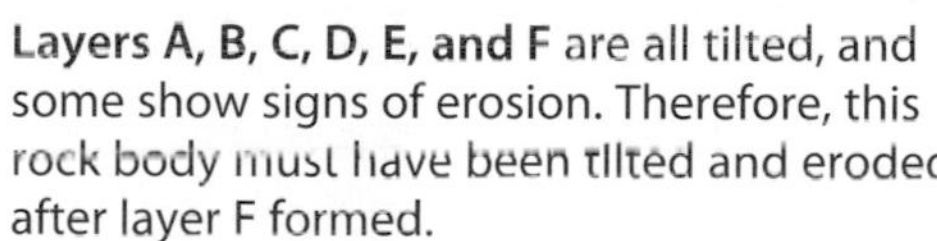
Layers A, B, C, D, E, and F are all tilted, and some show signs of erosion. Therefore, this rock body must have been tilted and eroded after layer F formed.

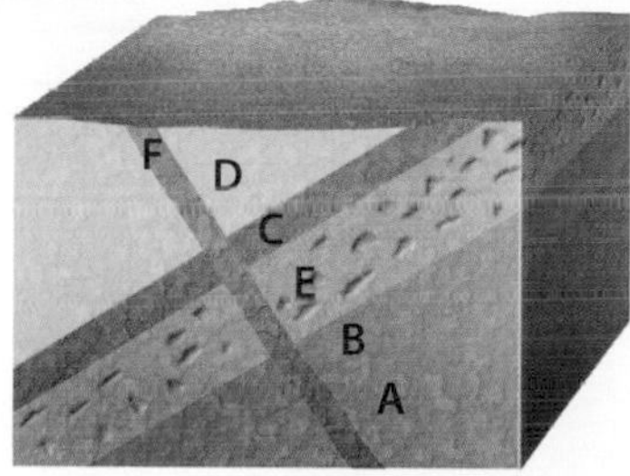

Layers G, H, I, J, and K are all horizontal, and are located above layers A, B, C, D, E, and F. Therefore, the uppermost layers in the rock body must have formed after the tilting and erosion of the lower layers.

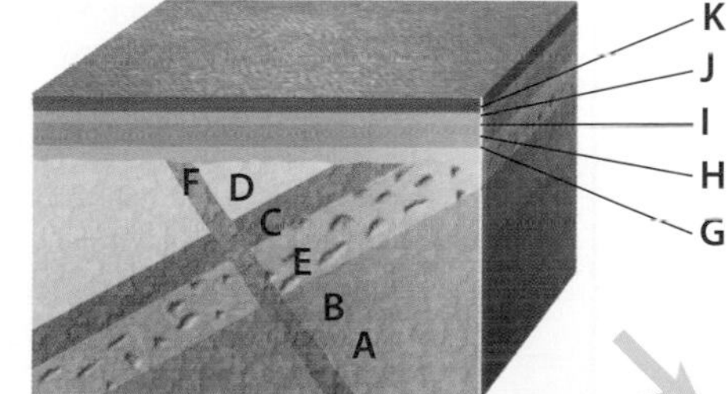

Layer K is the youngest layer in this rock body. It is uneven on top, which indicates that it has been eroded.

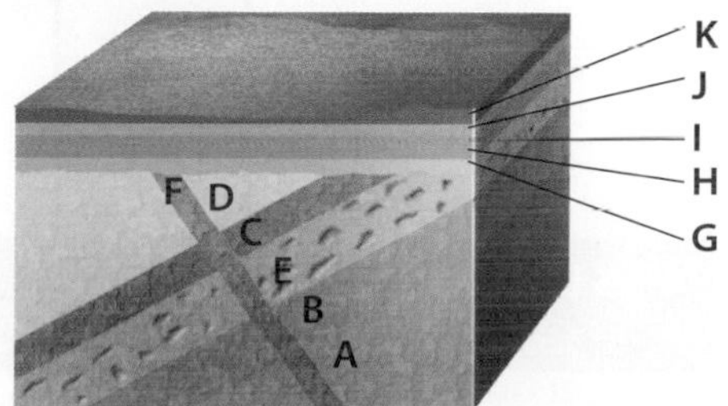

The surface and atmosphere of early earth were very different from what they are like today.

Over time, geologic and biologic processes have changed earth's atmosphere into its current form. Scientists study the earth's fossil record and other bodies in the solar system to estimate when changes occurred in earth's past.

Approximately 4.55 billion years ago, **earth began to form** from a cloud of dust and rocky materials orbiting the early sun. When the planet first took shape, it was molten. Dense materials, such as iron and nickel, sank to the center of the molten body to form the early core.

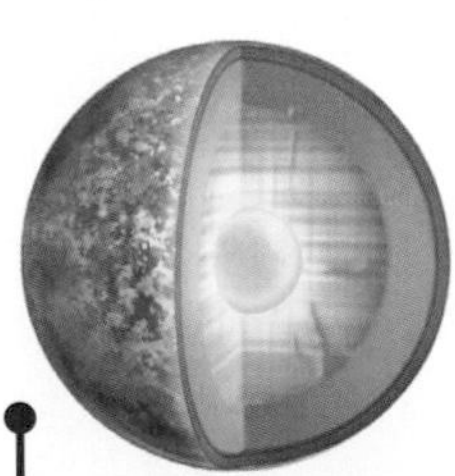

Scientists think the **moon formed** sometime between 500 million and 1 billion years after earth formed. The most well-supported theory explaining the formation of the moon states that a large object (roughly the size of Mars) collided with early earth. The collision produced a huge mass of material that was ejected from earth. Eventually, this material was drawn together by gravity to form the moon.

Earth's **earliest atmosphere** was composed mainly of hydrogen and helium. These gases have very low masses. As a result, earth's gravitational pull was not strong enough to keep them bound to the planet. The gases were blown away by the solar wind.

Like volcanoes today, **volcanoes** on early earth released many different gases into the atmosphere. These gases included carbon dioxide, water vapor, methane, and ammonia. They began to form earth's second atmosphere.

As earth cooled, water vapor in the atmosphere began to condense. Liquid water fell as rain on earth's surface. The impacts of comets brought more **water to earth**. The first oceans began to form.

4.55 billion years ago

The oldest known fossils are about 3.5 billion years old, suggesting that **life first evolved** on earth less than one billion years after earth formed. The earliest forms of life were microscopic single-cell organisms similar to some modern bacteria. Earth's atmosphere at this time contained almost no oxygen, so these organisms probably did not require oxygen to survive.

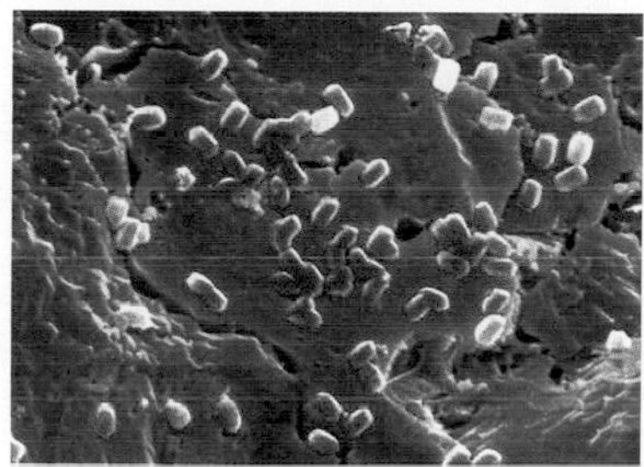

At first, the oxygen gas produced by photosynthesis reacted with the large amount of iron that was dissolved in earth's oceans. These reactions produced magnetite and hematite, which precipitated from the oceans to produce **banded iron formations**.

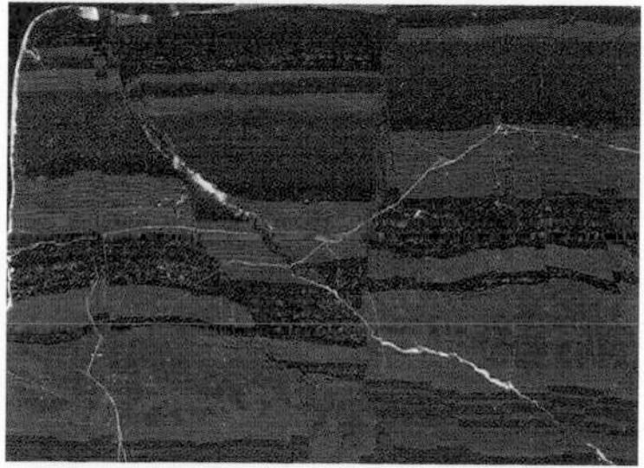

More than 2.3 billion years ago, **photosynthetic organisms** (such as algae and cyanobacteria) began to produce large amounts of oxygen gas.

By about 2 billion years ago, **oxygen** had begun to build up in earth's atmosphere. The ozone layer began to form as the amount of oxygen in the atmosphere increased. Today, earth's atmosphere is about 21 percent oxygen gas, but it took billions of years for so much oxygen to accumulate in the atmosphere.

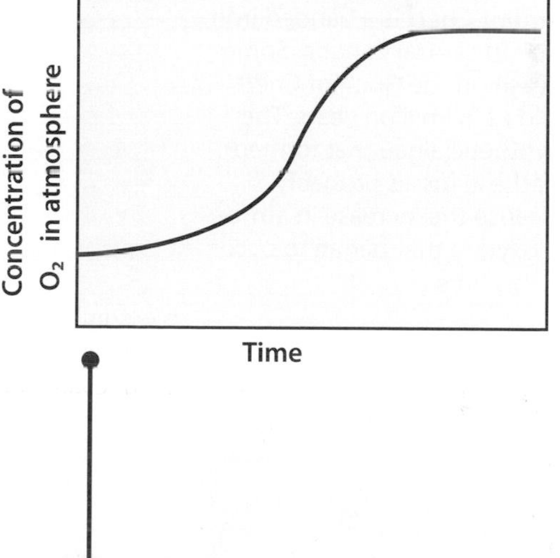

2 billion years ago

Fossils show that the earliest forms of life on earth existed 3.5 billion years ago, less than a billion years after earth formed.

Scientists have found fossils of early organisms in rocks that are more than 3.5 billion years old.

Today, these fossils are found in Precambrian shield rocks in the central portions of many continents. Precambrian shield rocks are the oldest rocks on earth's surface.

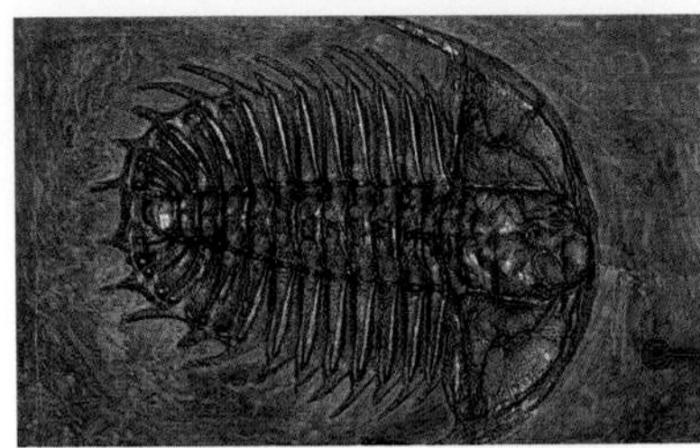

The **Burgess Shale formation** in Canada contains fossils that are more than 500 million years old.

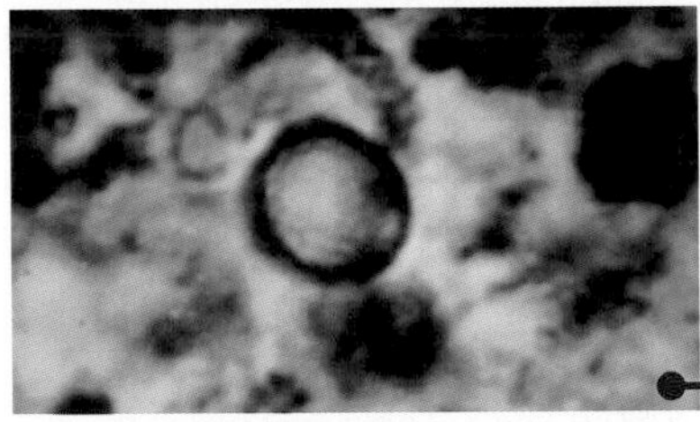

The **Gunflint Chert** is a banded iron formation in central Canada. Some of the fossils in the Gunflint Chert are as old as 2.3 billion years. The photosynthetic algae that formed some of these fossils probably helped cause the increase in atmospheric oxygen that began to occur around this time.

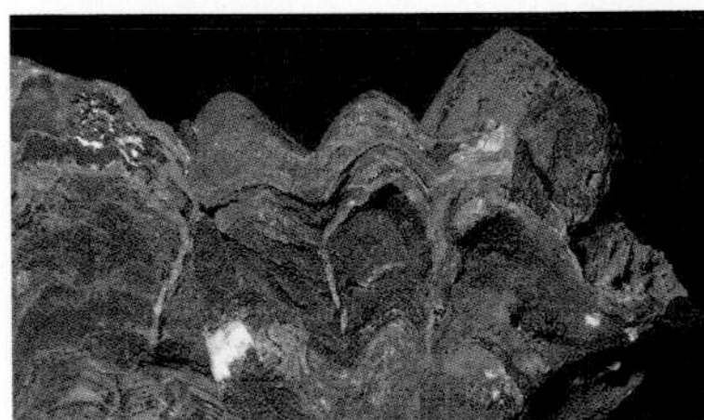

Fossilized **stromatolites**—mats of algae and bacteria layered with sediment—have been found in many parts of the world, including the central United States. Some of these stromatolites are more than 2.5 billion years old.

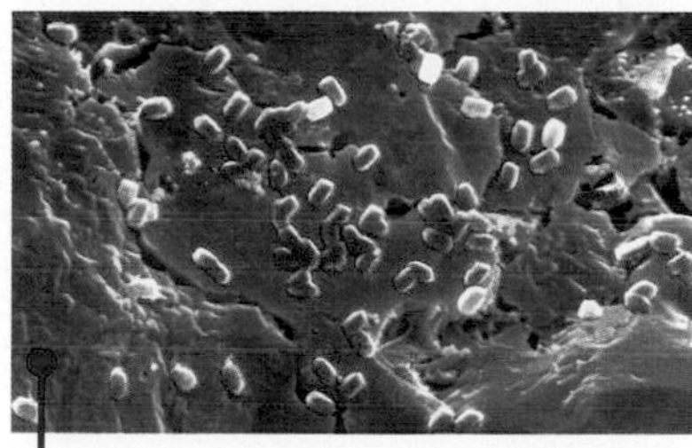

Scientists have found ancient **microfossils** in some geologic formations in South Africa. The tiny organisms that formed these fossils may have lived as long ago as 3.8 billion years.

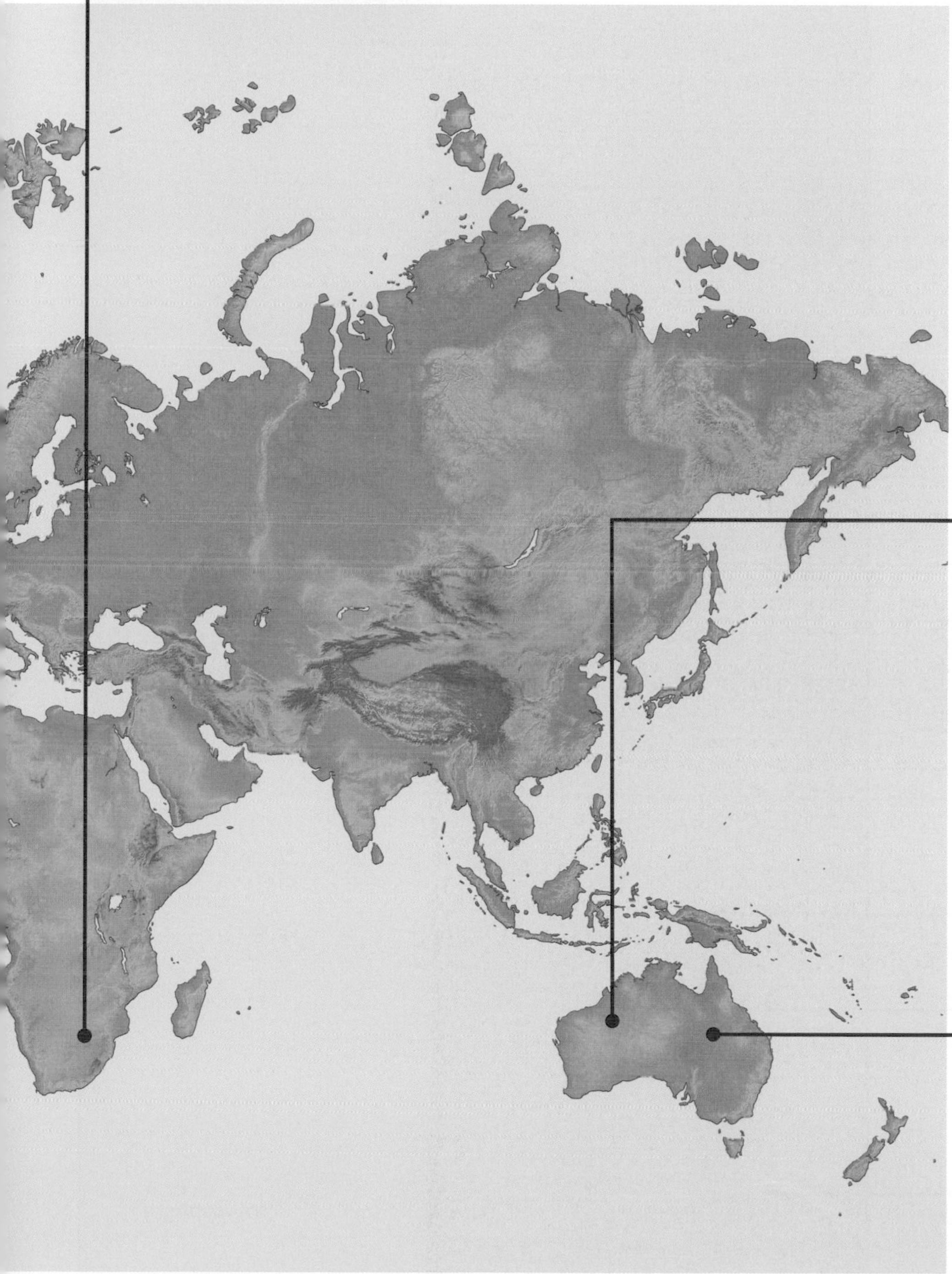

Dating back to the Proterozoic era (more than 850 million years ago), the **Bitter Springs formation** in central Australia contains fossils of many early life-forms. Although most of them are tiny, the fossils from the Bitter Springs formation are very well preserved.

The **Ediacaran fossils** of Canada, Russia, and Australia date back more than 550 million years. The organisms that formed these fossils were probably some of the first marine animals.

Our atmosphere is composed of many layers of gases held to earth by gravity.

The temperature, pressure, and composition of our atmosphere vary with altitude. The atmosphere protects life on earth by absorbing ultraviolet solar radiation and reducing temperature extremes between day and night.

The **ionosphere** is a layer of the atmosphere that contains charged particles, or ions. It includes the thermosphere and part of the mesosphere.

The **thermosphere** is the outermost layer of earth's atmosphere.

The **mesosphere** is the atmospheric layer located between the stratosphere and the thermosphere.

The **ozone layer** prevents ultraviolet light from hitting earth's surface. Ozone is a molecule with 3 oxygen atoms.

The **stratosphere** is located below the mesosphere. The ozone layer is part of the stratosphere.

The **troposphere** is the layer of earth's atmosphere closest to the surface. Nearly all weather occurs in the troposphere.

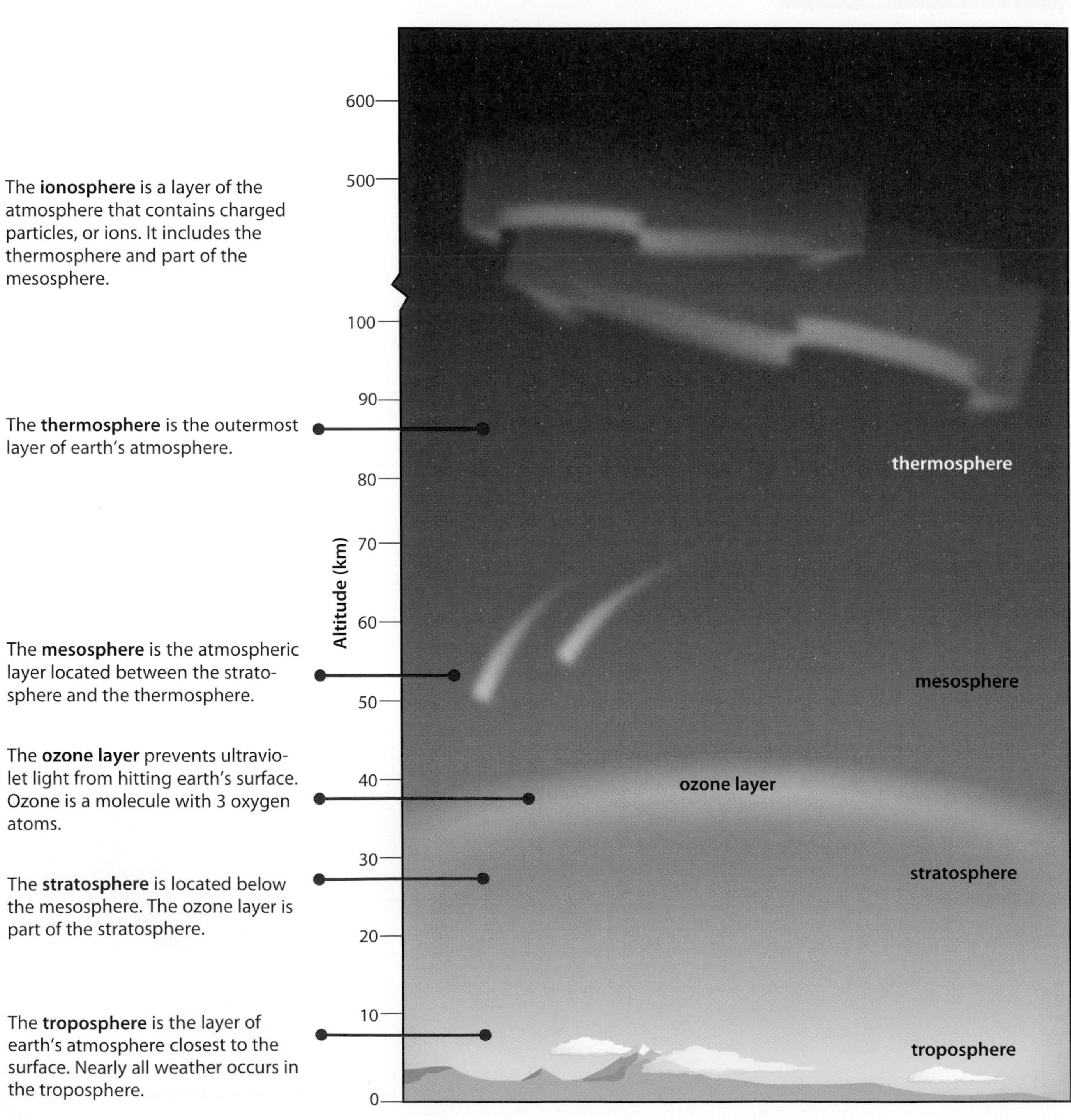

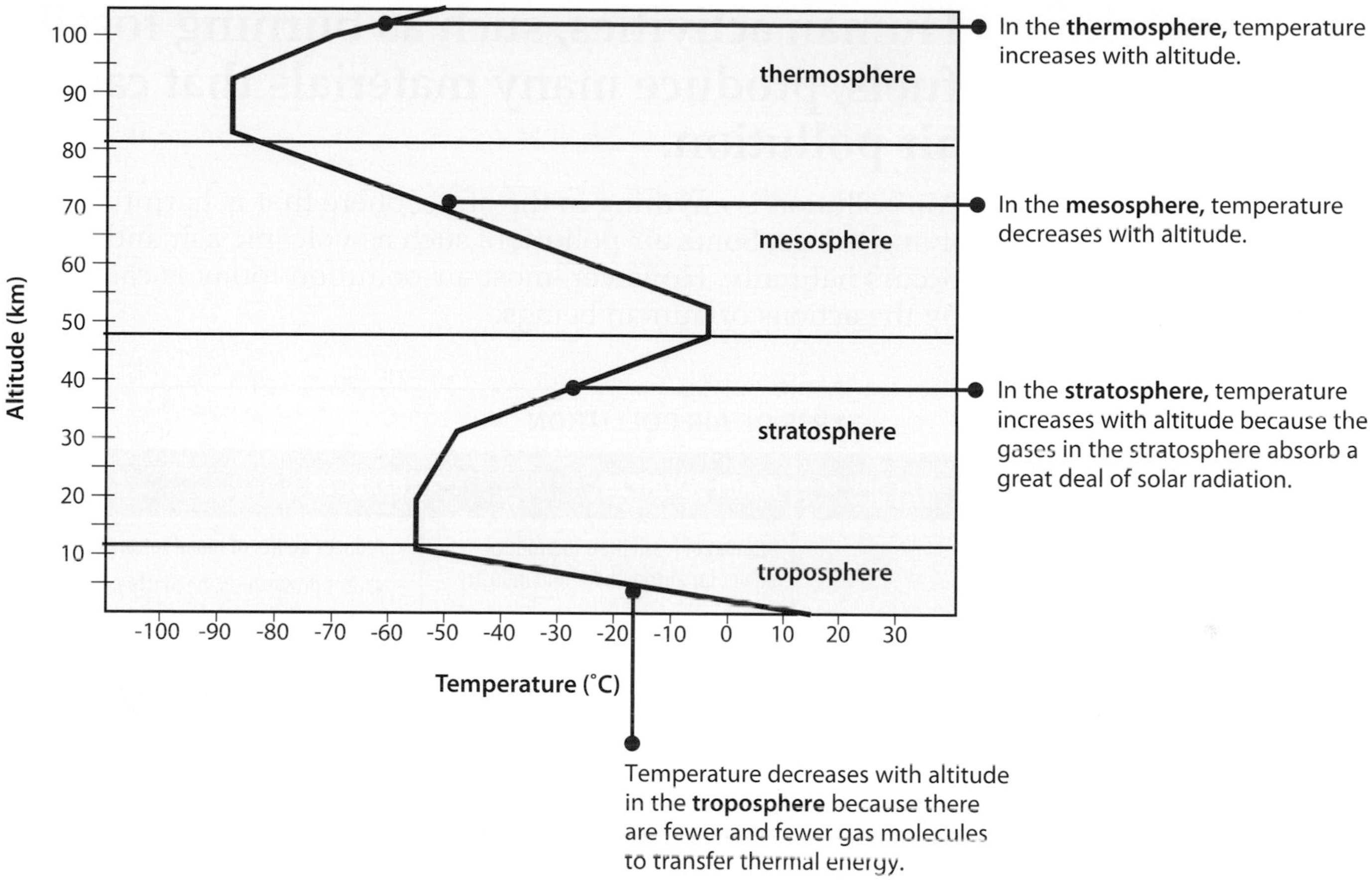

thermosphere

mesosphere

stratosphere

troposphere

Altitude (km)

100 90 80 70 60 50 40 30 20 10

-100 -90 -80 -70 -60 -50 -40 -30 -20 -10 0 10 20 30

Temperature (°C)

In the **thermosphere,** temperature increases with altitude.

In the **mesosphere,** temperature decreases with altitude.

In the **stratosphere,** temperature increases with altitude because the gases in the stratosphere absorb a great deal of solar radiation.

Temperature decreases with altitude in the **troposphere** because there are fewer and fewer gas molecules to transfer thermal energy.

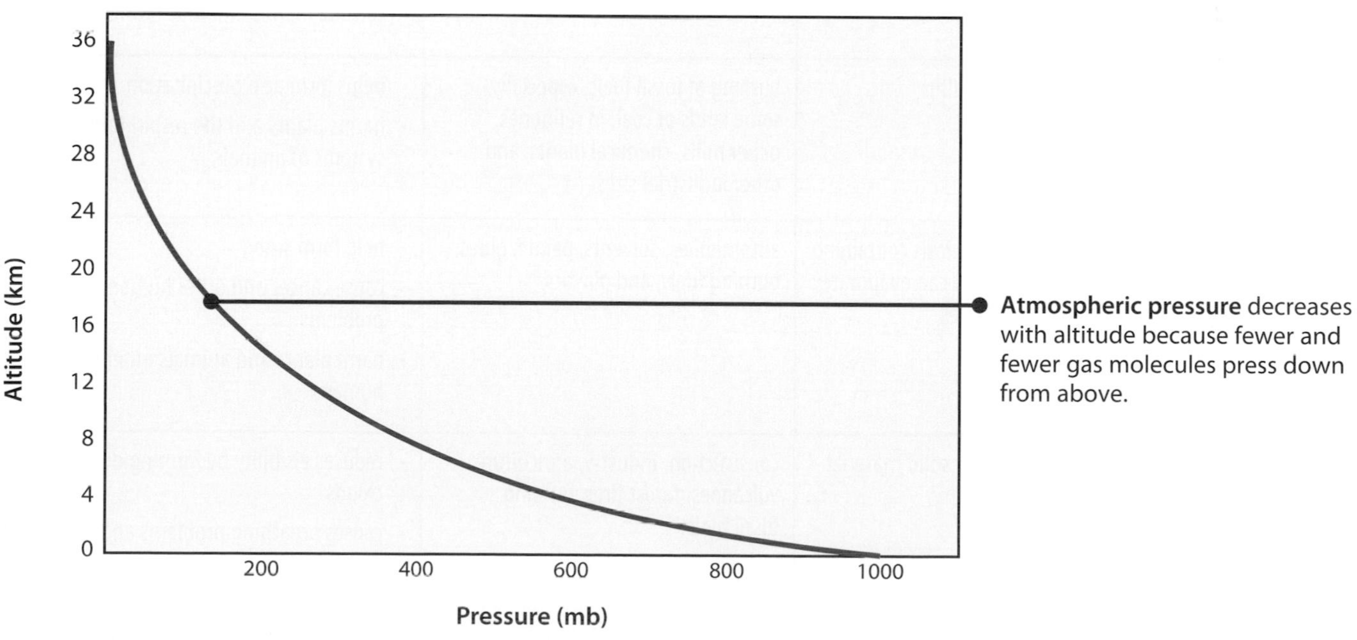

Atmospheric pressure decreases with altitude because fewer and fewer gas molecules press down from above.

Human activities, such as burning fossil fuels, produce many materials that cause air pollution.

Air pollution is anything in the atmosphere that is harmful to living things. Some air pollution, such as volcanic ash and gas, occurs naturally. However, most air pollution today is caused by the actions of human beings.

TYPES OF AIR POLLUTION

POLLUTANT	DESCRIPTION	MAIN SOURCES	EFFECTS
carbon monoxide (CO)	colorless, odorless gas	burning of fossil fuels to produce power for automobiles and industry	• reduces ability of blood to carry oxygen • causes drowsiness, headaches, heart stress, and slowed responses in low concentrations • causes unconsciousness and death in higher concentrations • harms developing fetuses
nitrogen oxides (NO_x)	compounds of nitrogen and oxygen (The *x* in the chemical formula indicates that different numbers of oxygen atoms can combine with nitrogen in these compounds.)	burning of fossil fuels in automobiles, power plants, and industry	• reduce resistance to lung infections, lung disease, and possibly cancer • contribute to smog and acid precipitation • corrode metals and fade fabrics
sulfur dioxide (SO_2)	gaseous compound of sulfur and oxygen	burning of fossil fuels, especially some kinds of coal, in refineries, paper mills, chemical plants, and other industrial sites	• helps form acid precipitation • harms plants and the respiratory systems of animals
volatile organic compounds	organic chemicals (chemicals containing carbon compounds) that can evaporate to produce harmful fumes	automobiles, solvents, paints, glues, burning fuels, and plastics	• help form smog • cause cancer and other human health problems • harm plants and animals other than humans
particulate matter	tiny particles of liquid or solid material	construction, industry, agriculture, volcanoes, forest fires, burning fossil fuels	• reduces visibility by forming dense clouds • causes breathing problems and lung disease • may cause cancers • corrodes metals, wears away building materials, and damages fabrics

A *temperature inversion* occurs when a body of warm air is located above a body of cool air. Because warm air is less dense than cool air, temperature inversions are very stable. The air does not tend to mix much. Therefore, temperature inversions can act as traps, keeping pollution near the ground.

Surface Inversion

One kind of temperature inversion, a **surface inversion,** forms when a cold land surface absorbs heat from the air near the ground. Air higher up remains warm. Surface inversions commonly form at night, when the air cools and the ground releases heat it absorbed during the day. Surface inversions can also form when cold air from high elevations flows downhill near the ground and collects in valleys.

Inversion Aloft

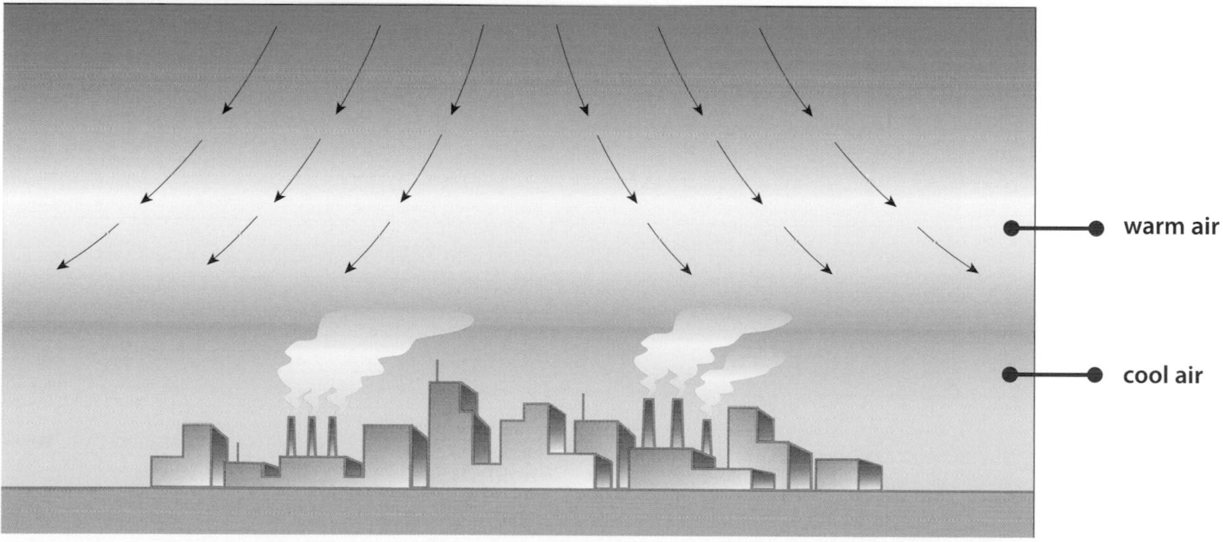

Another kind of temperature inversion, an **inversion aloft,** can form in areas of high pressure. The high-pressure areas cause air to sink. As the air sinks, its temperature rises because the pressure on it increases. The warm air can form a layer above the cooler air near the surface.

Most of the energy on earth's surface comes from the sun.

Sunlight contains primarily visible light, but it also contains other forms of electromagnetic radiation. Sunlight can interact with earth's atmosphere and surface in many different ways.

The **sun** makes up 99 percent of the mass in our solar system and provides most of the energy that strikes earth.

Earth's **atmosphere and clouds** absorb about 20 percent of the solar energy that strikes them. They eventally give off much of this energy as longer-wavelength radiation.

The **longer-wavelength energy** that is released by the atmosphere and by the clouds acts to warm earth. This radiation of heat helps keep the earth's temperatures moderate and stable.

The **atmosphere reflects** about 5 percent of the solar energy that strikes it. Clouds reflect an additional 20 percent of the solar radiation that enters earth's atmosphere. This reflected energy, along with energy reflected from the land and sea, travels back into space and has little effect on earth.

The **atmosphere absorbs** heat from land and water and helps keep earth's atmospheric temperatures stable.

About 50 percent of the solar energy that strikes earth is **absorbed by land and water**. Those materials can then reemit much of the energy as heat. In addition, about 5 percent of the solar radiation that strikes the land and sea is reflected back into space.

Different gases in a planet's atmosphere can absorb different amounts of sunlight.

Gases that readily absorb sunlight, such as carbon dioxide and methane, are called *greenhouse gases*. These gases reradiate the absorbed sunlight as heat. A planet with more greenhouse gases in its atmosphere will tend to have a higher surface temperature.

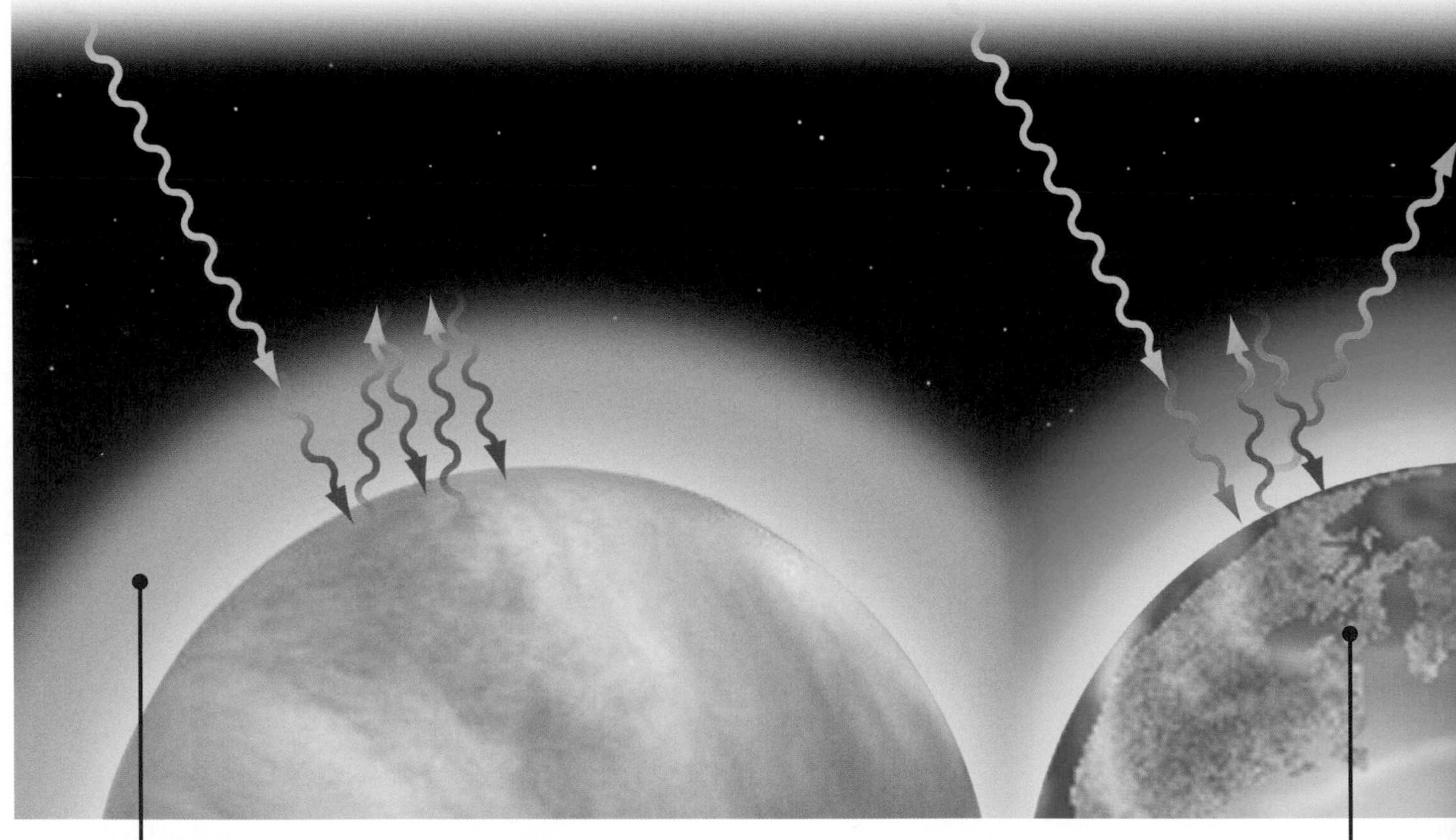

Venus's atmosphere is composed mainly of carbon dioxide and nitrogen. Its atmosphere is very thick, so the air pressure on the surface of Venus is 90 times that on Earth's surface. The composition and thickness of Venus's atmosphere lead to a pronounced greenhouse effect. As a result, the average temperature on Venus is about 400°C.

On **Earth**, the average temperature is about 14°C, and the surface pressure is 100 kPa.

kPa = kilopascals, a unit of pressure

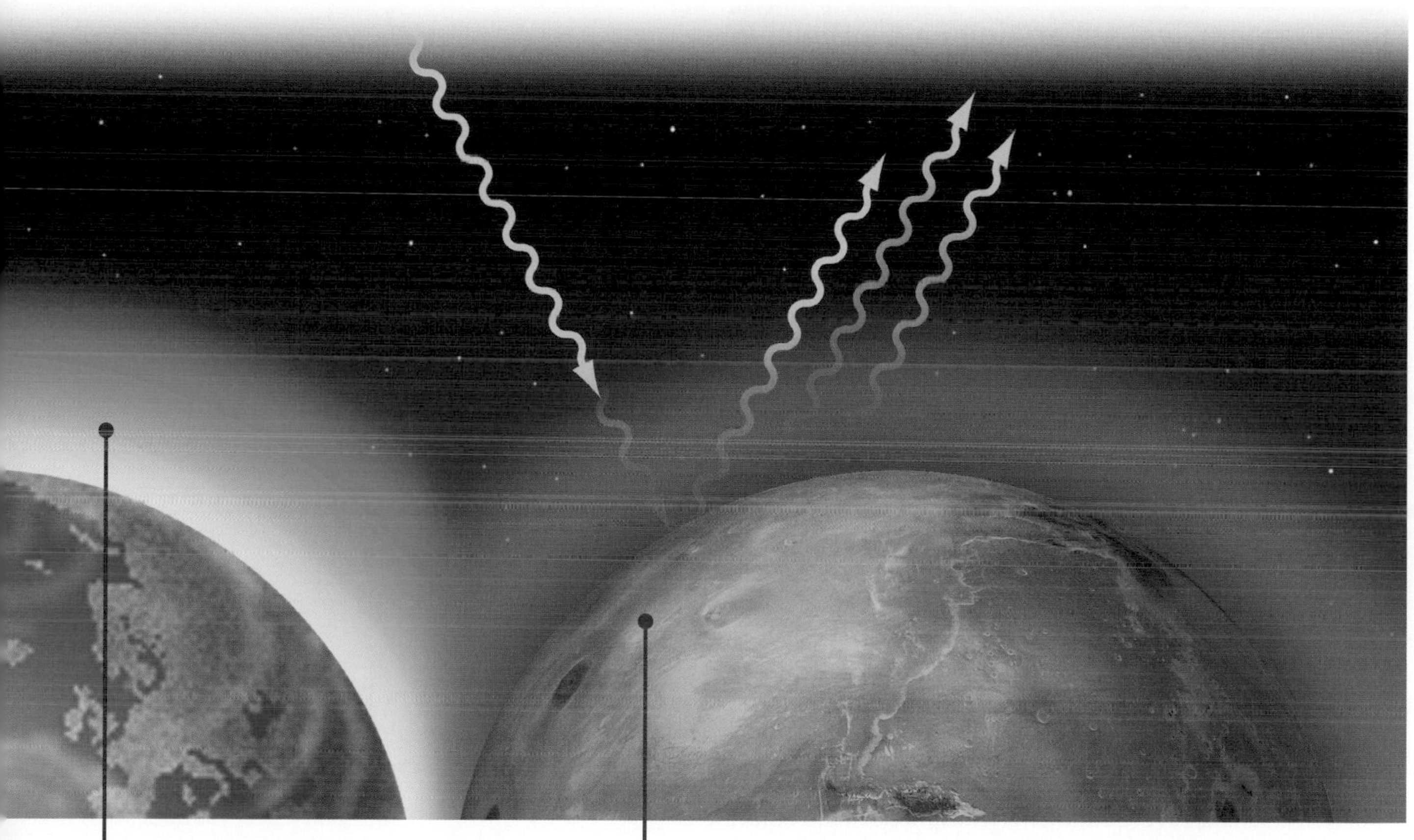

Earth's atmosphere contains mainly oxygen and nitrogen. It contains only a few hundredths of 1 percent of greenhouse gases.

The average temperature on **Mars** is about –63°C, and its atmosphere is very thin. The average surface pressure on Mars is less than 1 percent of the surface pressure on Earth. Very little radiant energy bounces back to the surface of Mars.

Air temperatures at earth's surface fluctuate throughout the year because of the way sunlight hits the planet's surface.

The tilt of earth's axis of rotation causes the Northern and Southern hemispheres to receive different amounts of sunlight at different times of the year. The amounts of sunlight a region receives has a significant effect on air temperatures.

January —Summer in the Southern Hemisphere

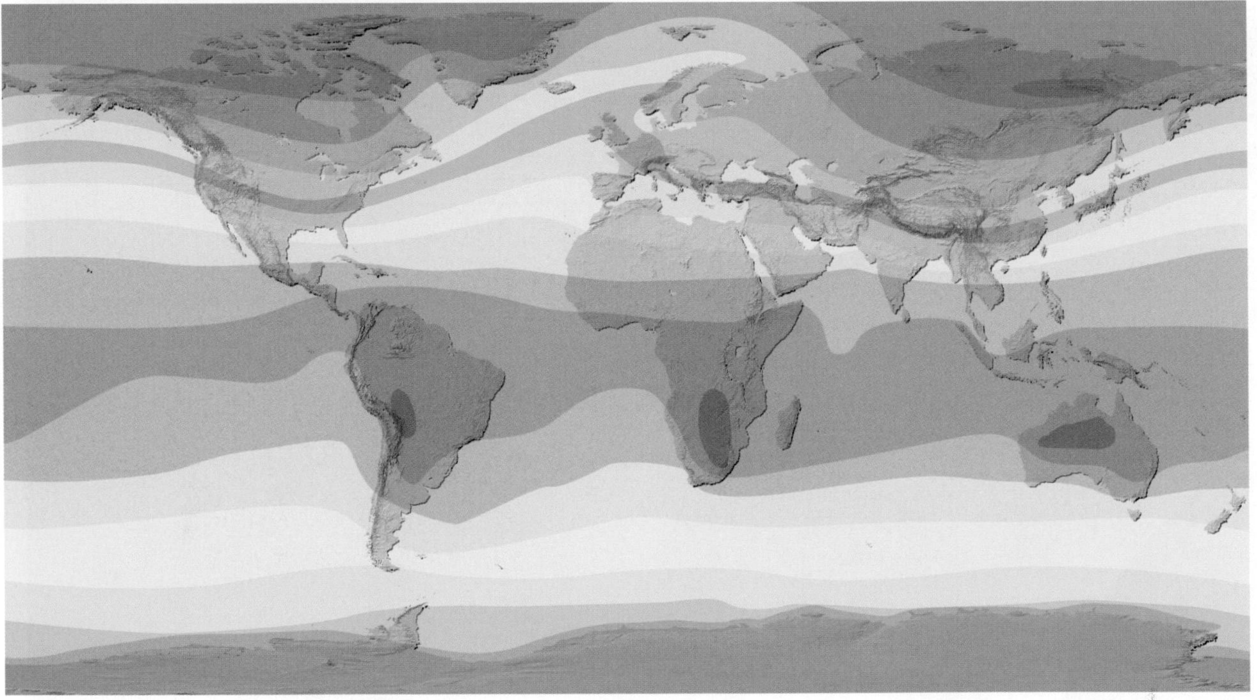

In January, it is summer in the **Southern Hemisphere**. At this time, the Southern Hemisphere is tilted toward the sun, so it receives more direct solar energy, and air temperatures are higher than in the winter.

Sea-Level Temperature	
< −40°C	5°C to 10°C
−40°C to −30°C	10°C to 15°C
−30°C to −20°C	15°C to 20°C
−20°C to −10°C	20°C to 25°C
−10°C to 0°C	25°C to 30°C
0°C to 5°C	> 30°C

July—Summer in the Northern Hemisphere

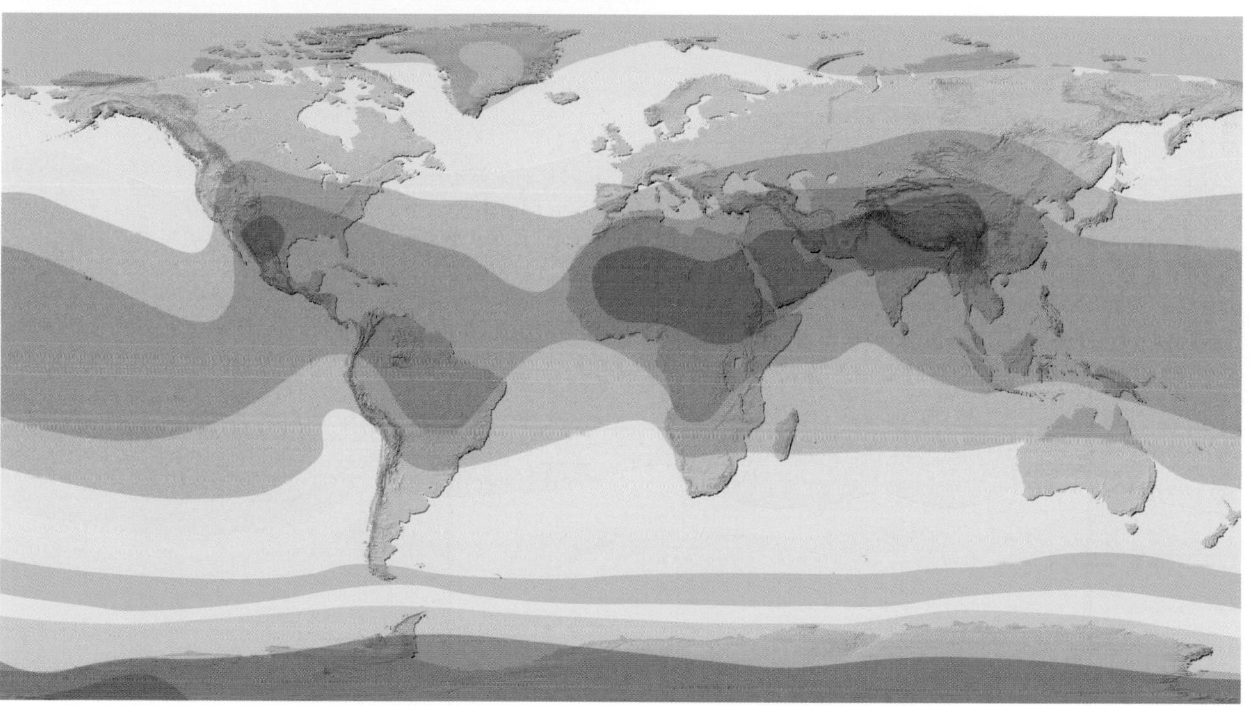

July is the middle of summer in the **Northern Hemisphere**. At this time, the Northern Hemisphere is tilted toward the sun and experiences warmer air temperatures.

Wind patterns result from the interaction of the sun's energy and earth's rotation.

Sunlight warms air near the equator, causing it to become less dense and then rise. As the air rises, an area of low pressure is produced. Cooler, denser air flows from areas of high pressure into areas of low pressure producing wind.

January

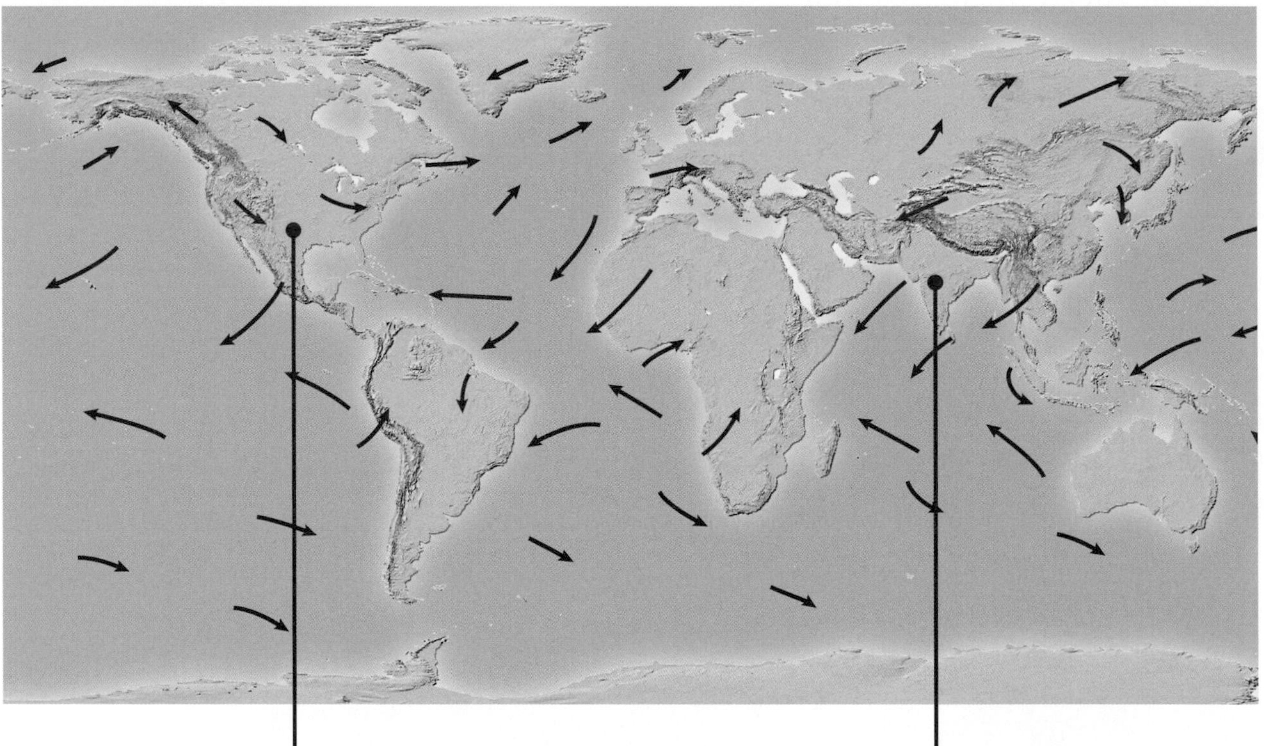

In January, most of **North America** experiences cold and moist northwesterly winds. Those areas experience colder temperatures because of the cooler air.

India's winters are dry and cool. *Monsoons* are winds that blow from the northeast, carrying cool, dry air from the Himalayas into the lowlands.

July

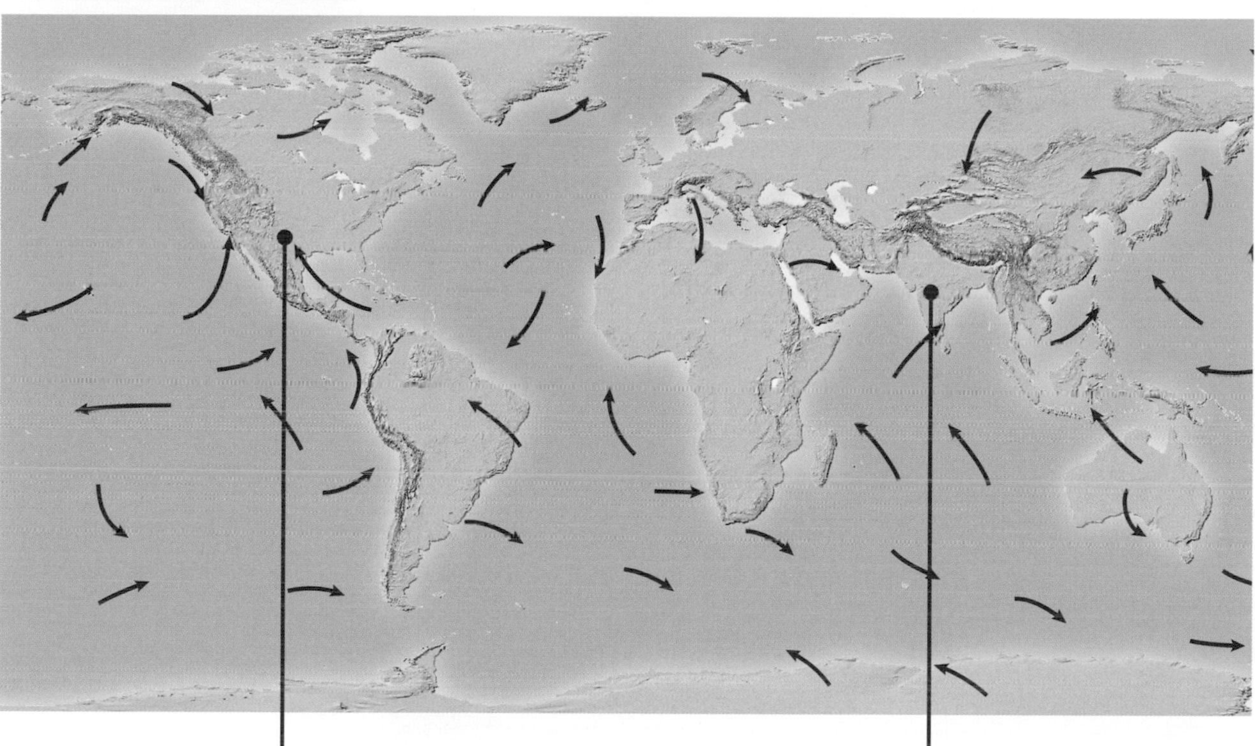

In July, eastern and central **North America** experience westerly winds that are dry and warm. Those winds can produce hot weather in the summer.

India's summers are warm and wet. Monsoons blow from the southwest, carrying warm, moist air from the ocean over the continent. This produces heavy rains.

Earth's rotation on its axis affects the movements of air in the atmosphere.

The rotation causes air in the atmosphere to travel along curved paths. This is called the *Coriolis effect*.

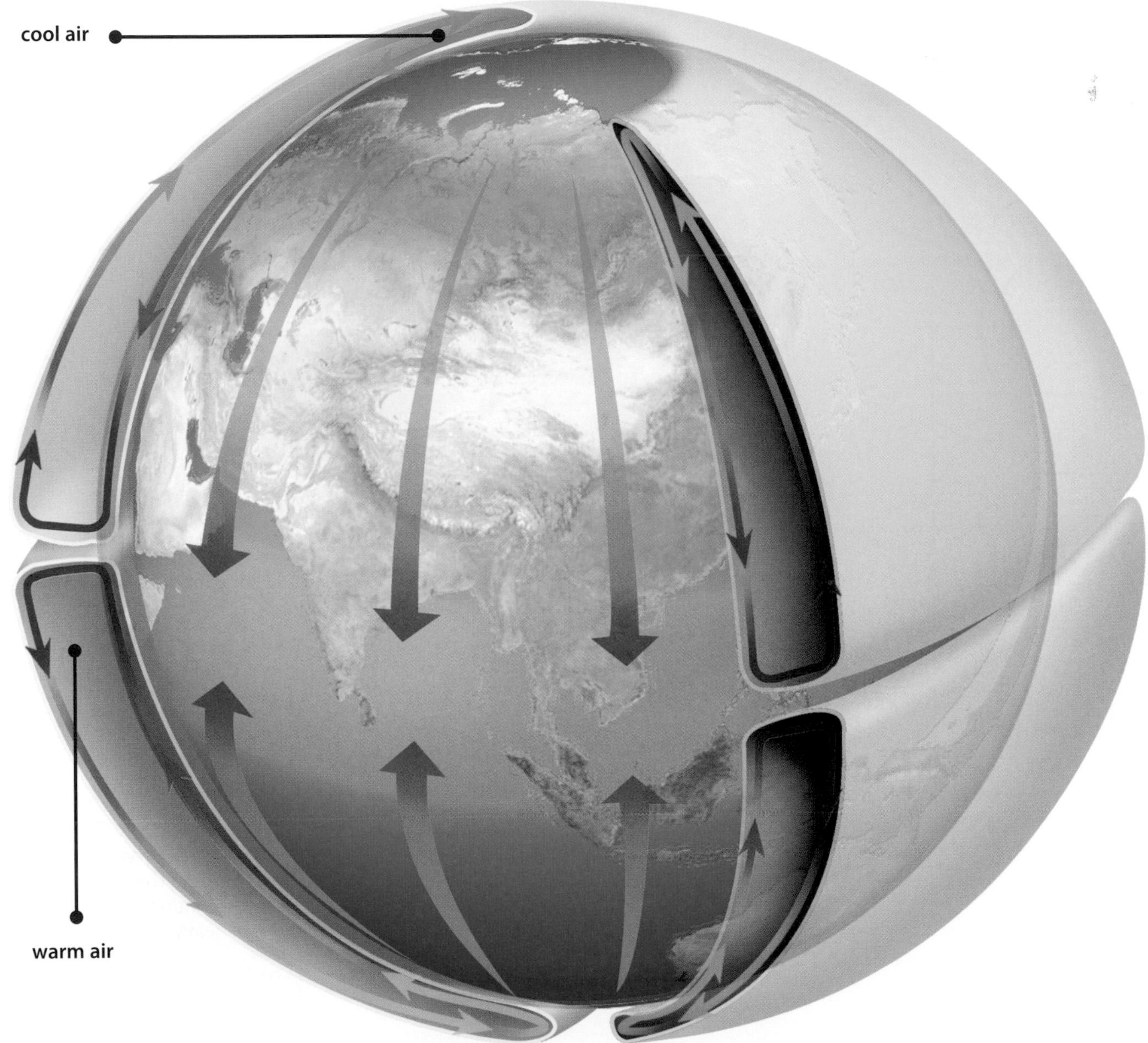

If earth did not rotate, air in the atmosphere would move along straight paths. Warm air at the equator would rise and flow toward the poles, while colder polar air would sink and flow toward the equator.

The differential heating of earth's atmosphere causes convection cells to form. Air near the equator receives direct sunlight, heats up, and becomes less dense. The less-dense air rises, producing an area of lower pressure near earth's surface. Colder, denser air from near the poles flows into the low-pressure area. The movement of the colder air away from the poles produces lower-pressure areas near the poles. The rising air from the equator moves into these low-pressure areas. In this way, warm air from the equator carries heat toward the poles.

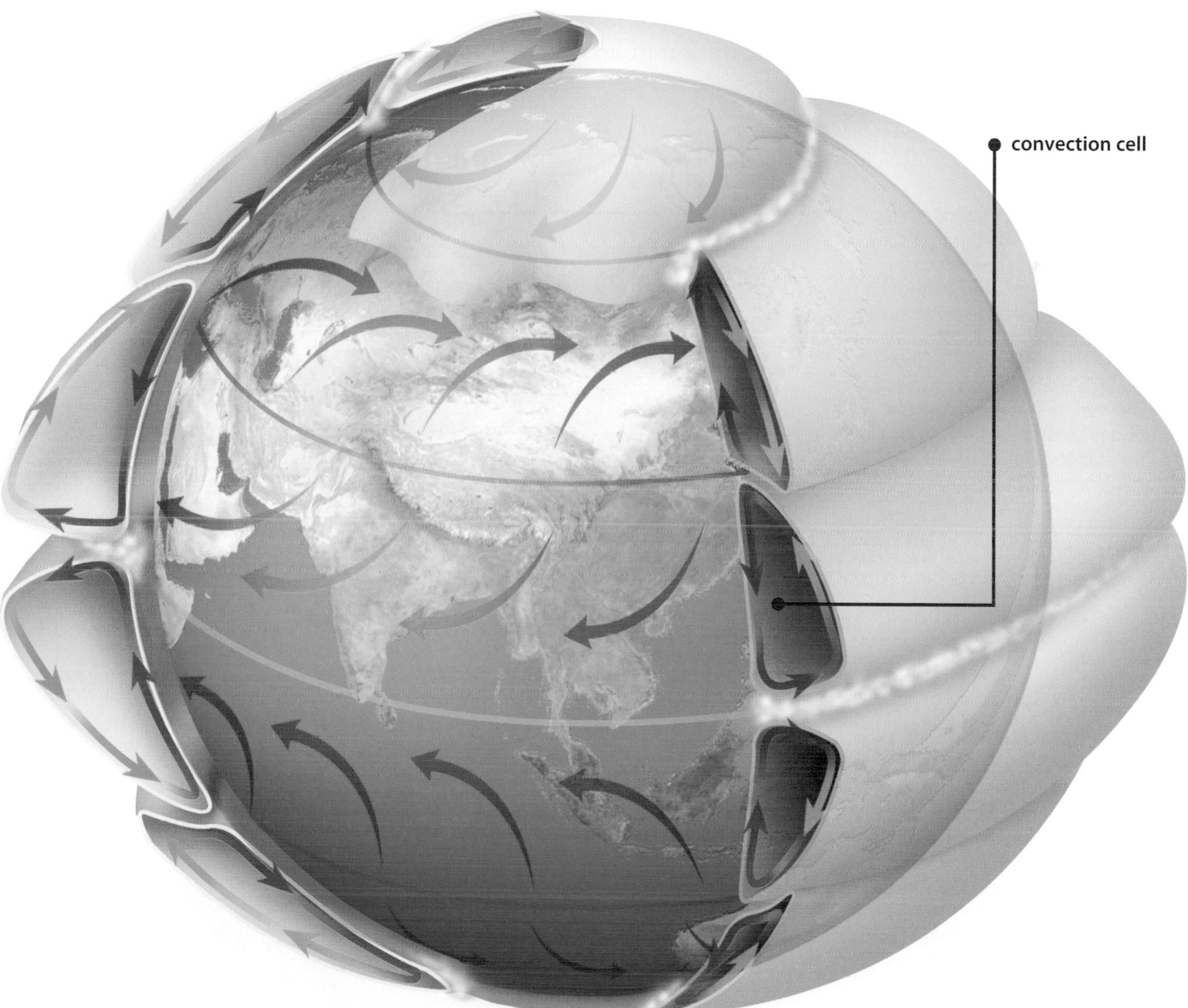

Because earth rotates on its axis, air in the atmosphere travels along curved paths. Air moving from the equator toward the poles is deflected to the east. Air moving from the poles toward the equator is deflected to the west.

Precipitation is part of the climate of an area. It affects the organisms that live there and the geologic processes that occur there.

Precipitation is rain, sleet, snow, hail, or any other form of water that falls to earth's surface. Topography, prevailing winds, and the locations of large bodies of water all affect the amount of precipitation in an area.

Many areas at high latitudes are classified as deserts because they receive very little precipitation. Unlike deserts at lower latitudes, these **cold deserts** may contain a great deal of water frozen in the form of ice and snow. Although very little precipitation falls each year, temperatures remain cold enough so that the little precipitation that does fall does not melt, run off, or evaporate.

Deserts are areas that receive less than 25 cm of precipitation each year. Many of the world's deserts are located in bands that run parallel to the equator. Deserts are more common in these regions because prevailing winds create areas of high pressure there. The high-pressure areas prevent moist air from rising and forming precipitation.

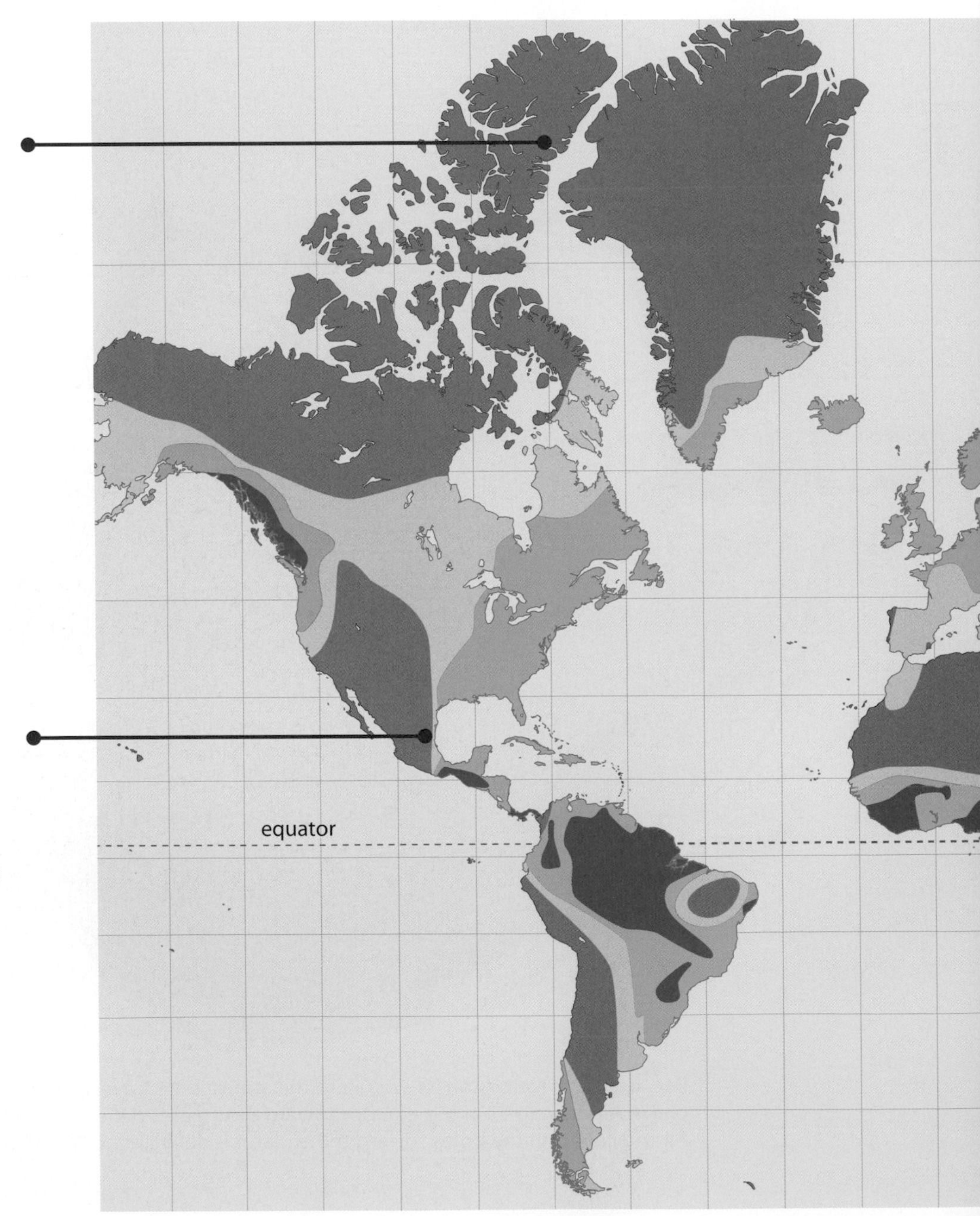

Average Annual Precipitation (cm)
< 40
40–80
80–160
> 160

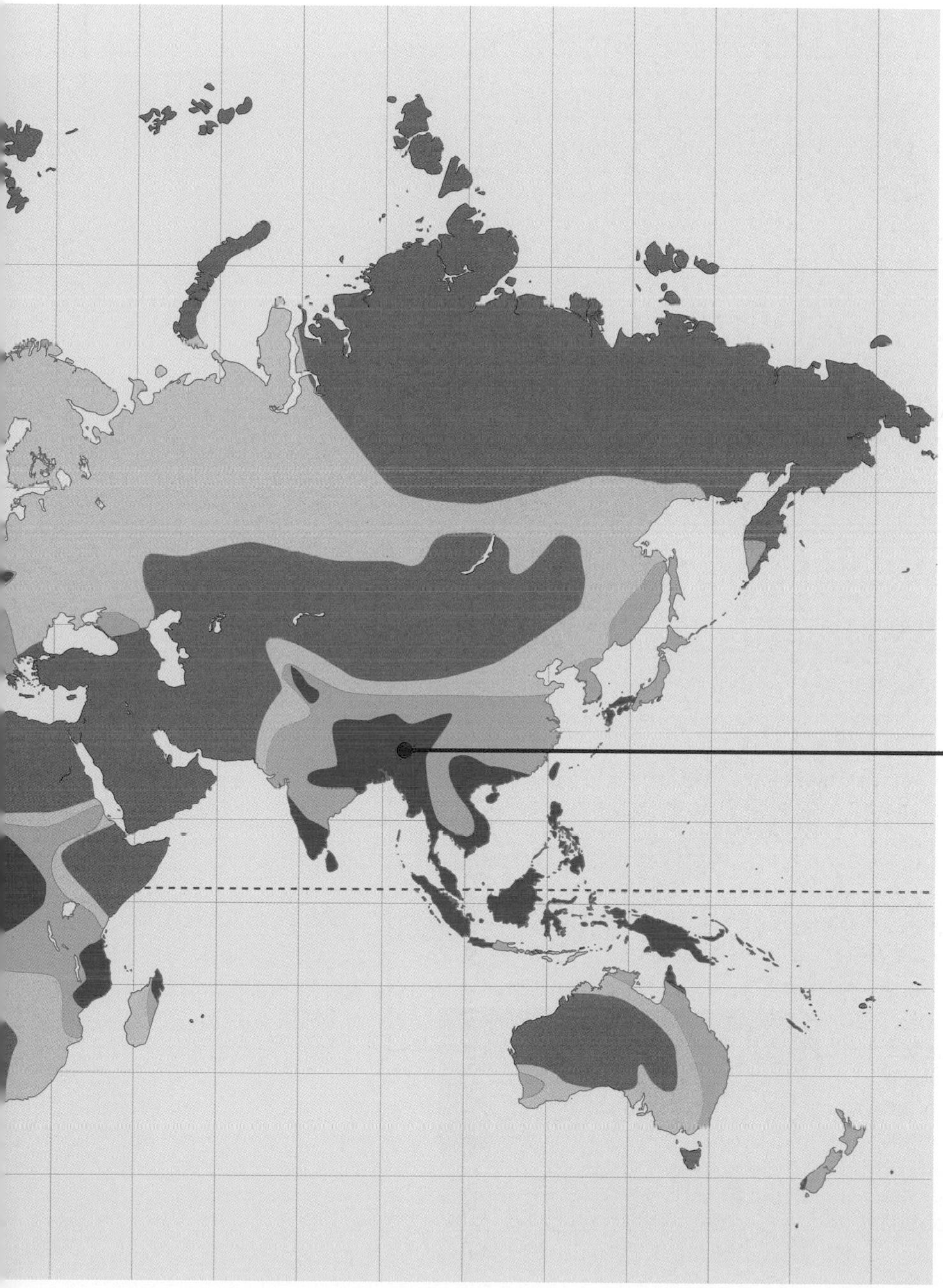

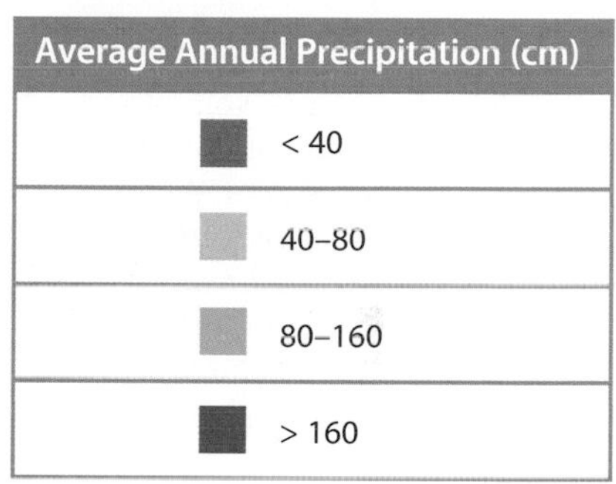

Rain forests and many other tropical areas receive significant precipitation each year. Tropical rain forests may receive more than 2.5 m of precipitation in a single year. Like low-latitude deserts, many of these high-precipitation regions are located in bands parallel to the equator. Rainfall may be high in these areas because prevailing winds create areas of low pressure, where moist air can rise and form heavy precipitation.

Mountains, lakes, plains, and ocean currents influence the weather in their local regions.

Regions located at about the same latitude, such as Maine, North Dakota, and Washington State, have very different climates because of those local influences.

In **northern California**, cold ocean currents cool the air and that air moves east over the land. When it meets the warmer air from the continent, dense fog forms. That fog provides an important source of water for the huge redwood forests in the region.

The **Sierra Nevada and the Rocky Mountains** produce a rain shadow in much of Nevada and Colorado. Regions on the leeward sides (those facing opposite the direction of prevailing winds) of these mountain ranges tend to have very dry climates.

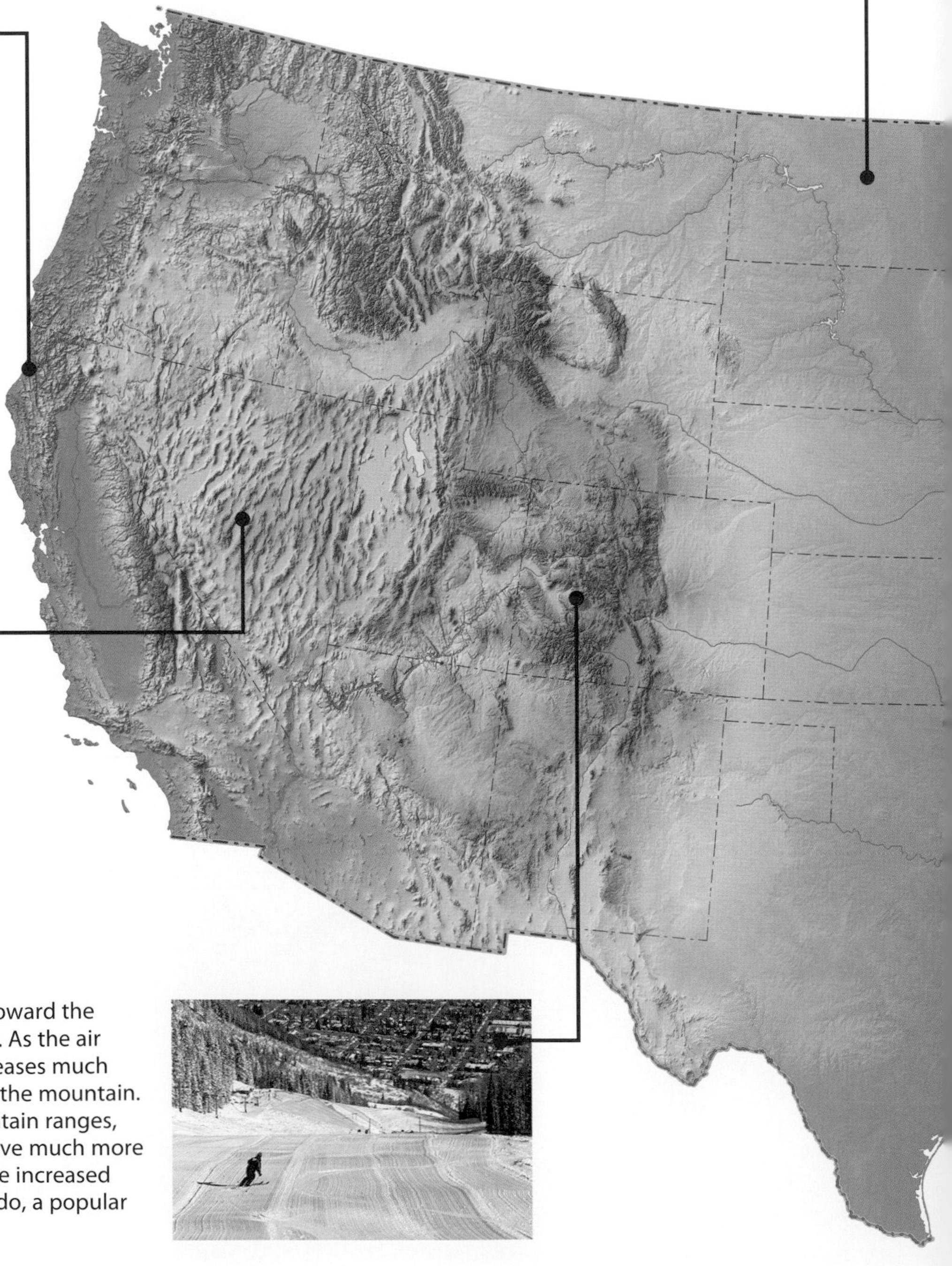

Winds blow moist air from the oceans toward the mountains in the western United States. As the air rises to move over the mountains, it releases much of its moisture on the windward side of the mountain. As a result, the windward sides of mountain ranges, such as those in **Colorado**, tend to receive much more precipitation than the leeward sides. The increased precipitation helps make Aspen, Colorado, a popular skiing destination.

North Dakota and other central plains states are far from any large bodies of water. The climate there can be extreme. In North Dakota, extremely cold and snowy winters contrast with hot, humid summers. Areas at similar latitudes that are near large water bodies experience much less extreme seasonal differences in weather.

In the winter, cold winds from the north pick up moisture from the surface of Lake Erie. As that air moves over central **New York**, the moisture produces heavy snowfall. Areas such as Buffalo, New York, may receive many feet of this lake-effect snow every winter.

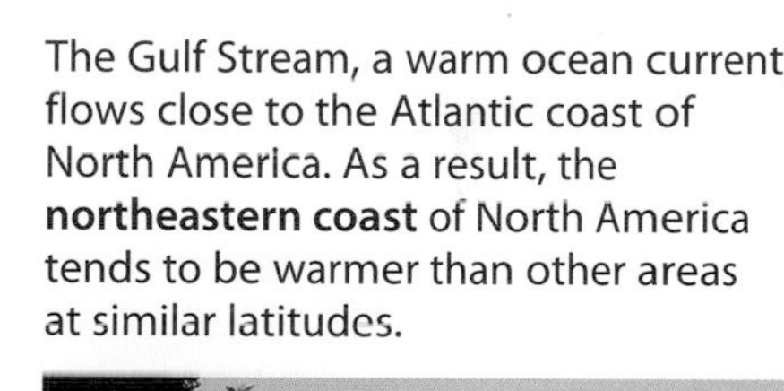

The Gulf Stream, a warm ocean current, flows close to the Atlantic coast of North America. As a result, the **northeastern coast** of North America tends to be warmer than other areas at similar latitudes.

Tornadoes are so common in parts of **Oklahoma** and other plains states that the region is known as Tornado Alley. The tornadoes form as a result of severe thunderstorms that move through the area. The thunderstorms form when warm, moist air from the Gulf of Mexico meets cooler, drier air moving south from Canada.

A biome is a large area of earth defined by its climate and the organisms that live there.

Latitude, elevation, terrain, temperature, and precipitation all affect the climate that defines each biome. The organisms that live in each biome have specific adaptations that allow them to survive there.

The **polar ice** biome includes the areas around earth's poles that are always covered in snow and ice. Almost no producers live at the top of the polar ice biome, though many producers are found in the waters here. Consumers in this biome, such as polar bears, rely on organisms that live in the oceans beneath the ice for their food.

Tundra is located in areas that are cold all year and receive little precipitation. Beneath the ground is a layer of *permafrost,* soil that remains frozen all year. During the summer, the top layer of soil may thaw, allowing a few plants to grow. A limited number of animal species and microorganisms live in the tundra, so it is one of the least diverse biomes on earth.

The **taiga,** or boreal forest, is characterized by long, cold winters and short, cool summers. Large numbers of coniferous trees, such as pines and cedars, are common in taiga biomes. Various consumers, such as mice, hares, finches, mink, and wolverines, also live in the taiga.

Mountains are characterized by many different climates and organisms.

Chaparral is a biome found in temperate climates that experiences periodic wildfires.

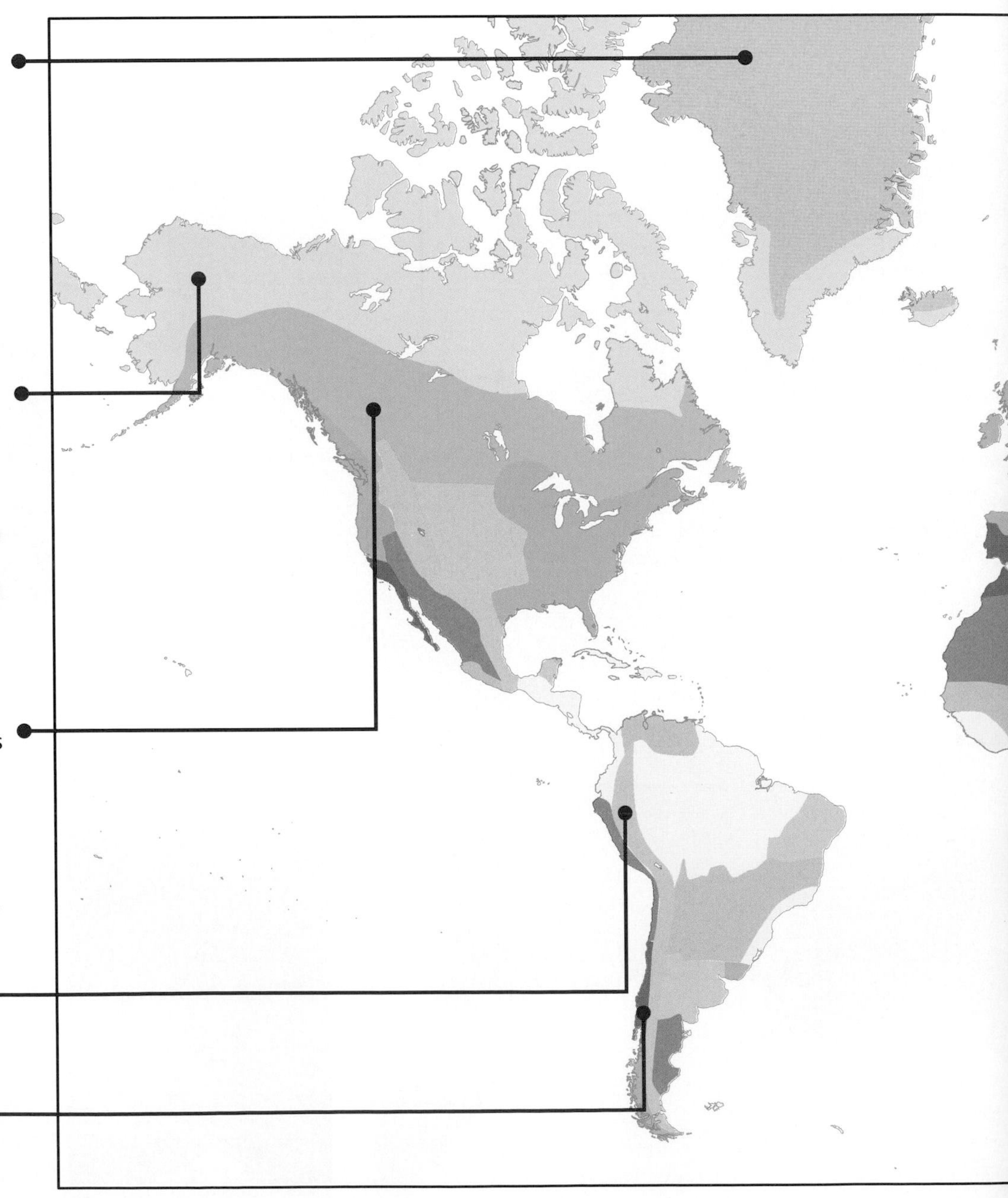

Temperate grasslands are located in areas that don't receive much rainfall and have distinct seasons. Because rainfall is limited, few trees grow in temperate grasslands. Grasses and shrubs are abundant, however, and the soil is very fertile. Many temperate grasslands have become centers of agriculture because of their highly fertile soil.

The **tropical rain forest** receives a large amount of precipitation and is warm all year. Many tropical forests experience two seasons: wet and dry. Tropical forests are some of the most diverse biomes on earth—they are home to thousands of species of plants, animals, and other organisms.

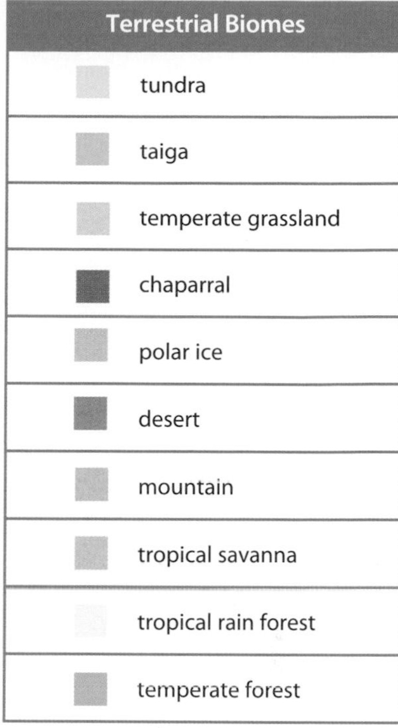

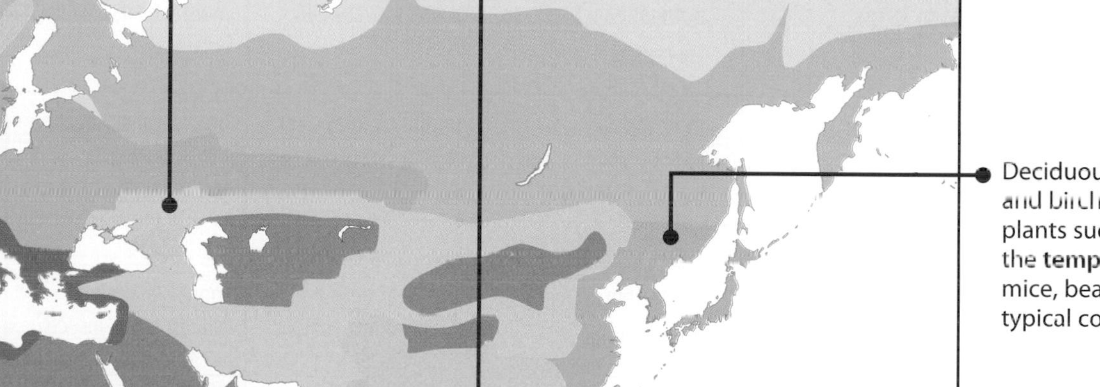

Deciduous trees such as maple, oak, and birch, as well as many smaller plants such as ferns, are common in the **temperate forest**. Deer, moose, mice, bears, owls, and foxes are typical consumers here.

Regions in the tropics that receive a moderate amount of rainfall are generally part of the **tropical savanna**. Tall grasses and shrubs are the most common plants, but some trees grow here as well.

Deserts cover about one-fifth of earth's surface. Deserts receive very little precipitation. Most deserts have very high temperatures during the day and low temperatures at night. Some regions of earth at high latitudes can also be classified as deserts because of their low rainfall, even though temperatures there are low most of the time.

Each biome has distinct temperature, precipitation, and soil characteristics.

These features affect the types of organisms that can live in a given biome. The organisms in a specific biome are adapted to that biome's characteristics.

BIOME CHARACTERISTICS

BIOME	AVERAGE YEARLY TEMPERATURE	AVERAGE YEARLY PRECIPITATION	SOIL CHARACTERISTICS	REPRESENTATIVE VEGETATION	REPRESENTATIVE ANIMAL LIFE
tundra	–26°C to 12°C	less than 25 cm	• thin, moist upper layer over permafrost • nutrient poor • slightly acidic	mosses, lichens, grasses, small woody plants	caribou, arctic foxes, lemmings, snowshoe hares
taiga	–10°C to 14°C	35 to 75 cm	• nutrient poor • very acidic	conifers (evergreen trees)	moose, bears, wolves, lynxes
temperate forest	6°C to 28°C	75 to 125 cm	• moist, fairly thick upper layer • moderate levels of nutrients	deciduous trees, shrubs, evergreen trees	bears, wolves, deer, foxes, raccoons, squirrels
tropical rain forest	20°C to 34°C	200 to 400 cm	• moist, thin upper layer • nutrient poor	broadleaf trees, shrubs, vines, moss	monkeys, snakes, lizards, birds, insects
temperate grassland	0°C to 25°C	25 to 75 cm	• thick upper layer • nutrient rich	dense grasses	buffalo, prairie dogs, coyotes
savanna	16°C to 34°C	75 to 150 cm	• dry, thin upper layer • nutrient poor	tall grasses, a few trees	zebras, wildebeests, giraffes, gazelles, lions, leopards
desert	7°C to 38°C	less than 25 cm	• dry, sandy • nutrient poor	succulent plants such as cacti, a few grasses	lizards, snakes, foxes

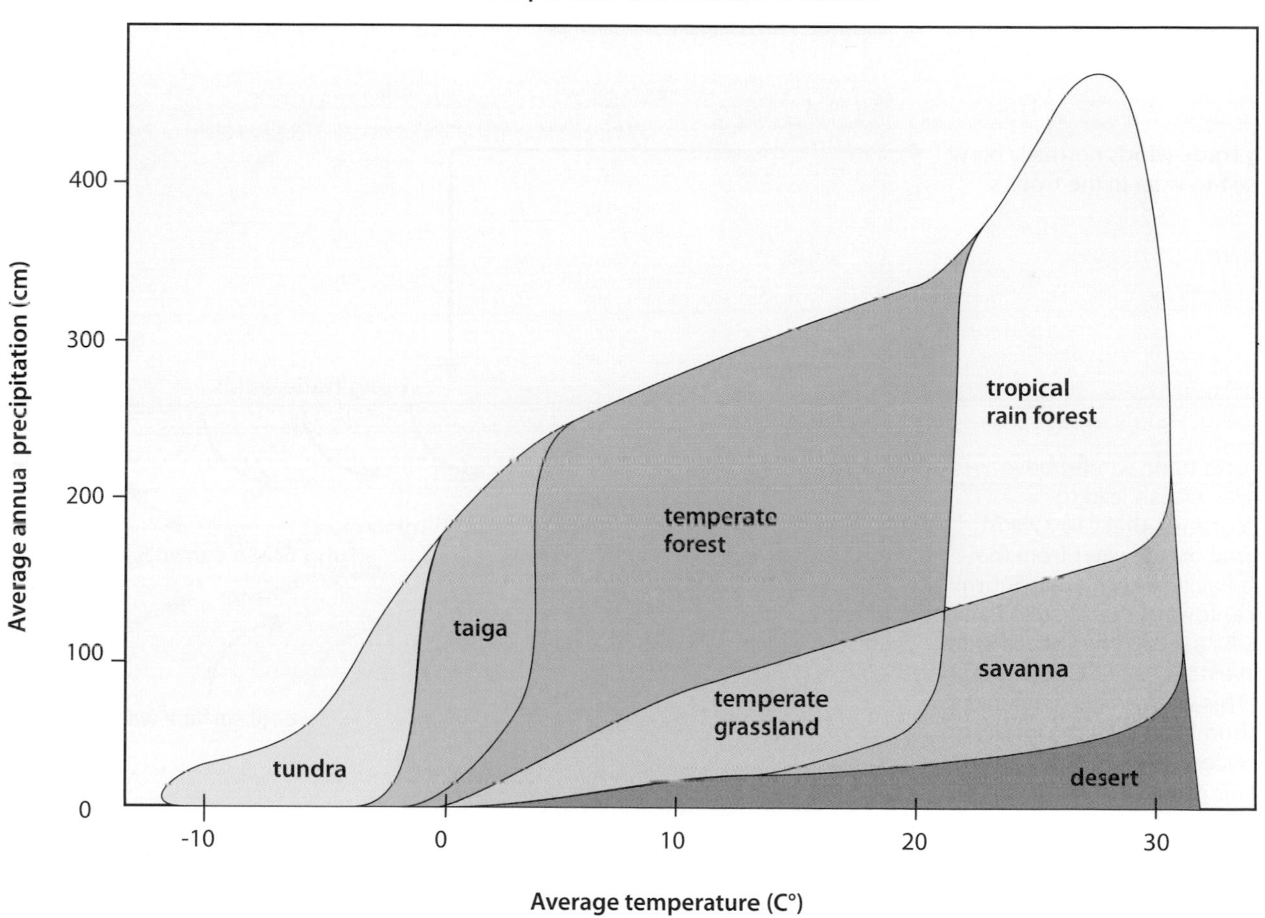

Refer to Biome Characteristics chart opposite page.

The El Niño–Southern Oscillation (ENSO) is a phenomenon that affects the world's oceans and atmosphere.

Strong ENSO events occur every 2 to 12 years and are the main cause of short-term temperature and climate fluctuations in the world. They cause heavy rainfall in some areas and droughts in others.

Non–ENSO Year

Strong trade winds normally blow from east to west in the tropics.

The strong trade winds above the Pacific Ocean lead to **ocean currents** that carry warm, equatorial ocean water from the eastern Pacific Ocean (near South America) toward the western Pacific Ocean. As a result, the surface water in the western Pacific Ocean is fairly warm. This warm water produces warm, humid air over the western Pacific Ocean islands. The warm, humid air causes increased rainfall in these areas.

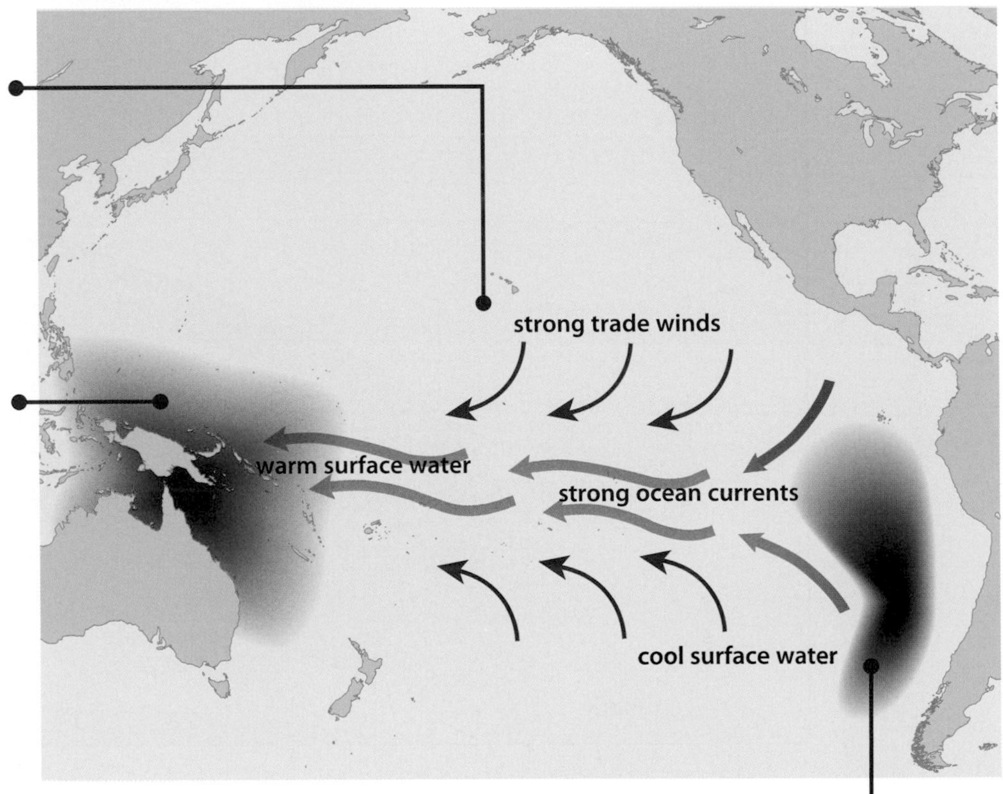

Normally, the water off the coast of South America is cold because the trade winds blow warmer water toward the western Pacific Ocean. Cooler, deeper water rises toward the surface to replace the warmer **surface waters**. This cooler water is nutrient rich and supports many diverse marine ecosystems.

Continental polar (cP) air masses form over northern Canada and bring dry, cold air to the northern United States. In the winter, these air masses cause heavy snowfall when they pass over the Great Lakes. In the summer, they bring cool, dry weather.

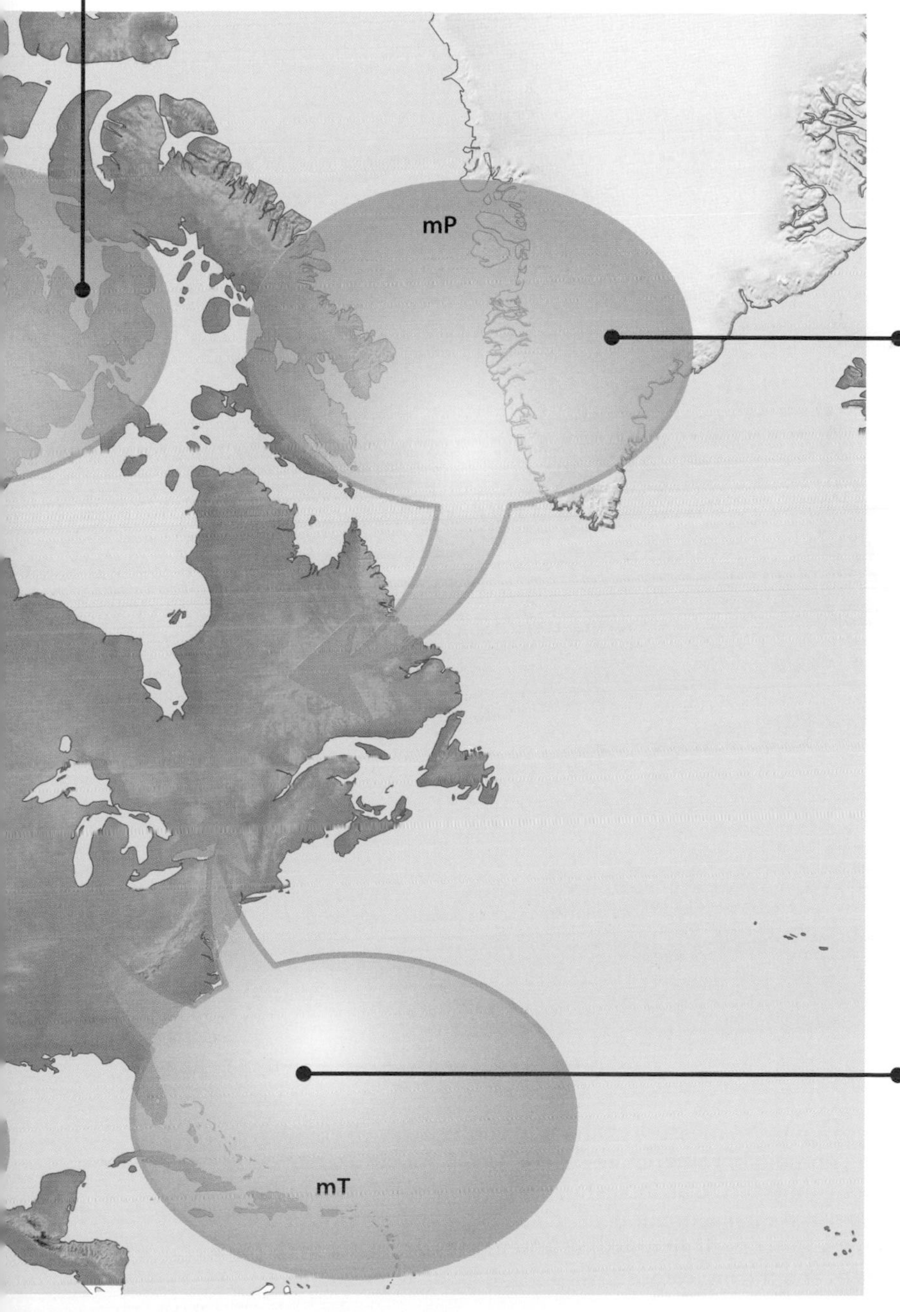

Maritime polar (mP) air masses form over the northern Atlantic Ocean and bring moist, cold air to the Atlantic coast of North America. They cause serious winter storms in the northeastern United States in the winter.

Maritime tropical (mT) air masses that form over the southern Atlantic Ocean bring warm, moist air to the Atlantic coast of North America. These air masses are largely responsible for the heat and humidity experienced by the east coast of the United States in the summer.

Meteorologists use specialized tools to measure weather conditions.

For example, a meteorologist may use a barometer to monitor air pressure and a rain gauge to measure rainfall.

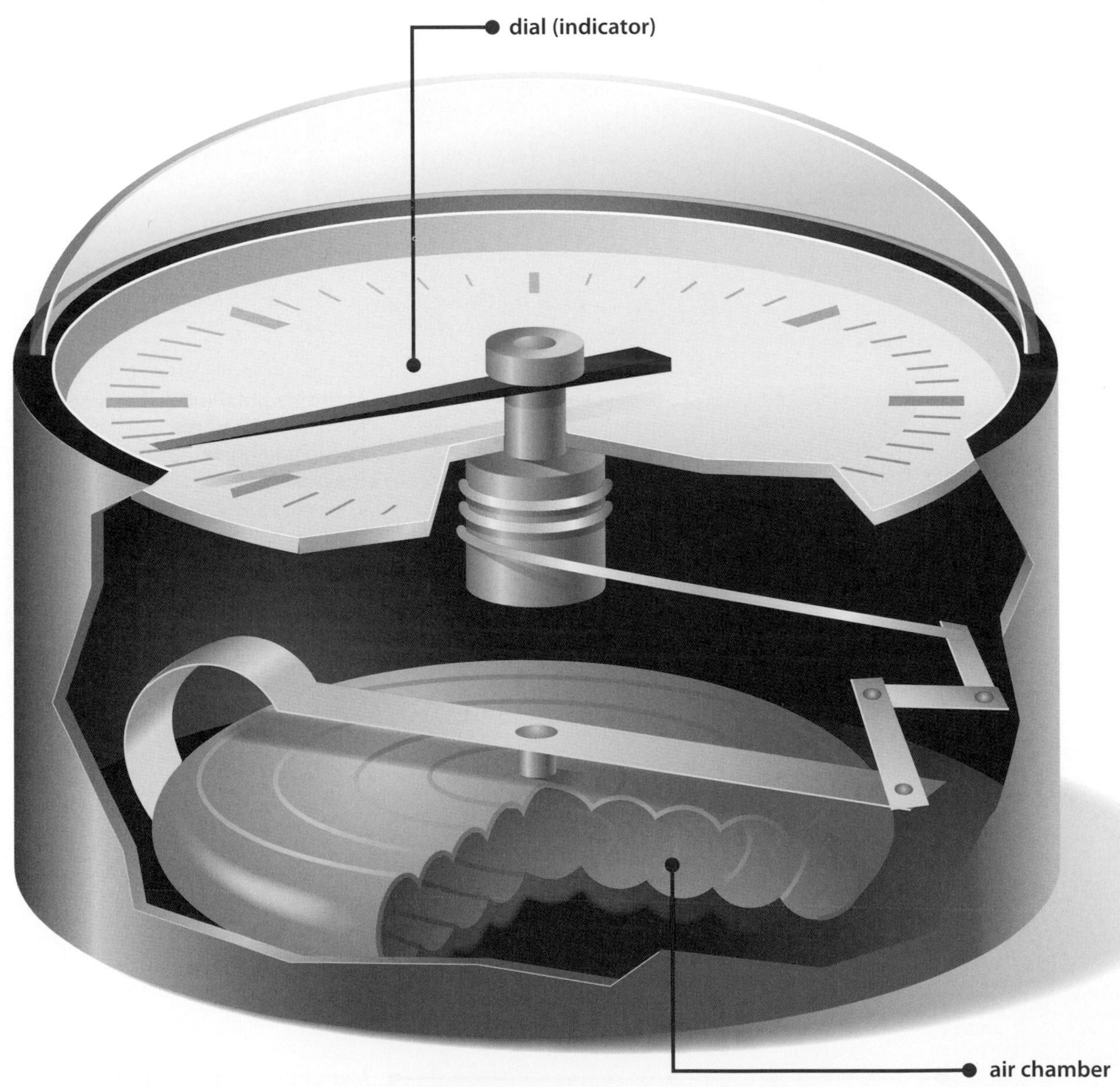

Although some barometers use liquids to measure changes in air pressure, an aneroid barometer does not. Within an **aneroid barometer** are several small, hollow disks or chambers. These chambers are connected to an indicator, such as a dial. If air pressure around the barometer increases, the small amount of air in each chamber contracts, and the indicator shows an increase in air pressure. If air pressure around the barometer decreases, the air in the chambers expands and the indicator shows a decrease in air pressure.

You can create your own **rain gauge** to measure the amount of rain that falls in your area. Use a ruler to make 1-cm marks on the outside of a clear glass or plastic bottle. The opening at the top of the tube should be 2 cm in diameter. You can place a funnel inside the mouth of the bottle to prevent larger objects from falling in. The mouth of the funnel should be 20 cm wide. Place the rain gauge in a flat area (such as on top of a railing or fence), away from trees, awnings, or other overhanging objects.

If you do not use a funnel, measure the rain directly on the scale. If you do use a funnel, be sure to divide the rain amount by 10.

Fronts form when different air masses collide. Fronts are accompanied by a variety of weather phenomena.

For example, when a warm air mass meets a cold air mass, precipitation may form. The four main kinds of fronts are *warm, cold, occluded,* and *stationary*.

A **warm front** forms where a warm air mass moves in and replaces a cold air mass. Warm fronts often produce light precipitation over large areas. As the warm air mass replaces the cold air mass, temperatures increase. Therefore, warmer weather often follows the passage of a warm front.

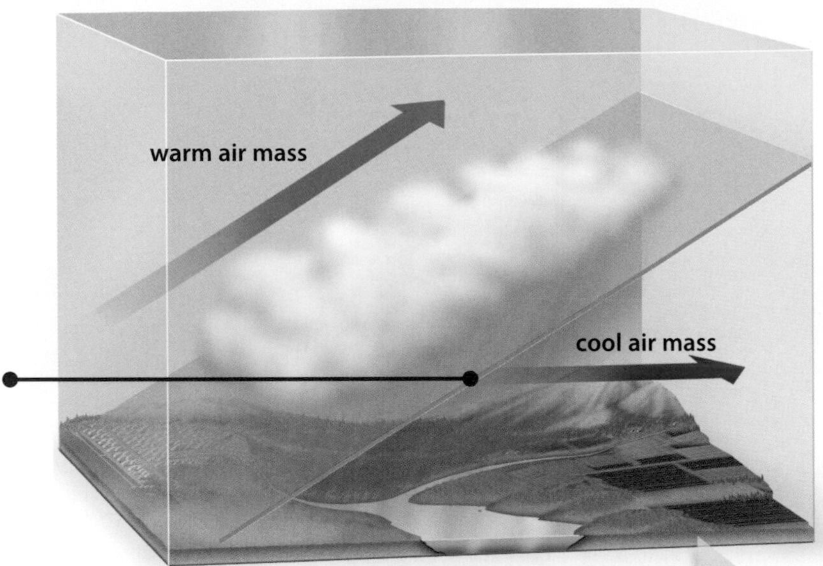

A **cold front** forms where a cold air mass moves in and replaces a warm air mass. Heavy precipitation and severe storms form along many cold fronts. Cooler, drier weather often follows a cold front.

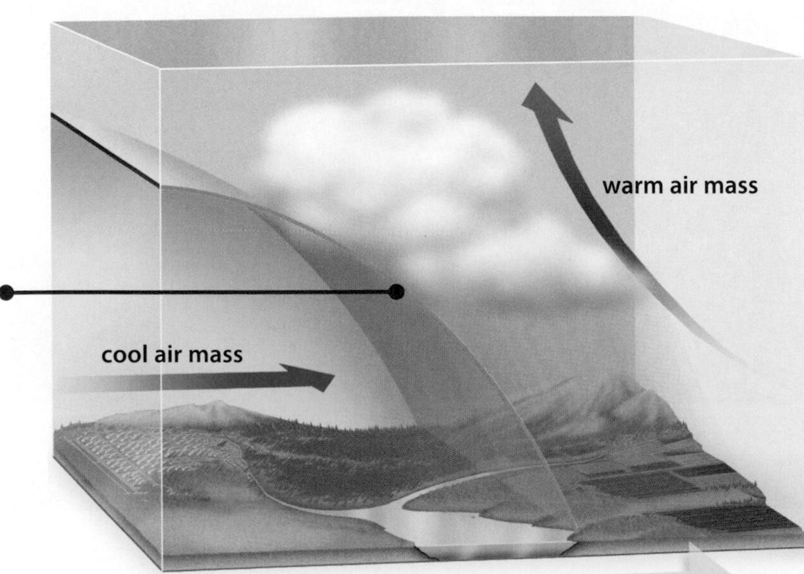

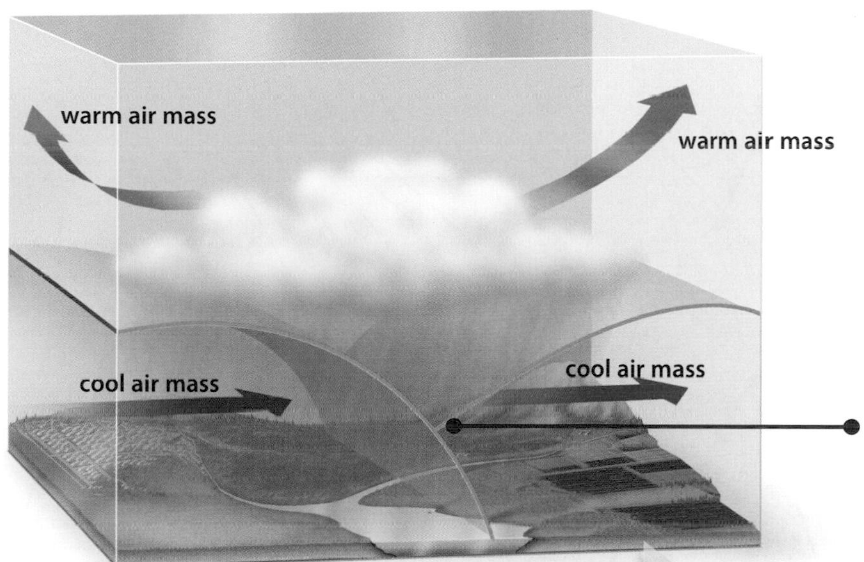

An **occluded front** forms when a warm air mass is caught between two colder air masses. Occluded fronts can produce heavy precipitation and are usually followed by lower temperatures.

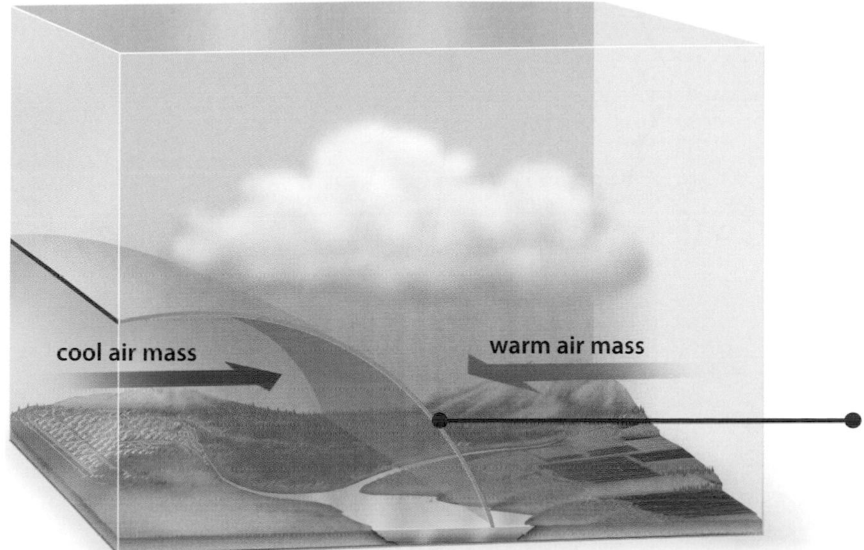

A **stationary front** forms when two air masses meet, but neither one replaces the other. Precipitation may form along the front. Because the front does not move, the precipitation may fall for long periods of time in the same place.

Weather maps show fronts, air pressure, air temperature, wind speed and direction, dew point, and precipitation in different areas.

Each piece of information on a weather map is represented by a specific symbol.

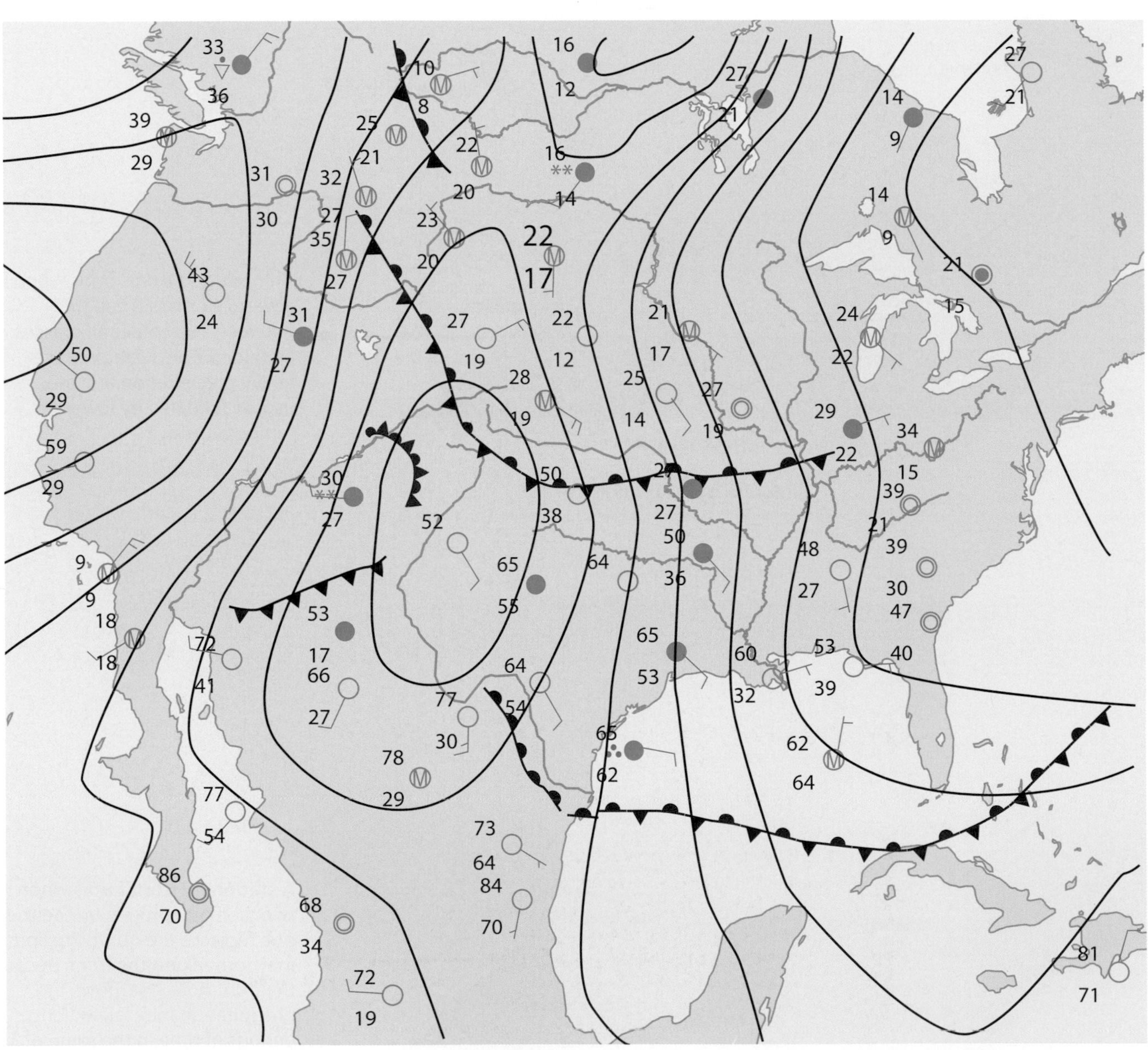

Weather Symbol Key	
——	isobar
Winds	
◎	calm winds
50 + 10 + 10 + 5	winds blowing west at 75 knots
Precipitation	
,,	drizzle
= ≡	fog (light, heavy)
	freezing rain
∞	haze
	rain (light, moderate, heavy)
	shower (rain, snow)
	ice pellets (sleet)
	snow (light, moderate, heavy)
	thunder (with rain, snow, no precipitation)
Cloud Cover	
	no clouds
	scattered clouds
	partly cloudy
	mostly cloudy
	overcast
⊗	sky obscured
Ⓜ	no data available
Fronts	
	cold front
	occluded front
	stationary front
	warm front

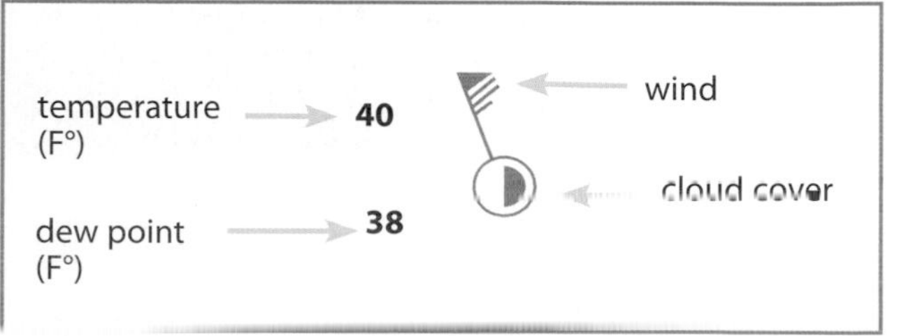

Clouds form when water vapor in the atmosphere condenses on tiny particles of solid material, such as dust.

The condensed water vapor forms tiny droplets. Clouds may contain millions of these droplets. Meteorologists classify clouds by shape and by altitude.

High clouds form more than 6,000 m above earth's surface. Temperatures at this height are very low, and water freezes as soon as it condenses. Therefore, high clouds are made of tiny crystals of ice. High clouds are denoted by the prefix *cirro-*.

Middle clouds form between 2,000 m and 6,000 m above earth's surface. They may contain water droplets, ice crystals, or both. Middle clouds are denoted by the prefix *alto-*.

Low clouds form less than 2,000 m above earth's surface. Because temperatures at this altitude tend to be fairly high, low clouds rarely contain ice crystals. There is no prefix used to denote low clouds.

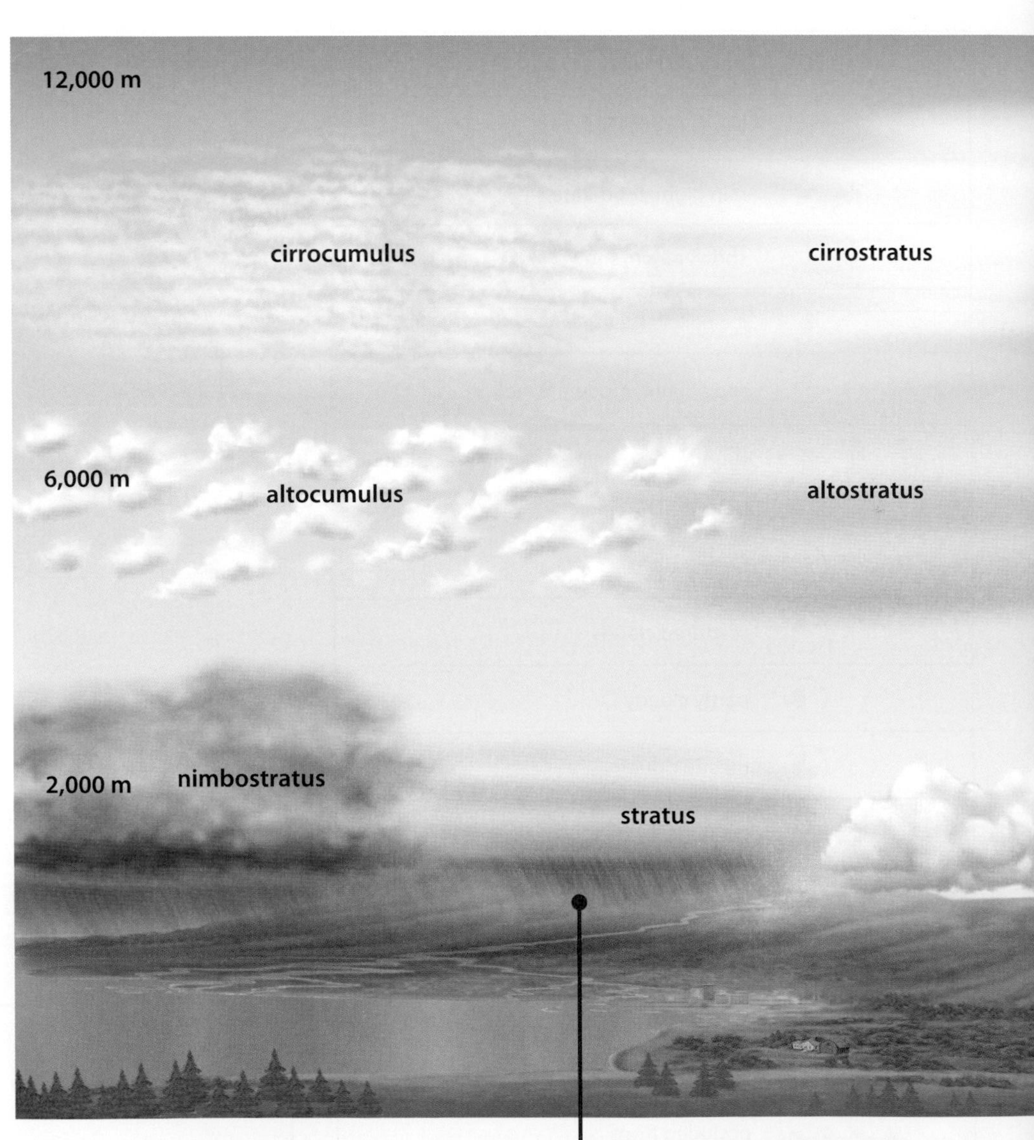

Stratus clouds are horizontal, layered, sometimes dark-gray clouds that can cover large areas of the sky. The name *stratus* comes from a Latin word that means stretched or extended.

Cirrus clouds are thin, wispy-looking clouds that are usually very high. The name for these clouds comes from the Latin word *cirrus,* which means curl of hair.

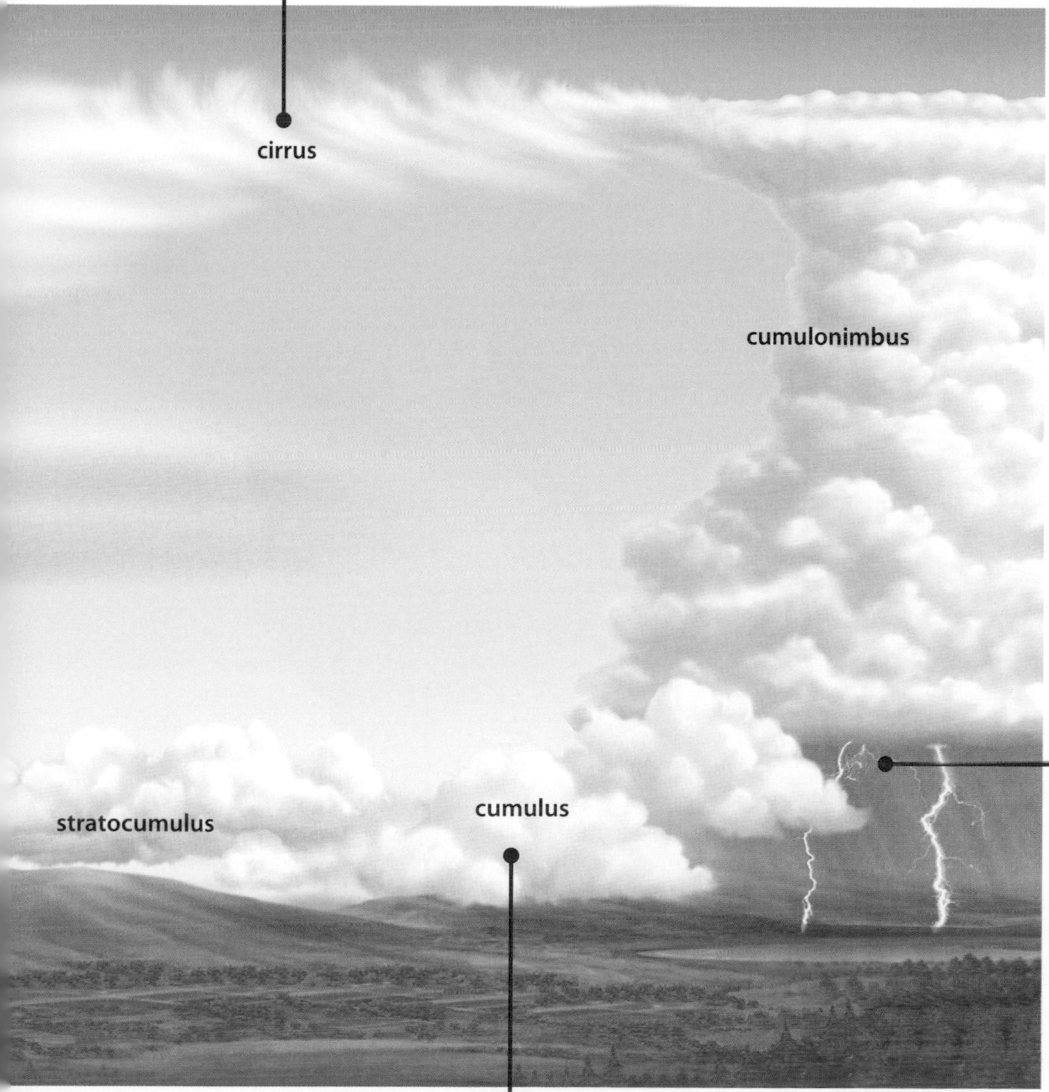

Clouds that are producing precipitation are denoted by the prefix *nimbo-* or the suffix *–nimbus*. For example, cumulonimbus clouds are cumulus clouds that are producing precipitation. Nimbostratus clouds are stratus clouds that are producing precipitation.

Cumulus clouds are large, fluffy-looking clouds. They are often white, but they may turn dark gray or black during severe weather. The Latin word *cumulus*, from which these clouds get their name, means heap.

Tornadoes and hurricanes are some of the most destructive storms that affect North America.

Both hurricanes and tornadoes form as a result of rising warm air that begins to spin. Tornadoes generally form over land, when large air masses combine. Hurricanes initially form over warm ocean water, but they often move toward land.

Frequency of Tornado Occurrence in the Contiguous United States

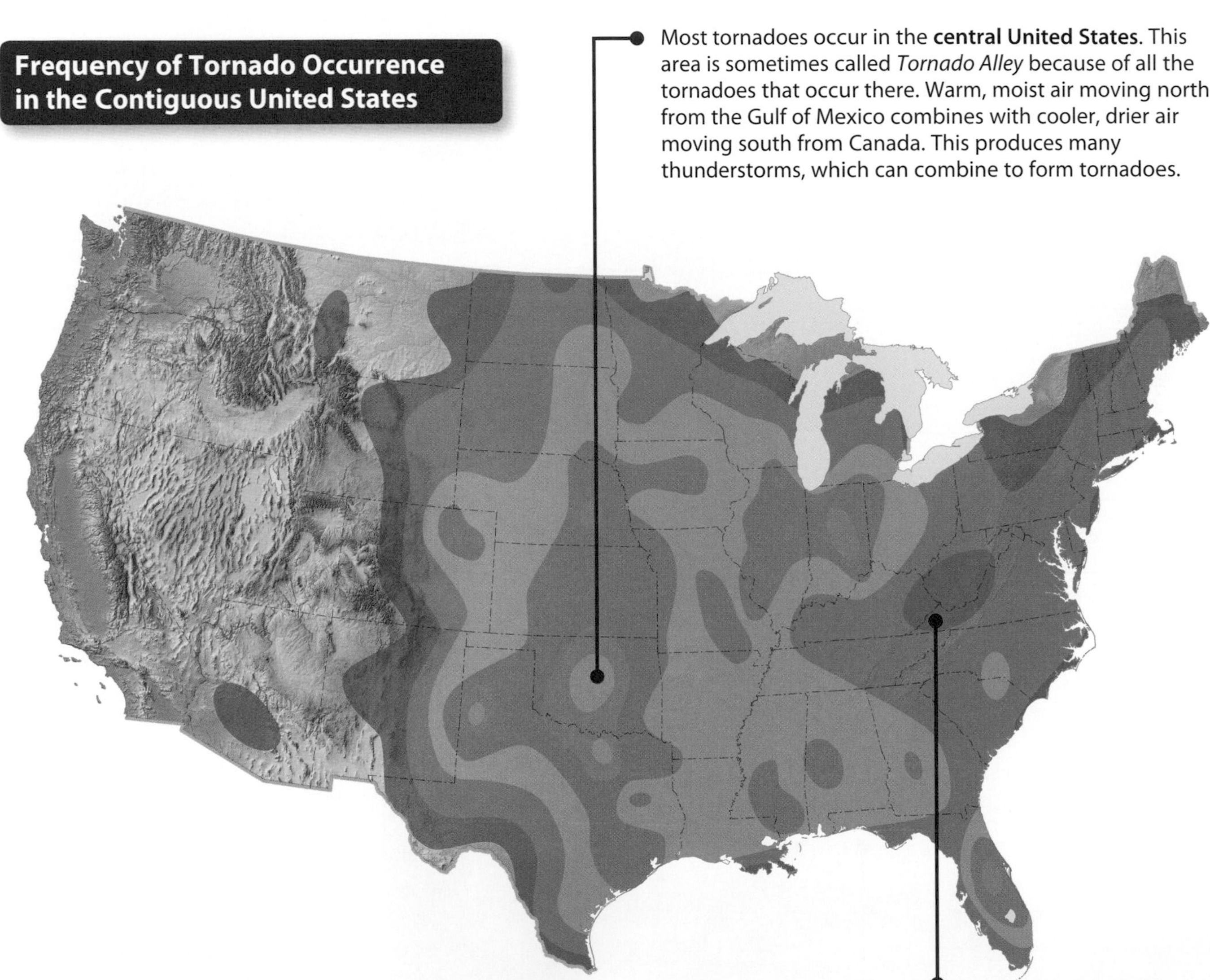

Most tornadoes occur in the **central United States**. This area is sometimes called *Tornado Alley* because of all the tornadoes that occur there. Warm, moist air moving north from the Gulf of Mexico combines with cooler, drier air moving south from Canada. This produces many thunderstorms, which can combine to form tornadoes.

Tornadoes do occur in **places other than the Great Plains.** The thunderstorms that can form tornadoes can happen anywhere. Therefore, tornadoes can also occur anywhere.

Average Number of Tornadoes per Year per 10,000 km²		
9.0	5.0	1.0
7.0	3.0	0.5

Most hurricanes that strike North America form in the **eastern Atlantic Ocean.** These hurricanes travel north and west. Some of them eventually move over North America, bringing heavy rains, strong winds, and coastal flooding. Most Atlantic Ocean hurricanes happen between June 1 and December 15.

Historic North American Hurricane Paths

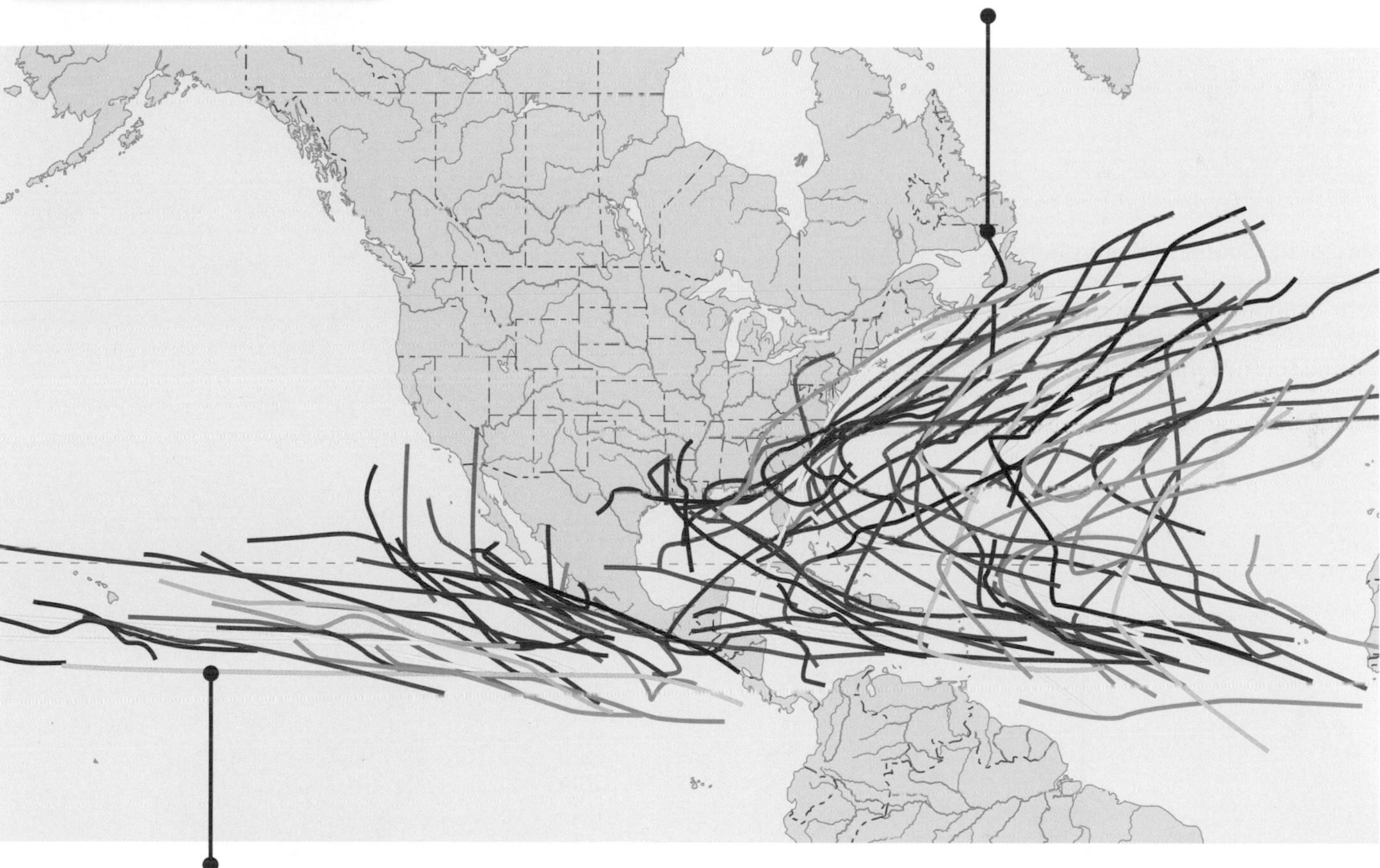

Some hurricanes form in the **eastern Pacific Ocean**. These hurricanes travel north and west. Most Pacific Ocean hurricanes occur between May 15 and November 15.

Oceans cover 70 percent of earth's surface.

Earth's oceans are home to a huge number of organisms. The oceans also help regulate earth's temperatures, and they significantly affect weather and climate.

Most of the **Southern Hemisphere** is ocean. Australia, Antarctica, and parts of South America and Africa are the only major landmasses located south of the equator.

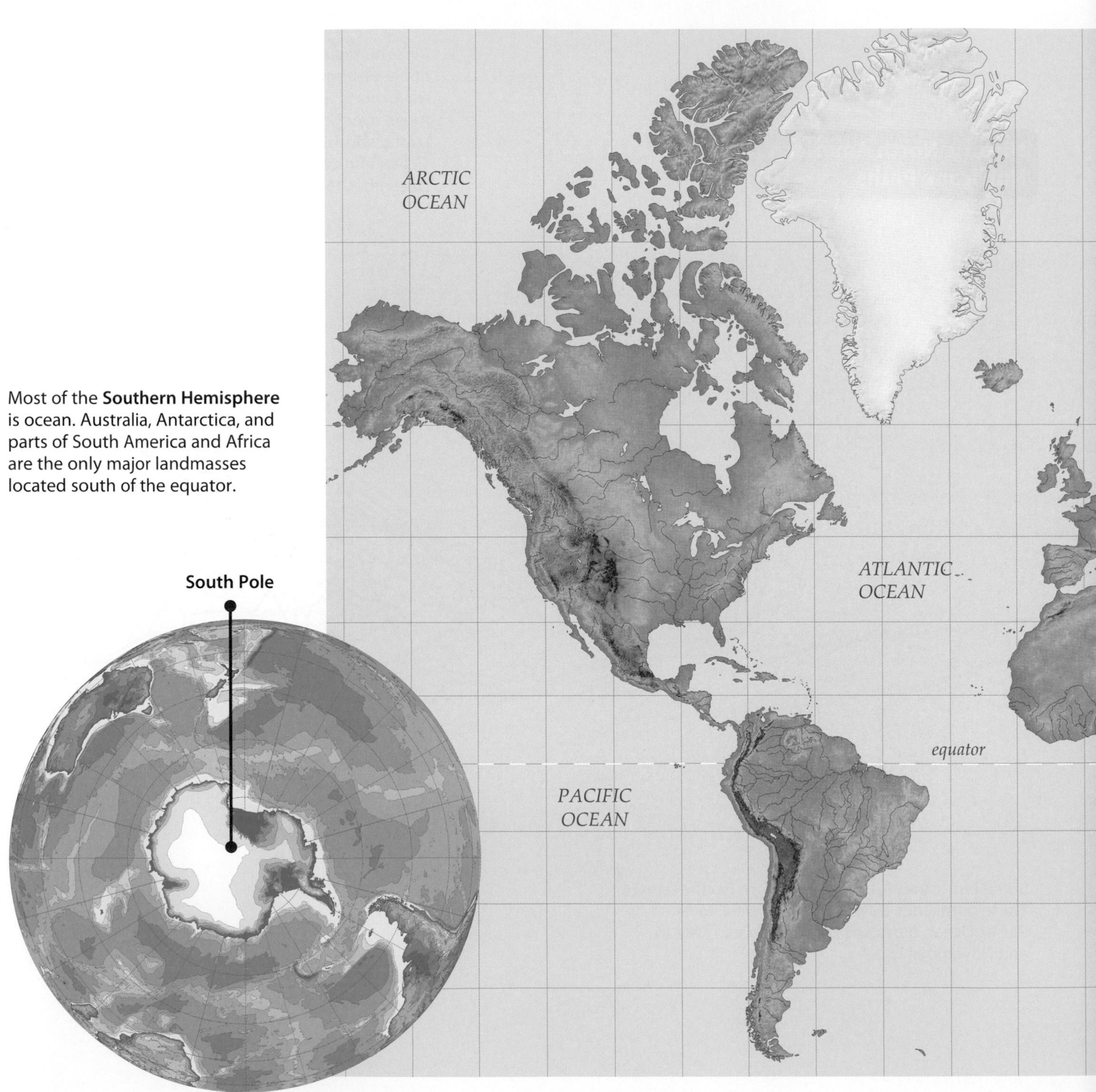

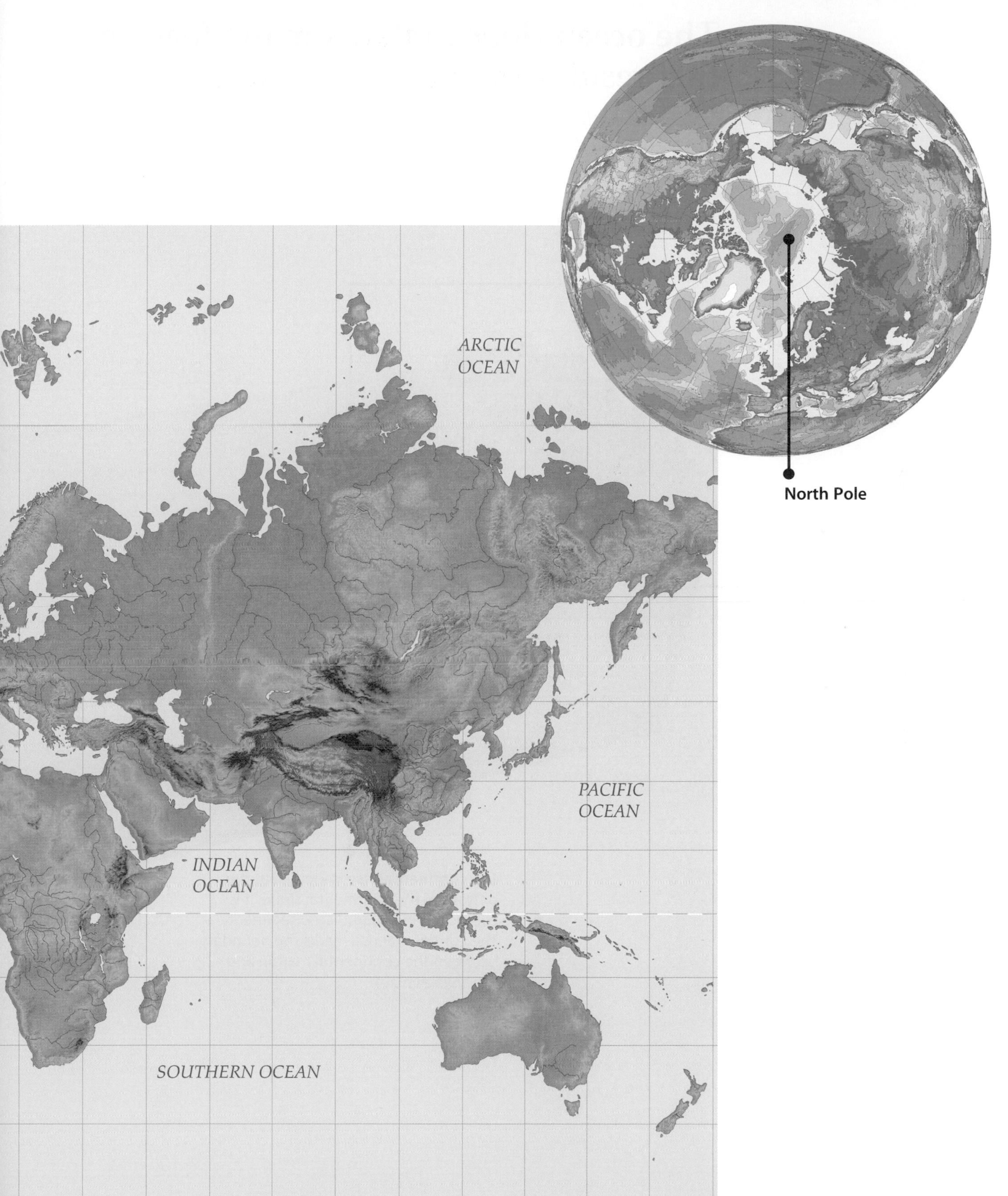
ARCTIC
OCEAN
PACIFIC
OCEAN
INDIAN
OCEAN
SOUTHERN OCEAN
North Pole

The ocean floor contains many features that result from geologic processes.

The *continental margin* is the area of the ocean floor that extends from the coast of a continent to the deepest part of the ocean floor. The *abyssal plain* is the deepest part of the ocean floor.

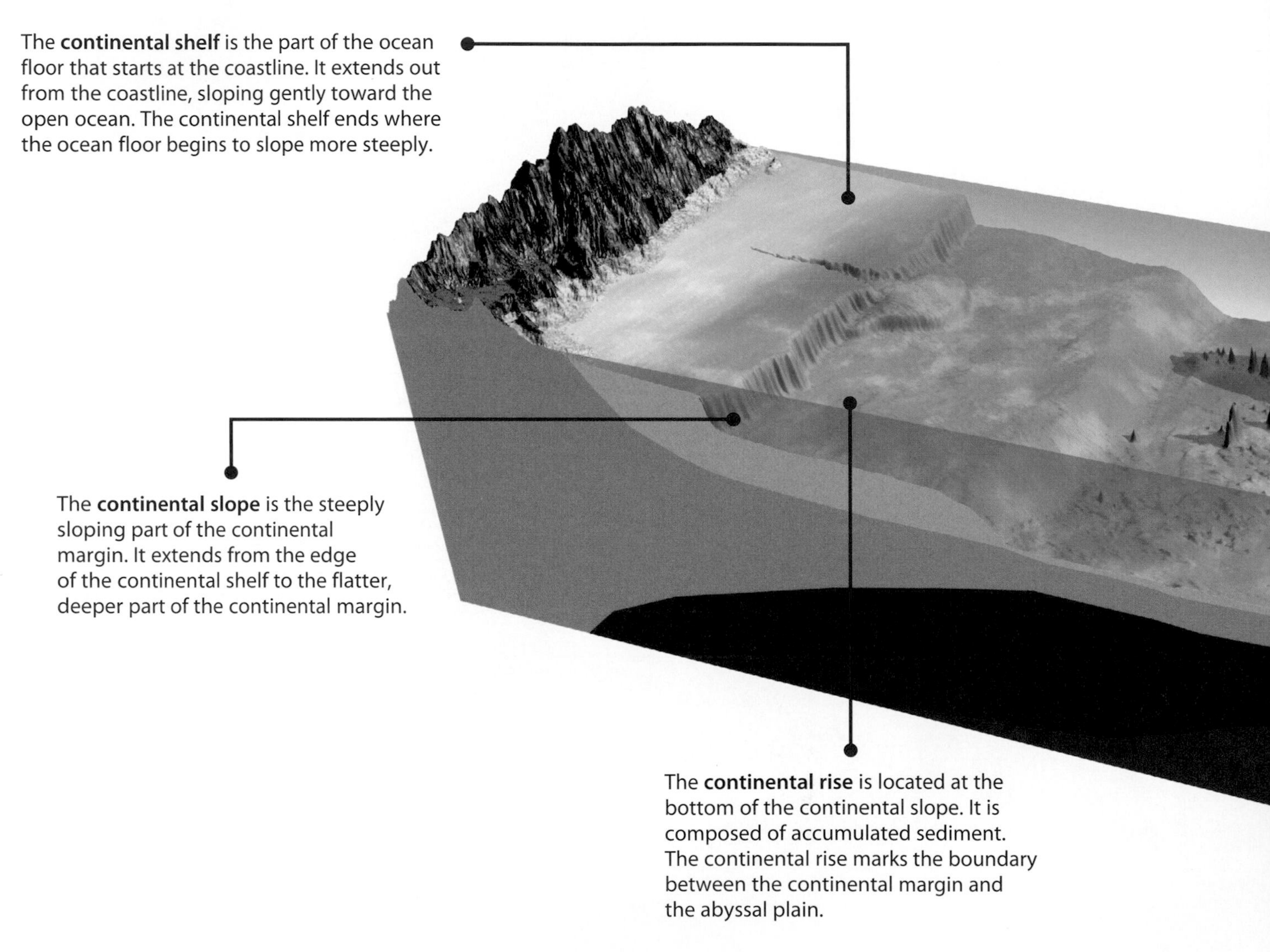

The vast, mostly flat plains that make up much of the ocean floor are **abyssal plains**. They are covered with layers of sediment that fall to the ocean floor from the water above.

Seamounts form where volcanoes erupt on the ocean floor.

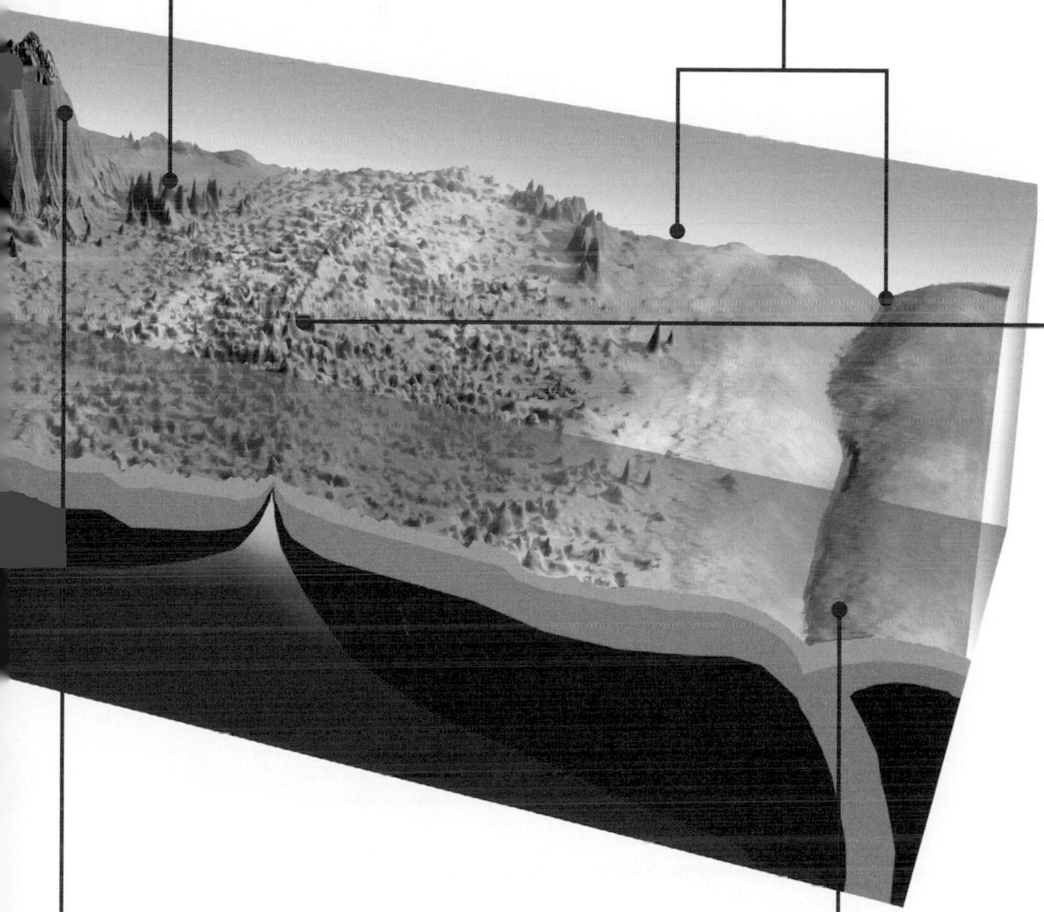

Mid-ocean ridges are long, underwater chains of volcanoes that form where two tectonic plates move apart. As the plates separate, cracks form in the lithosphere. Magma rises through the cracks, erupts, and solidifies to form a new layer of the oceanic lithosphere.

Volcanoes that erupt on the ocean floor build up tall mountains. If the volcanoes erupt enough lava, the mountains may grow tall enough to rise above sea level and form an **island**.

Ocean trenches form where one plate subducts, or sinks, beneath another plate. Trenches can be thousands of meters deep.

The world's tallest mountains, deepest valleys, and longest mountain ranges are all found on the ocean floor.

Most of the features of the ocean floor are the result of plate tectonics and processes within earth's mantle.

The **islands of Hawaii** formed through a combination of processes. A mantle plume causes volcanoes to erupt through the ocean floor. At the same time, the Pacific plate is moving across earth's surface. As the plate moves, the volcano is eventually carried away from the mantle plume. The plume then pushes through another point in the lithosphere. As a result, a long, straight chain of volcanoes forms.

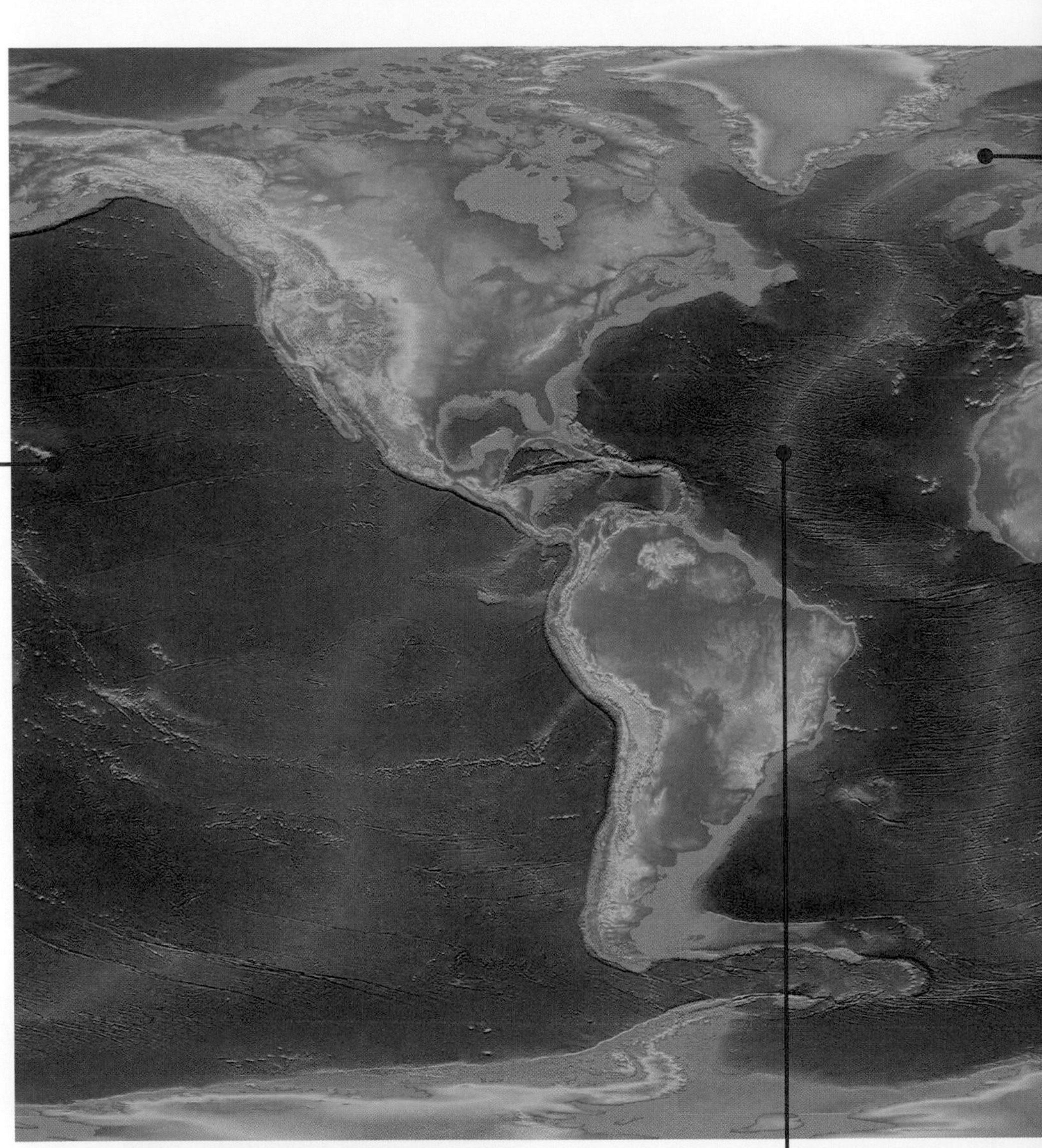

The **Mid-Atlantic Ridge** is a long chain of volcanoes that runs through the center of the Atlantic Ocean. It grows as the North American and South American plates move away from the Eurasian and African plates. These plates are moving apart at about 2.5 cm per year.

Iceland is one of the only places on earth's surface where a divergent boundary rises above the surface of the ocean. The island is located directly along the Mid-Atlantic Ridge. Most scientists agree that a mantle plume—a column of hot, solid rock rising toward the surface from deep within the mantle—is what causes Iceland to protrude above sea level.

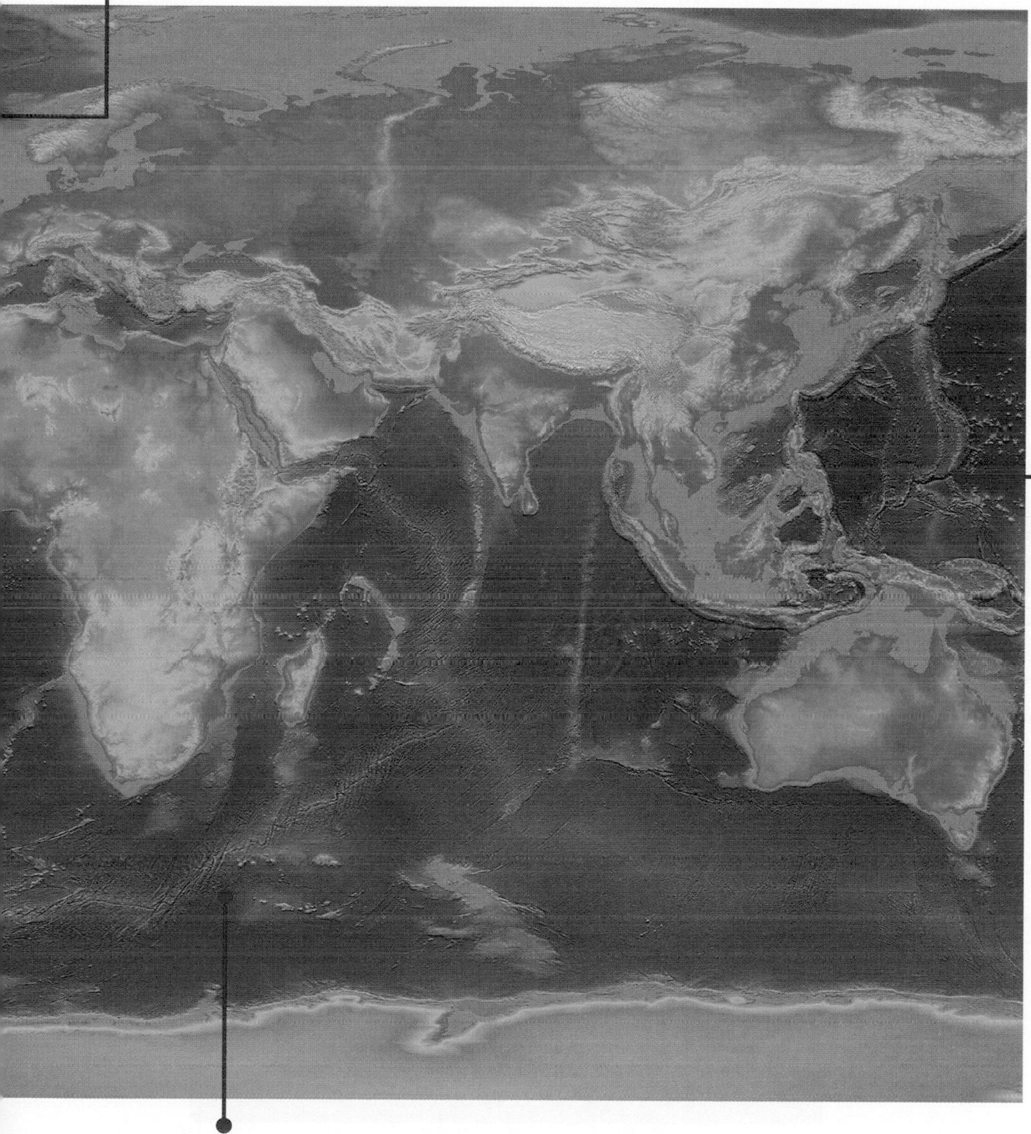

The **Mariana Trench** is the result of the collision of the Pacific plate with the Philippine plate. A deep trench has formed where one plate is sinking into the mantle. The deepest region of the Mariana Trench, the Challenger Deep, is about 11 km deep. This is farther below the ocean's surface than the peak of Mount Everest is above sea level.

As two oceanic plates pull apart, a mid-ocean ridge forms. However, the ridge is not a straight, even line. Instead, sections of the ridge are offset from one another. Transform faults—places where plates move past each other horizontally—connect each short section of mid-ocean ridge to the next. The **Southwest Indian Ridge** is an example of a mid-ocean ridge that is offset by transform faults.

Ocean temperatures vary around the world and according to the season.

Ocean water temperature varies from less than 1°C to more than 37°C. The latitude of an ocean surface affects the temperature of the ocean water. The water temperature is also related to water depth.

Average Ocean Surface Temperatures in August

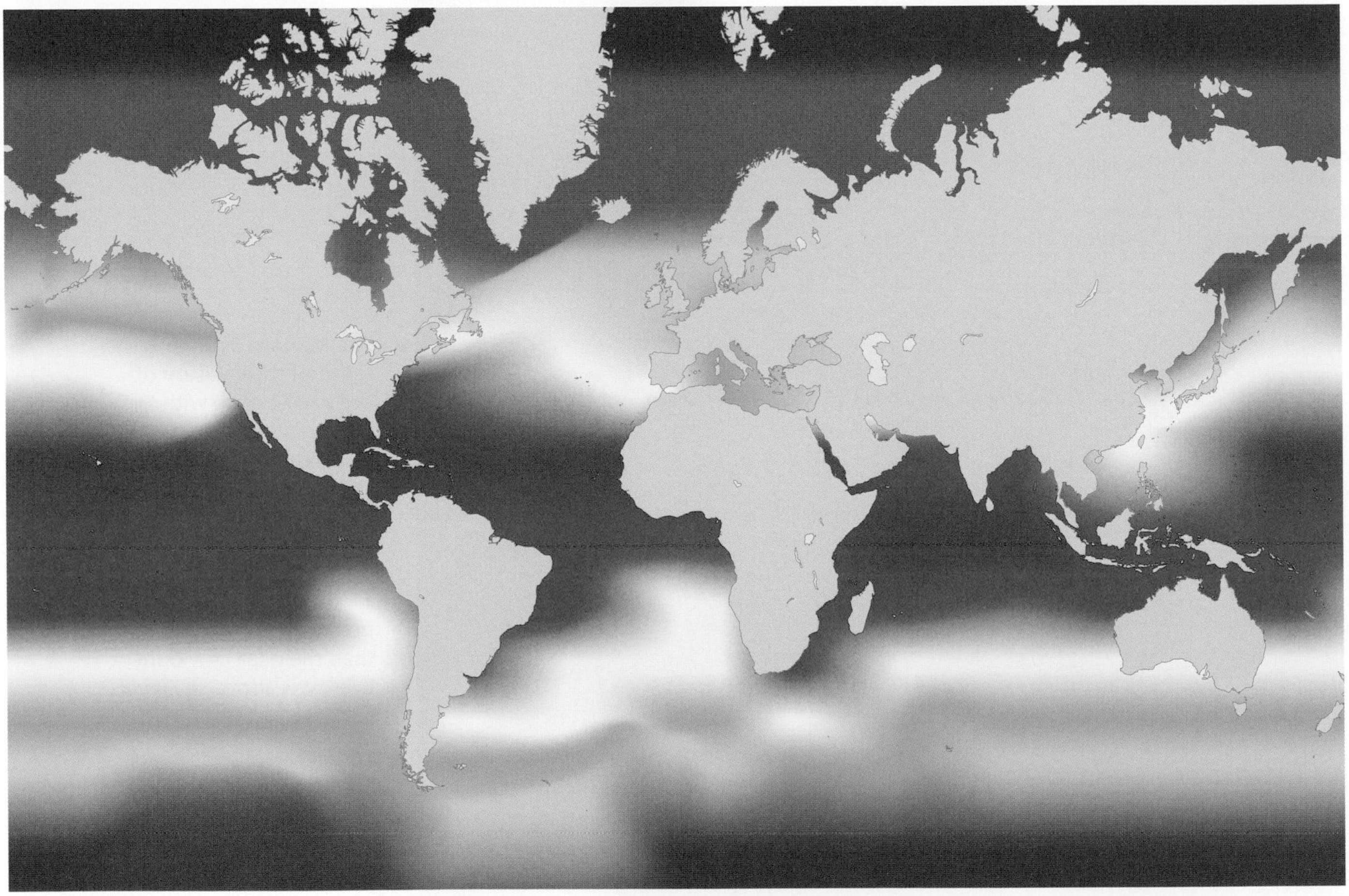

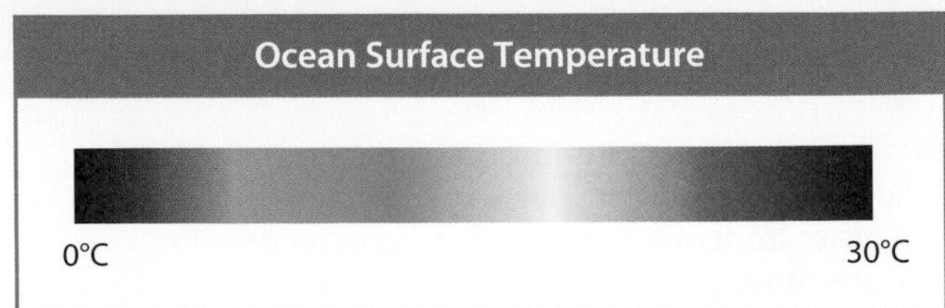

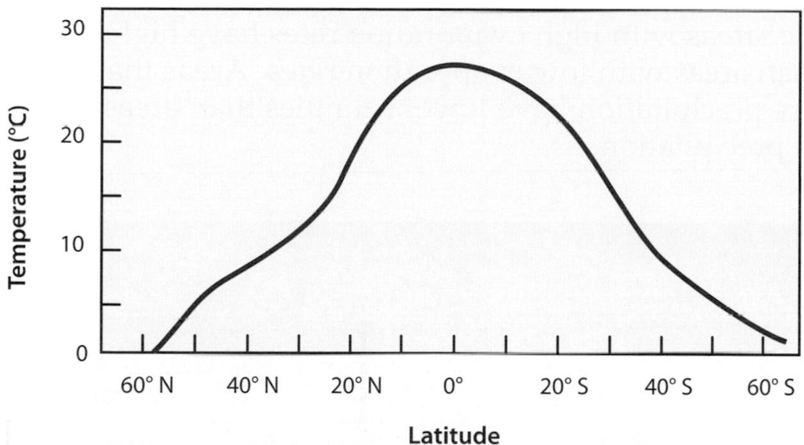

Temperatures and Depth

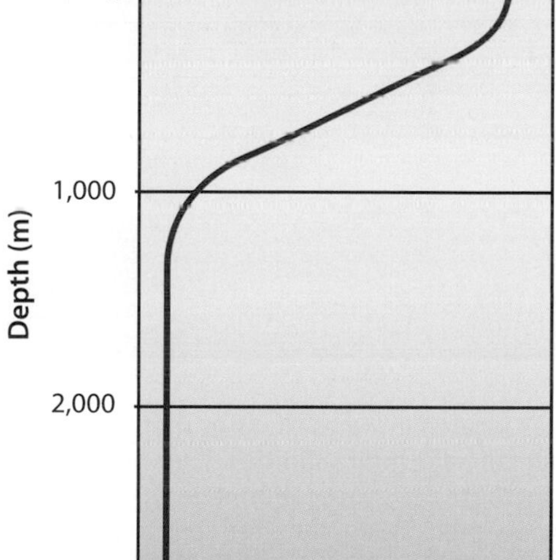

The salinity, or saltiness, of ocean water varies from place to place.

For example, areas with high evaporation rates have higher salinities than areas with low evaporation rates. Areas that receive more precipitation have lower salinities than areas that receive less precipitation.

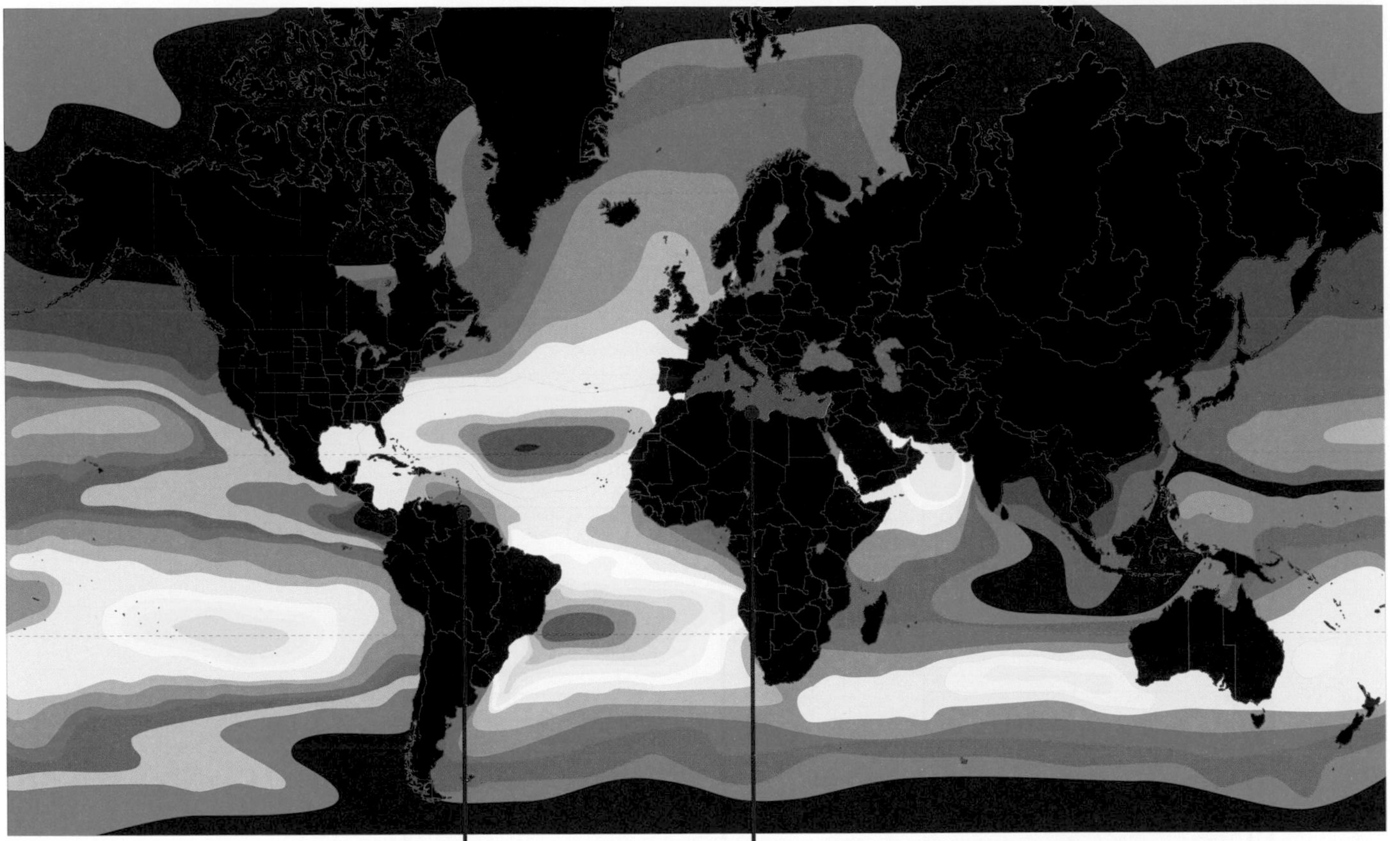

Areas where rivers flow into the ocean have **lower salinities** than other parts of the ocean. This is because the freshwater from the river dilutes the salty ocean water.

Oceans and seas located in warm climates that receive little rainfall have **high salinities**. The warm temperatures increase the evaporation rate of water, and the low precipitation rate reduces the amount of freshwater entering the sea. The salts remain in the liquid water, increasing its salinity.

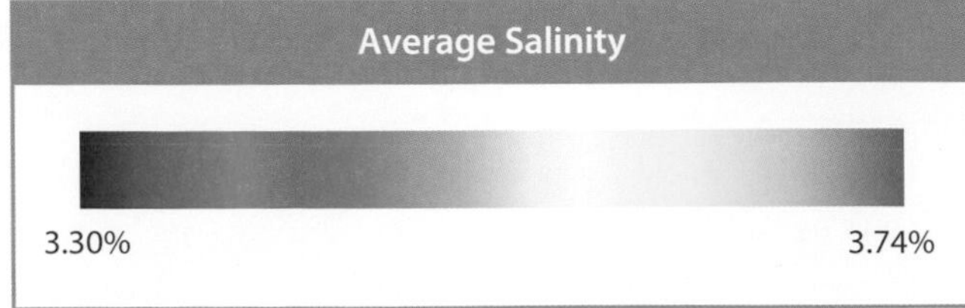

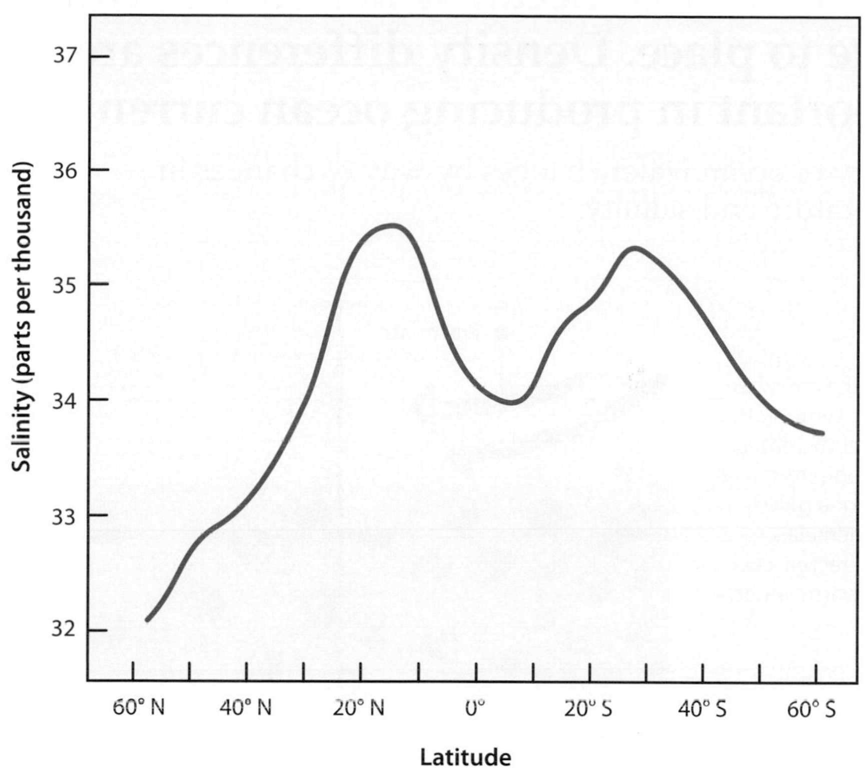

AVERAGE COMPOSITION OF SEAWATER

COMPONENT	*AVERAGE MASS OF COMPONENT IN 1,000 g OF SEAWATER*
sodium chloride (NaCl)	23.48 g
magnesium chloride (MgCl)	4.98 g
sodium sulfate (Na_2SO_4)	3.92 g
calcium chloride ($CaCl_2$)	1.10 g
potassium chloride (KCl)	0.66 g
sodium bicarbonate ($NaHCO_3$)	0.192 g
potassium bromide (KBr)	0.096 g
boric acid (H_3BO_3)	0.026 g
strontium chloride ($SrCl_2$)	0.024 g
sodium fluoride (NaF)	0.003 g

The density of ocean water varies from place to place. Density differences are important in producing ocean currents.

Density of ocean water changes by way of changes in temperature and salinity.

When warm air blows across cool water, the water absorbs heat from the air. The water's temperature increases. As the temperature of water increases, water molecules move faster. They begin to move farther apart. As a result, the volume of a given mass of water increases. As volume increases and mass remains the same, **density** decreases. Therefore, cold water is denser than warm water.

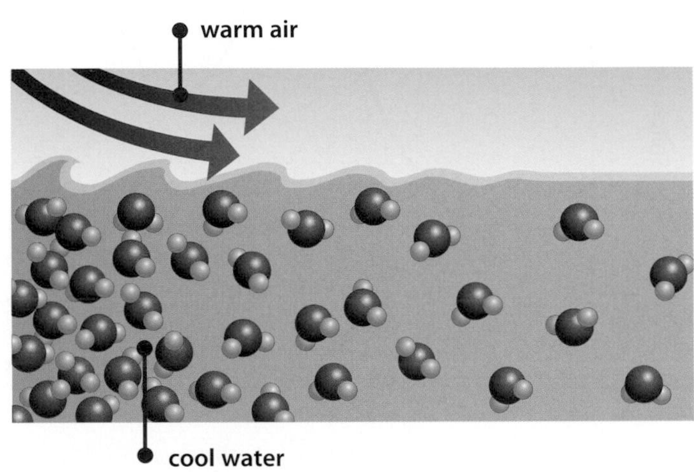

The salinity of ocean water can increase through **freezing**. The ice in the polar ice caps forms as water freezes from the oceans. However, only water molecules freeze to form the ice; materials that are dissolved in the water, such as salts, do not become part of the ice. These materials stay dissolved in the remaining water. Therefore, the concentration of salt in the remaining ocean water increases. This increases the density of the ocean water. Therefore, salty water is denser than less-salty water.

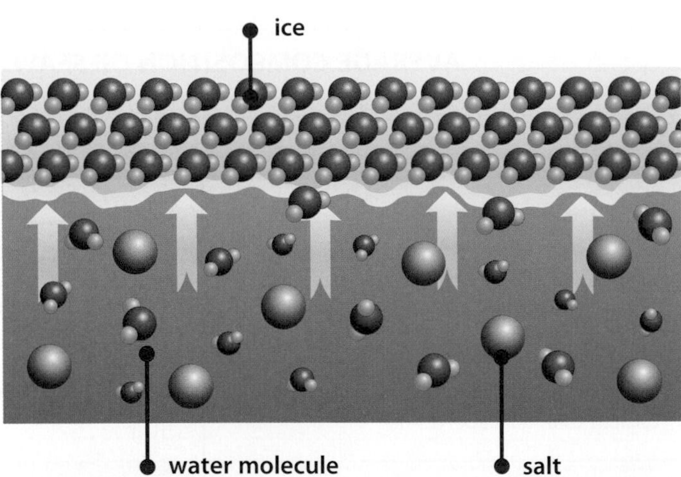

Another way that the salinity of ocean water increases is through **evaporation**. Water molecules are constantly evaporating from the ocean surface. In some areas, less liquid water enters the ocean than is evaporated. As a result, the concentration of dissolved materials in the water increases, causing the density of the water to increase.

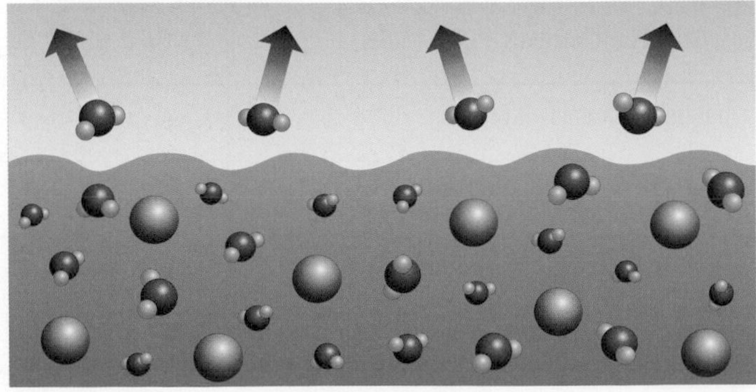

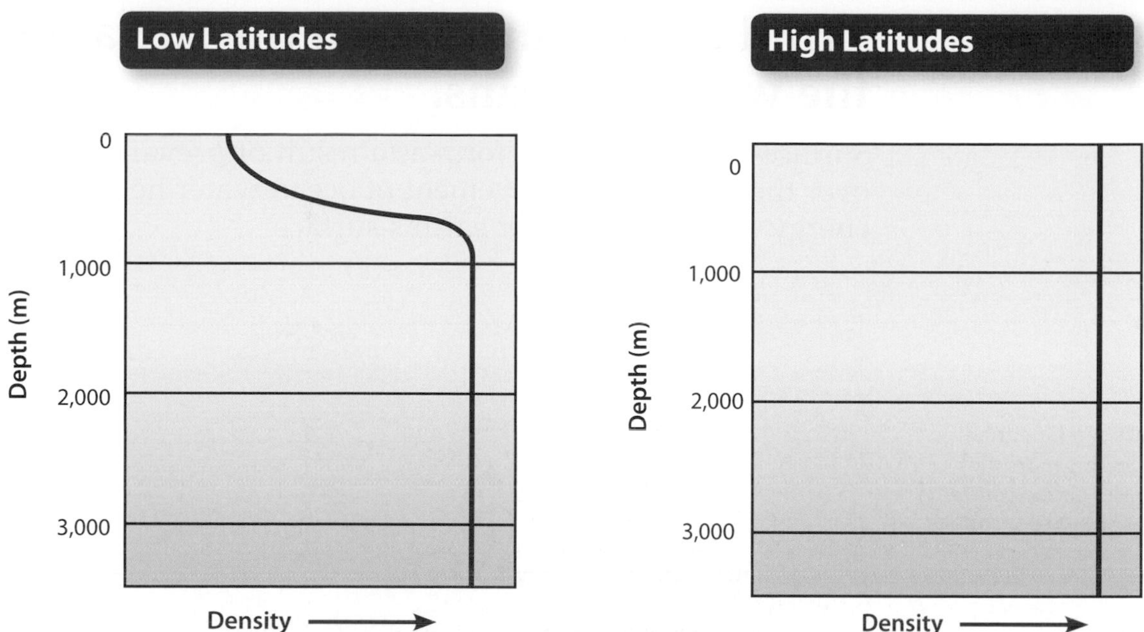

Differences in ocean water density are greater at low latitudes (near the equator) than at higher latitudes (near the poles). This is because there is a greater variation in water temperature at the equator than near the poles.

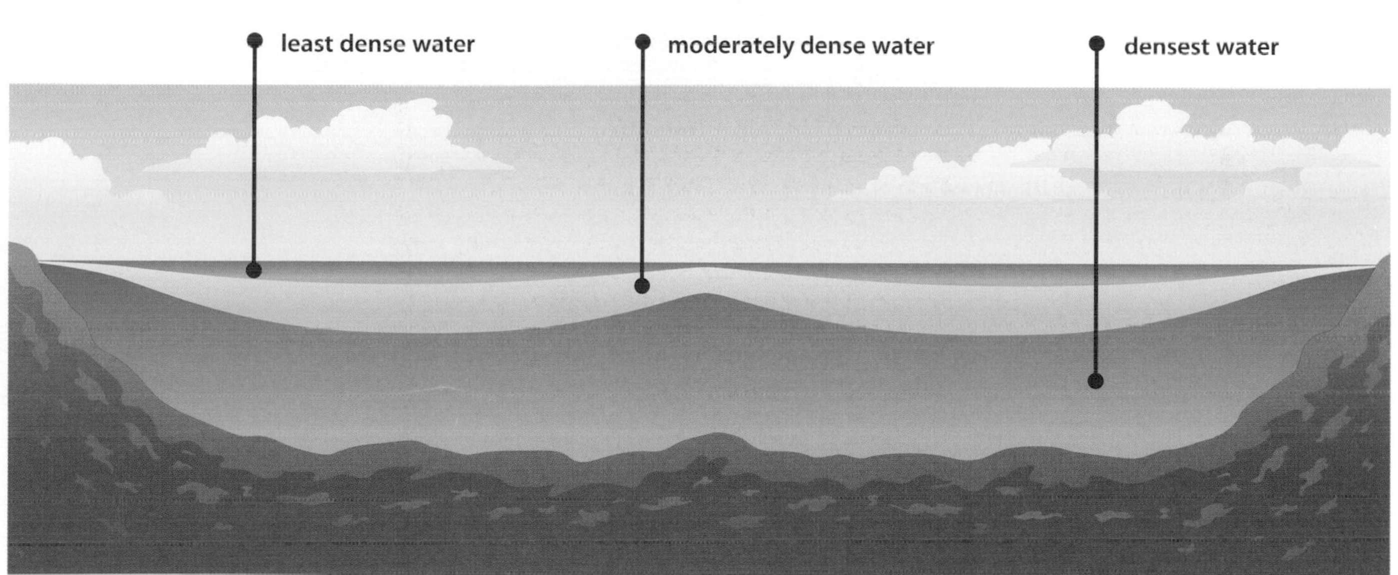

Scientists can divide the water in the oceans into **three layers based on density**. Deep, cold water is the densest. Warm, shallow water is the least dense.

Wind drives currents on the surface of the world's oceans.

Surface-current patterns form as a result of prevailing winds over the oceans. The movement of ocean water helps distribute energy from the sun over earth's surface.

Surface currents can be either warm or cool. On this map, pink arrows show warm surface currents and blue arrows show cool surface currents.

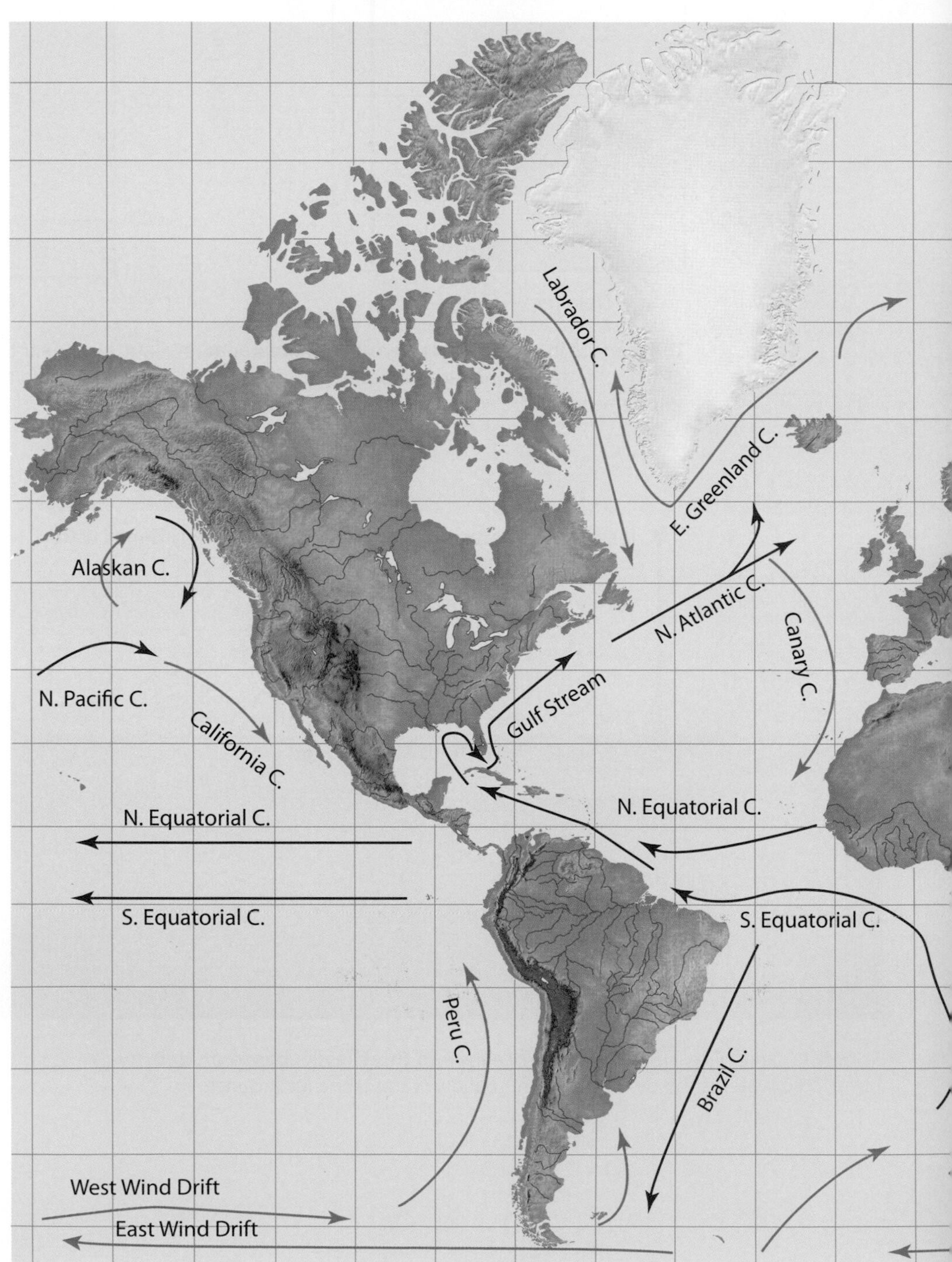

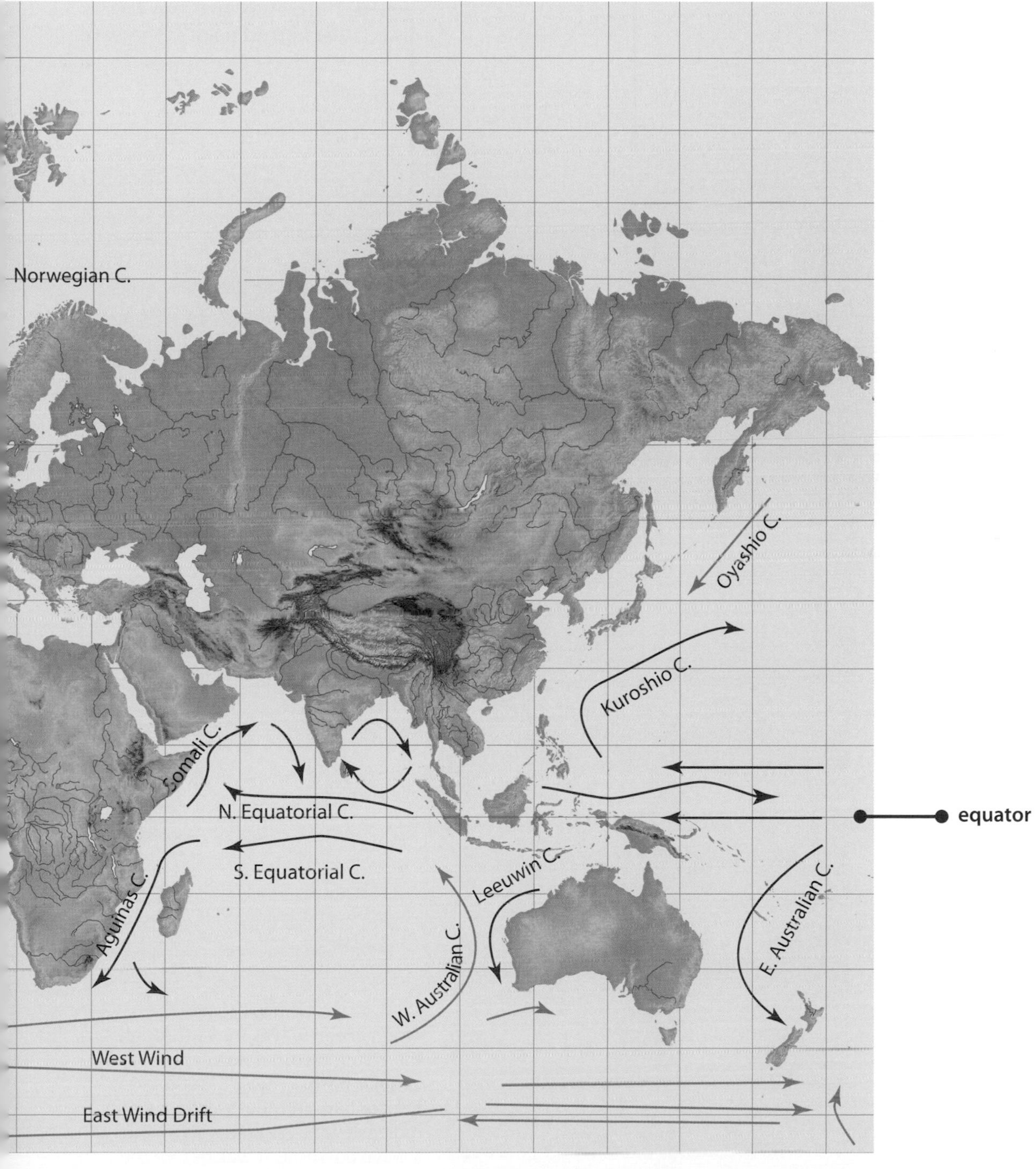
Norwegian C.
Oyashio C.
Kuroshio C.
Somali C.
N. Equatorial C.
S. Equatorial C.
Leeuwin C.
Agulhas C.
W. Australian C.
E. Australian C.
West Wind
East Wind Drift
equator

Ocean waves form when wind blows over the ocean's surface.

Ocean waves transfer energy from the air (wind) to the land. Although ocean waves transfer energy, they do not transfer matter.

The **wavelength** of an ocean wave is the horizontal distance between one crest or trough and the next.

The **crest** of an ocean wave is the highest point of the wave.

Like all waves, an ocean wave has an **amplitude.** The amplitude of an ocean wave is the difference in height between the middle of the wave and the crest or the trough.

The **trough** of an ocean wave is the lowest point of the wave.

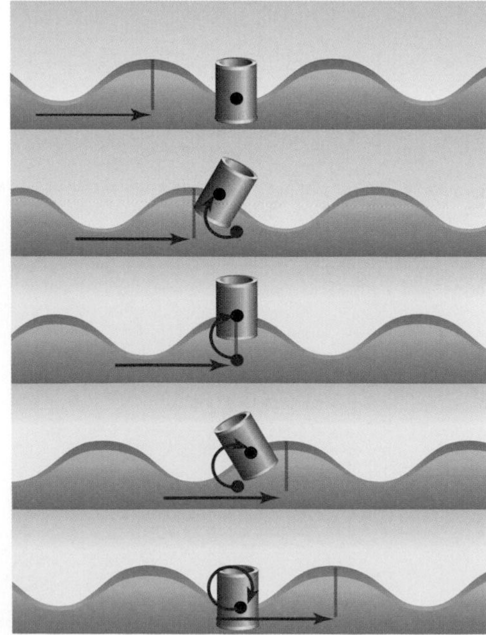

When ocean waves move through the water, water molecules move in **circular paths**. They do not move horizontally. In other words, a water molecule is in the same place after a wave passes as it was before the wave passed.

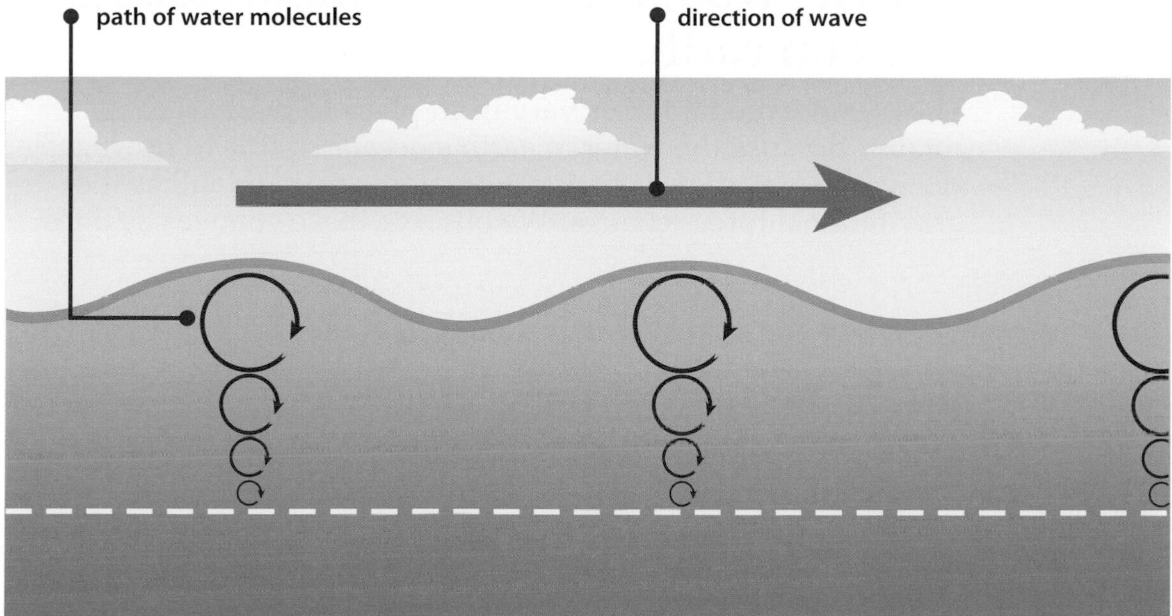

In the open ocean, **particles move in smaller and smaller circular paths** as water depth increases. Particles at the ocean surface move in larger circles than do particles deeper in the water. Water that is deeper than about one-half of the wavelength of a wave at the surface is not moved by the wave at all.

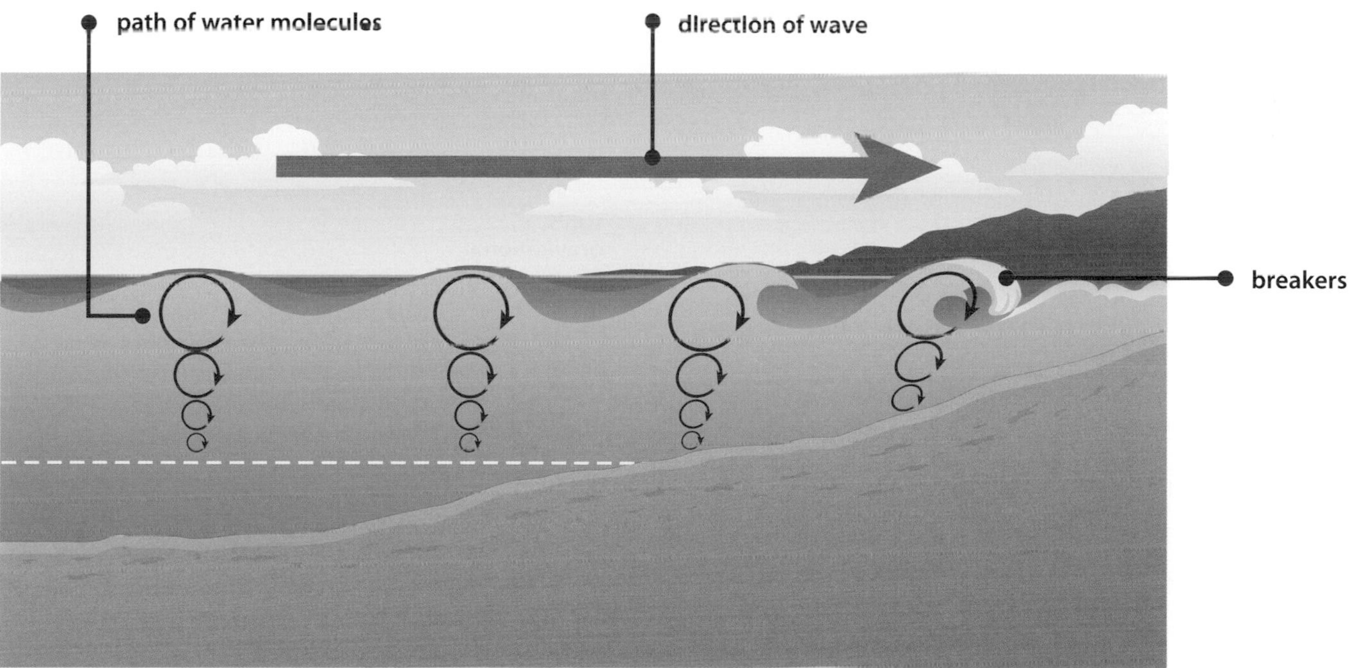

As waves move closer to shore, moving **particles begin to drag along the ocean floor**. The bottom of the wave moves more slowly than the top of the wave. The top of the wave piles up and the wave gets taller. Eventually, the wave gets too tall to support itself. When this happens, the water collapses downward to form breakers, which crash onto the shore.

Gravity from the sun and moon causes tides on earth.

These gravitational forces vary from place to place as earth rotates. Because the water in earth's oceans is able to flow easily, the water levels in different places rise and fall as the gravitational forces change. We observe these changes as tides.

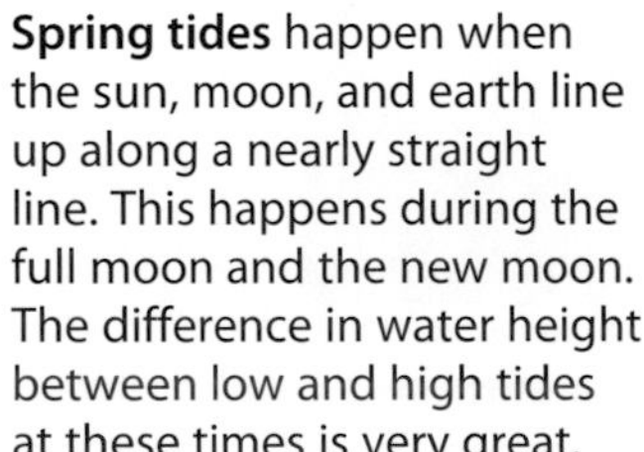

Spring tides happen when the sun, moon, and earth line up along a nearly straight line. This happens during the full moon and the new moon. The difference in water height between low and high tides at these times is very great.

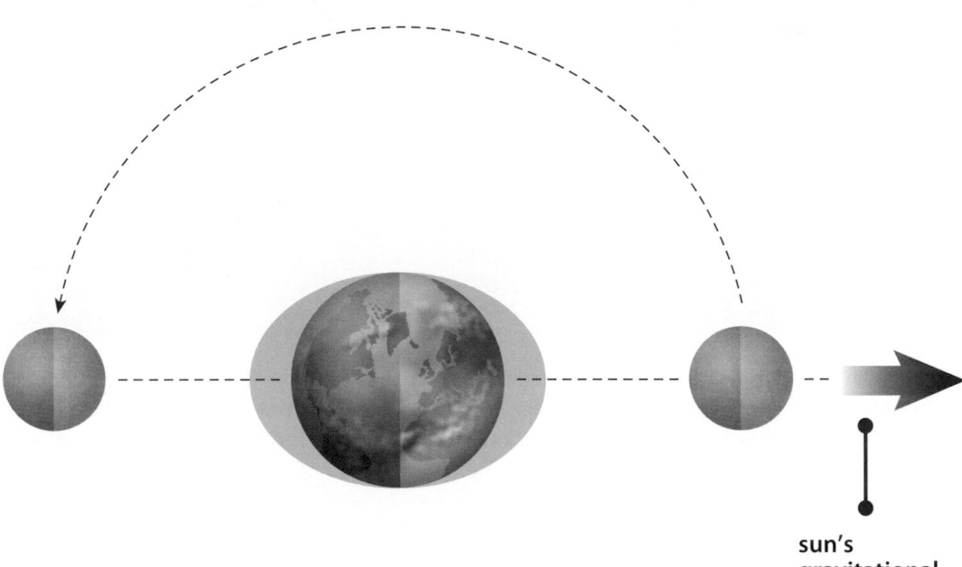

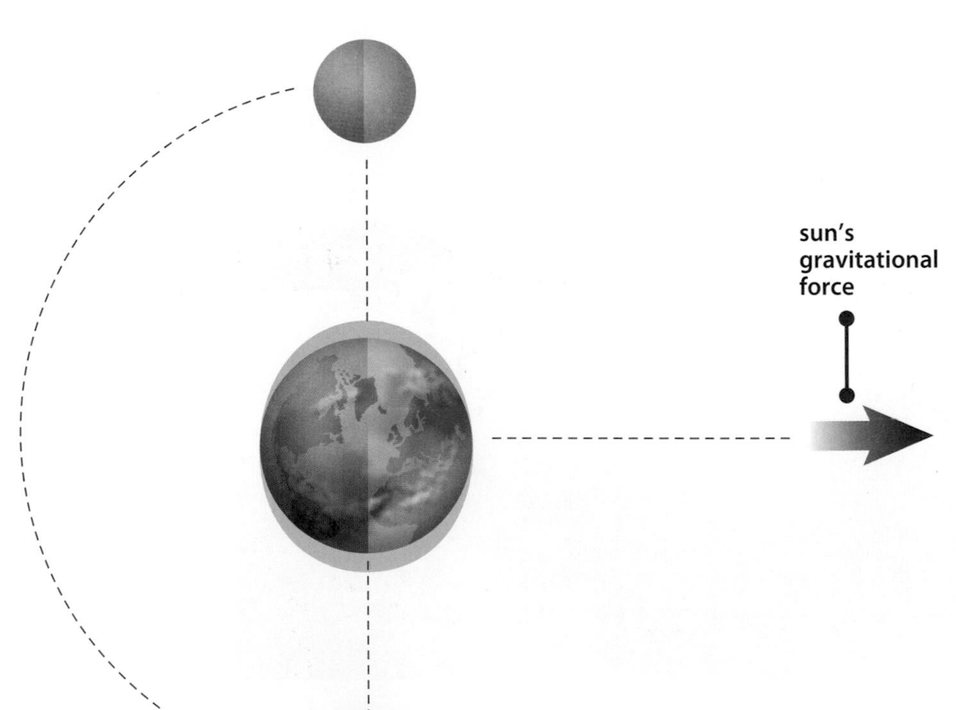

Neap tides happen when the sun, moon, and earth form a right angle. This happens during the first and third quarter moons. At these times, the sun's gravity and the moon's gravity pull the water in different directions. This causes the difference in water height between the low tide and the high tide to be smaller than it is at other times.

Some coastal areas (marked in yellow on this map) have one **low tide** and one **high tide** every day. Some areas (marked in green on this map) have two high tides that are about the same height and two low tides that are about the same height. Other areas (marked in red on this map) have two high tides and two low tides each day, but each tide is a different height.

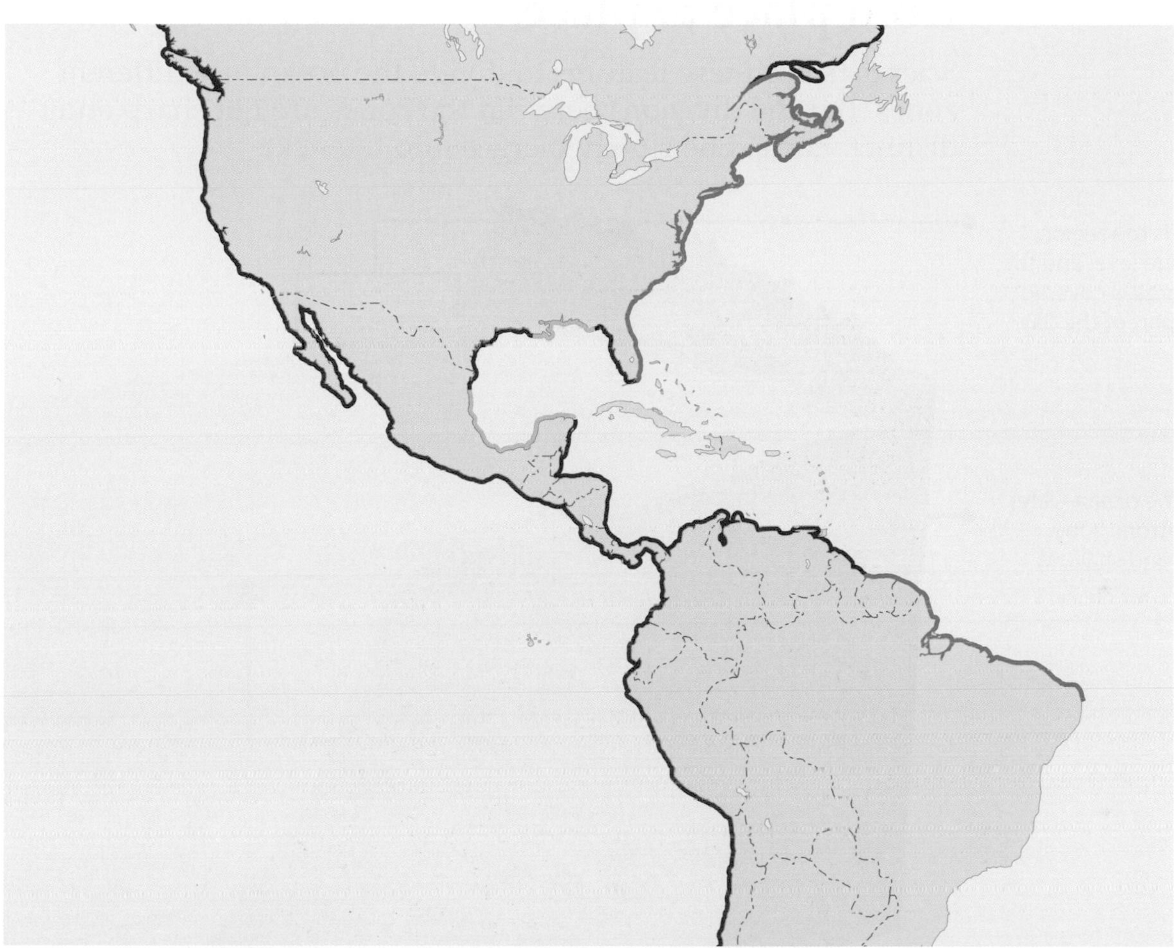

TIDE TABLE FOR BALTIMORE, MD, JUNE 4, 2006: FIRST QUARTER MOON			
FIRST HIGH TIDE	***FIRST HIGH TIDE HEIGHT***	***SECOND HIGH TIDE***	***SECOND HIGH TIDE HEIGHT***
9:27 a.m.	1.3 ft	10:27 p.m.	1.4 ft
FIRST LOW TIDE	***FIRST LOW TIDE HEIGHT***	***SECOND LOW TIDE***	***SECOND LOW TIDE HEIGHT***
3:44 a.m.	0.5 ft	4:06 p.m.	0.5 ft
SUNRISE	***SUNSET***	***MOONRISE***	***MOONSET***
5:41 a.m.	8:29 p.m.	1:35 p.m.	1:50 a.m.

The depth, temperature, and other features of the ocean vary greatly from place to place.

Scientists use these features to divide the ocean into different zones. But the divisions between the zones are not sharp and distinct. Each zone slowly merges into the next.

The **intertidal zone** is the region between the high-tide level and the low-tide level. This area is covered with water for only part of the day.

The **neritic zone** is the ocean water that covers the sublittoral zone. It is generally warm and shallow.

The **sublittoral zone** is the area of the ocean floor that extends from the low-tide level onshore to about 200 m from shore. Water in this region is well lit and generally much warmer than in deeper regions.

The **bathyal zone** is the part of the ocean floor that extends from the sublittoral zone to the abyssal zone. The bathyal zone ranges from about 200 m below sea level to about 4,000 m below sea level.

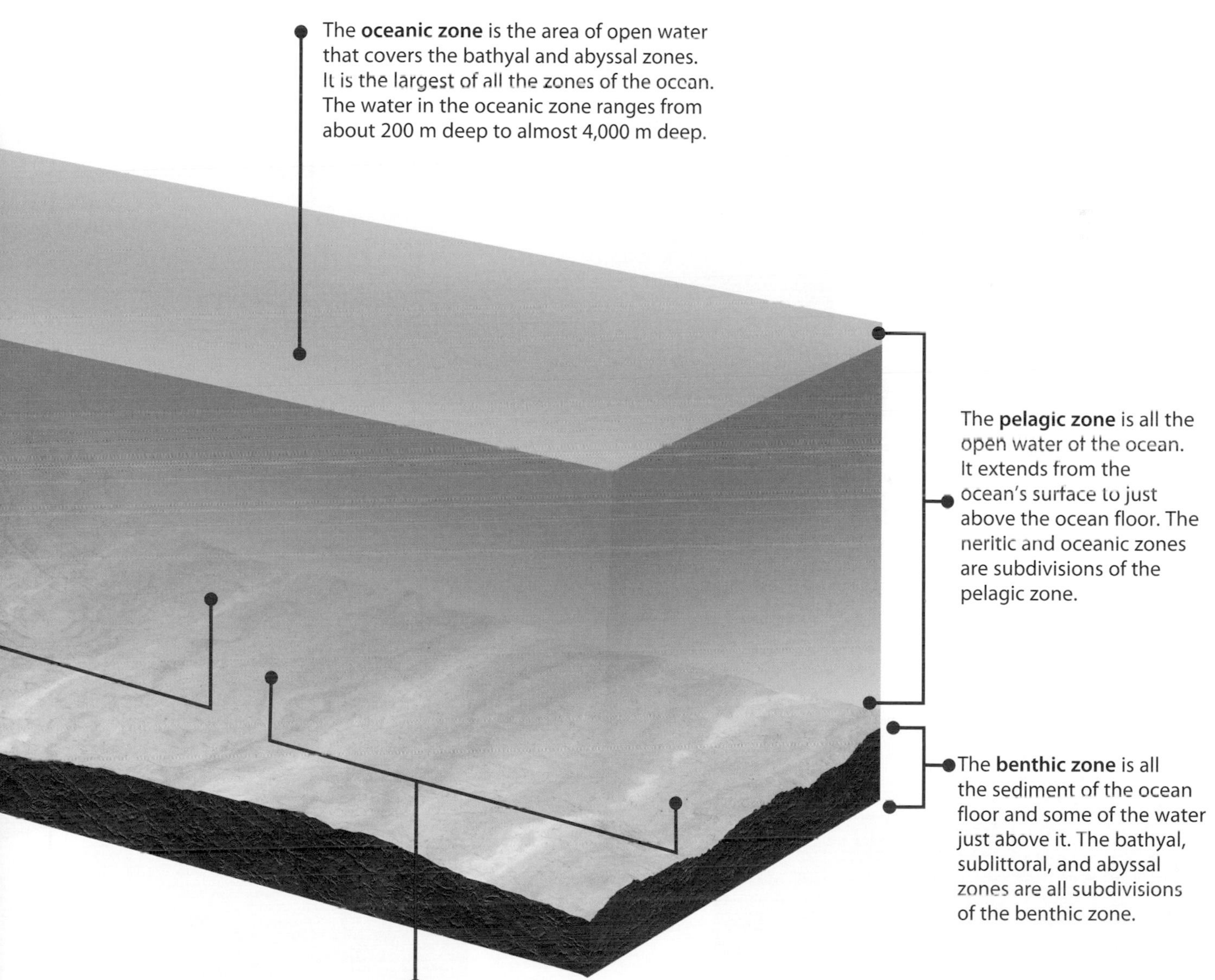

The **oceanic zone** is the area of open water that covers the bathyal and abyssal zones. It is the largest of all the zones of the ocean. The water in the oceanic zone ranges from about 200 m deep to almost 4,000 m deep.

The **pelagic zone** is all the open water of the ocean. It extends from the ocean's surface to just above the ocean floor. The neritic and oceanic zones are subdivisions of the pelagic zone.

The **benthic zone** is all the sediment of the ocean floor and some of the water just above it. The bathyal, sublittoral, and abyssal zones are all subdivisions of the benthic zone.

The **abyssal zone** is the deepest part of the ocean floor. It can be more than 4,000 m below sea level. This part of the ocean floor is very cold and dark because no sunlight can reach it. However, small parts of the abyssal zone near mid-ocean ridges and hot spots may be heated by the hot rock and magma below the surface.

Marine organisms have adaptations that help them survive the conditions that exist where they live.

For example, organisms that live in the intertidal zone have adaptations that allow them to survive crashing waves. Organisms that live in the abyssal zone are adapted to extremely dark, cold, high-pressure conditions.

Intertidal zone: ghost crab

Sublittoral zone: coral reef

Bathyal zone: octopus

Oceanic zone (shallow): manatee

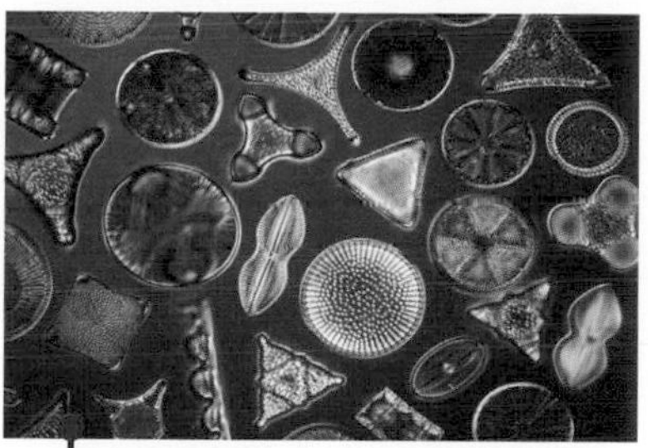

Oceanic zone (shallow): plankton

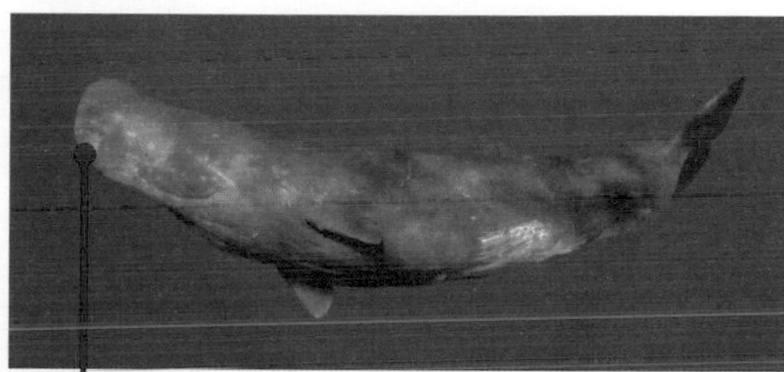

Oceanic zone (moderate depth): sperm whale

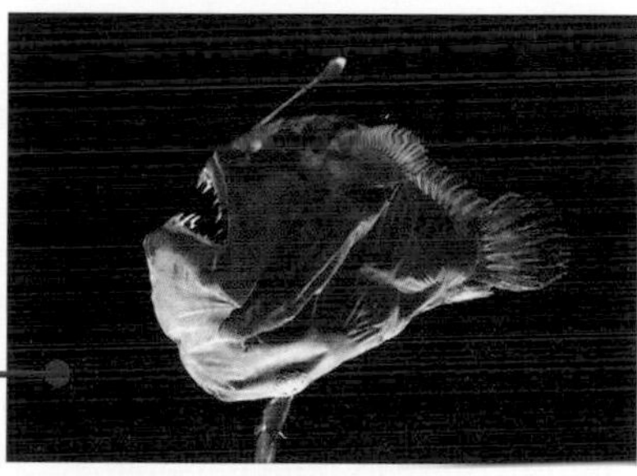

Oceanic zone (deep): anglerfish

Abyssal zone: tube worm

Nitrogen on earth moves continuously between reservoirs.

The reservoirs of nitrogen and the processes that cause nitrogen to move between them make up the nitrogen cycle.

The **atmosphere** is the largest nitrogen reservoir on earth. Nearly all nitrogen in the atmosphere is in the form of nitrogen gas (N_2).

Soil bacteria known as *denitrifying bacteria* convert nitrate into **nitrogen gas**, which moves from the soil into the atmosphere.

All living things require **nitrogen** to survive, but most cannot use nitrogen gas directly from the atmosphere. *Nitrogen-fixing bacteria* convert nitrogen gas into other compounds, such as ammonia and nitrates, that plants can use. Animals obtain usable nitrogen by eating plants or other organisms. When organisms decay, the nitrogen in their bodies returns to the soil.

Human activities—such as the use of fertilizers and the burning of fossil fuels—can change the amount of nitrates and other nitrogen compounds in earth's system. In fact, human actions have nearly doubled the amount of atmospheric nitrogen converted into ammonia and nitrate. This increase has a significant impact on the global nitrogen cycle.

Ammonia, nitrates, and other nitrogen compounds dissolve easily in rainwater. **Runoff** can carry these compounds into the oceans.

Marine organisms rely on nitrogen-fixing bacteria to convert atmospheric nitrogen gas into ammonia and nitrate. Phytoplankton and other producers in the oceans can use these nitrogen compounds. Marine animals obtain nitrogen by feeding on other organisms. Decomposers also help mobilize nitrogen in marine ecosystems. Denitrifying bacteria convert nitrate into nitrogen gas, which flows back into the atmosphere.

Carbon is the basis of all life on earth.

The carbon on earth moves through various reservoirs, including living things, the atmosphere, the oceans, and even rocks. Carbon exists in different forms in each reservoir, and various processes act to move carbon between reservoirs. Without these processes, life on earth would not be possible.

When **fossil fuels and organic matter** burn, carbon dioxide is released into the atmosphere.

Plants and other photosynthetic organisms remove carbon dioxide from the atmosphere. During **photosynthesis**, the carbon dioxide is converted into glucose and other materials that the organisms use for growth and development.

Plants, animals, and most other organisms on earth release carbon dioxide into the atmosphere as part of **cellular respiration**. During celullar respiration, sugars combine with oxygen to produce carbon dioxide, water, and energy that the organisms can use.

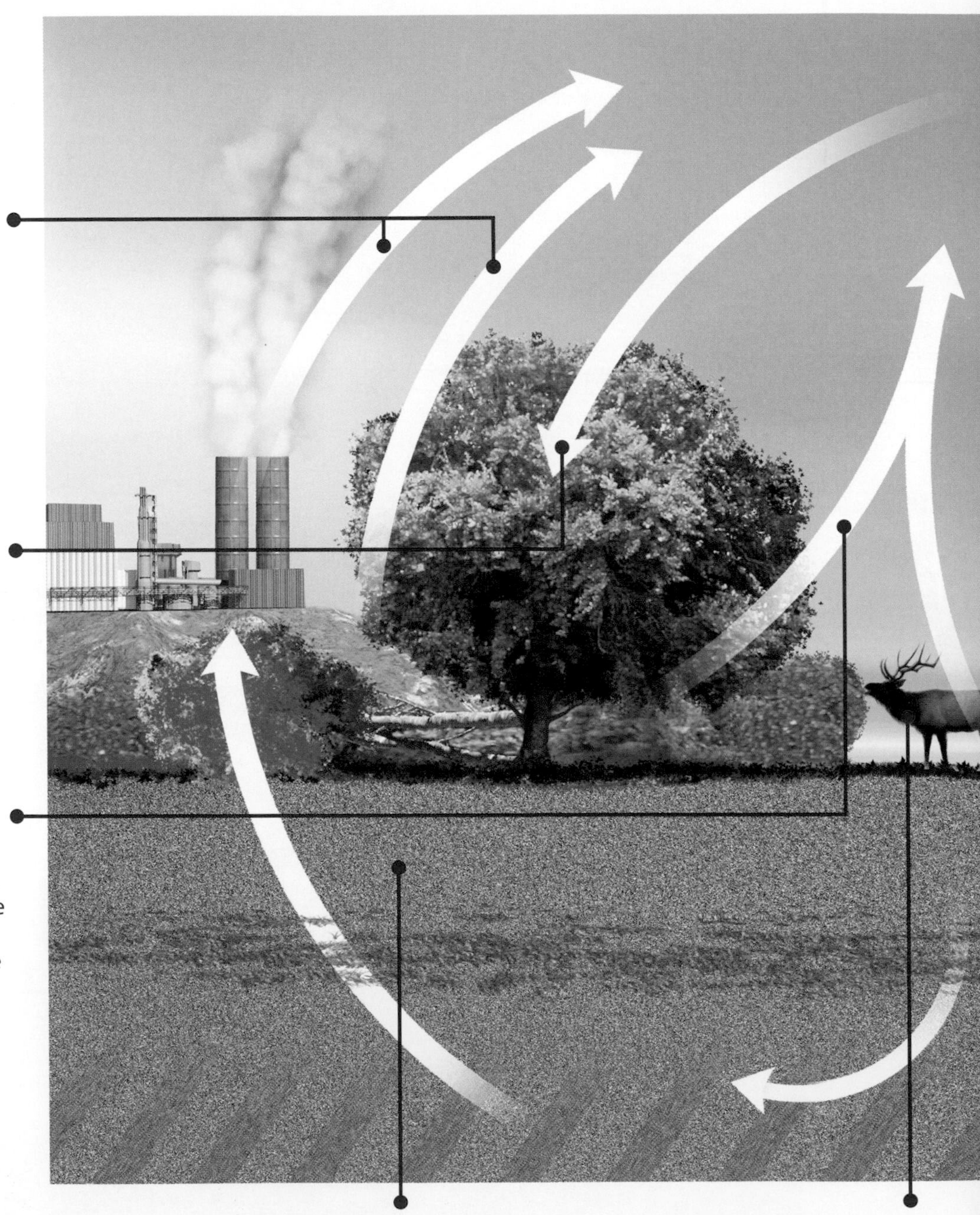

The **organic matter and rock fragments** in soil contain carbon. Rocks below or at earth's surface may contain carbon in the form of minerals or fossil fuels.

The **living things** on earth's surface contain carbon in their cells.

Most of the carbon in the atmosphere is in the form of carbon dioxide. Methane and other gaseous hydrocarbons in the atmosphere also contain carbon. These compounds act as **greenhouse gases**, absorbing light and radiating it back as heat. This helps keep the earth warm.

Carbon dioxide dissolved in **water** diffuses into the atmosphere. In addition, carbon dioxide in the atmosphere dissolves in surface water.

The **oceans** are the largest carbon reservoir on earth. Ocean-dwelling organisms contain carbon in their cells. Some, such as corals and shellfish, use carbon to build hard skeletons out of carbonate minerals. Carbon dioxide is also dissolved in ocean water, but in a less significant amount.

When organisms in the oceans die and decompose, one by-product is carbon dioxde, which can dissolve in ocean water or move into the atmosphere. The remains of some **ocean organisms** do not completely decay, or release all their carbon dioxide. Instead, they can be buried and compressed. After millions of years, they may become fossil fuels such as oil and natural gas.

When dead organisms break down, carbon dioxide is released into the atmosphere. Some of the carbon is also incorporated into soil organic matter. The **remains of some organisms** can be buried and compressed to produce fossil fuels, such as coal.

Phosphorus constantly recycles through different processes.

These processes and the reservoirs that store phosphorus make up the phosphorus cycle.

Rocks represent a major phosphorus reservoir. They may be uplifted and brought close to the surface by geologic processes such as mountain building. Weathering by water and mining by people can break down the rock, allowing phosphorus to be released into the soil. In addition, people add fertilizers that may contain phosphorus to the soil.

Plants and soil microorganisms take up phosphorus from soil and from rainwater. Animals obtain phosphorus by eating plants and other organisms. When organisms die, decomposers, such as bacteria and fungi, break down the remains and return the phosphorus to the soil.

Runoff carries phosphorus from the land to the oceans.

Sewage treatment plants release **wastewater** that contains significant amounts of phosphorus, which ends up in the oceans.

The **oceans** represent a major phosphorus reservoir on earth. Most of this phosphorus is dissolved in the ocean water.

Marine organisms obtain phosphorus from ocean water and by eating other organisms.

When marine organisms die, their **remains** may settle to the ocean floor and become buried in sediment. Phosphorus buried in marine sediment is not available for organisms to use.

Water does not enter or exit the earth system.

But the water on earth does change form, moving from one reservoir to another, such as from the oceans to the atmosphere. The processes that move water between reservoirs make up the water cycle.

Water that moves over the land into the oceans is called **runoff.** Runoff may flow directly into the oceans or it may flow into other water bodies, such as ponds, lakes, and rivers, which then drain into the oceans.

Water in the **atmosphere** exists primarily as water vapor, a gas. However, the water in clouds is in the form of tiny liquid water droplets.

Living things, especially plants, give off water vapor during **respiration** and **transpiration.** When animals breathe out, they give off water vapor. During transpiration, water passes out of plants by way of tiny pores in the surfaces of their leaves and then evaporates into the atmosphere.

All **living things** contain water in their cells. Without water, organisms on earth could not survive.

Only a tiny fraction of earth's water is **surface water:** lakes, rivers, ponds, swamps, and streams.

Water moves from earth's surface into the ground through **percolation.** Percolation is the primary way that water enters aquifers (underground rock bodies that store water).

Groundwater is the second largest freshwater reservoir on earth. Groundwater is water that is located below earth's surface, stored in layers of rock.

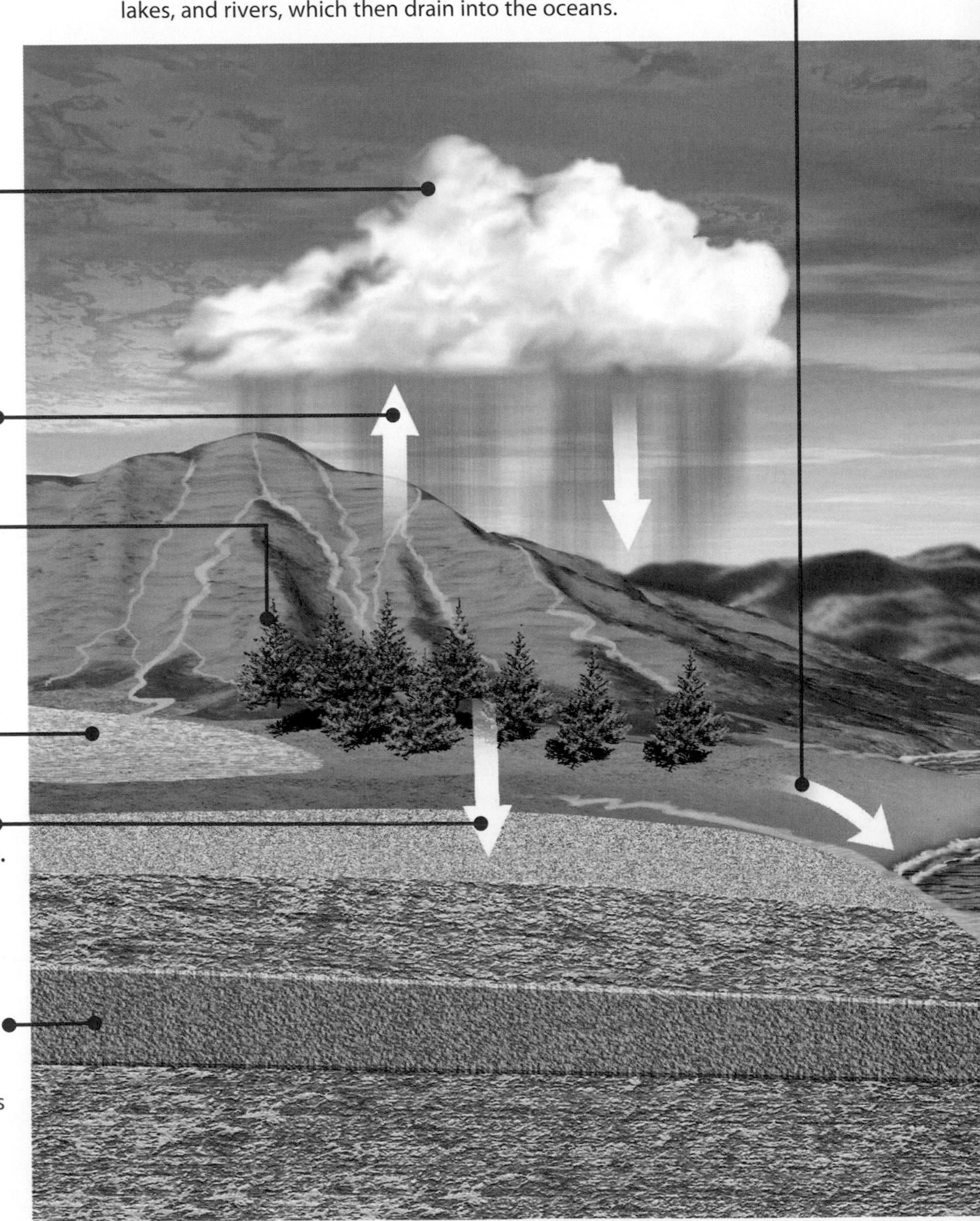

Water moves from oceans and surface waters into the atmosphere through **evaporation.** The main source of energy for evaporation is sunlight.

Water moves from the atmosphere to earth's surface through **precipitation.** Most precipitation falls on the oceans

The **oceans** are the largest water reservoirs on earth. About 97 percent of the water in the earth system is located in the oceans.

Approximately 2 percent of all water in the earth system is located in **ice** in glaciers and icebergs. This ice makes up the largest freshwater reservoir on earth.

Liquid water turns into ice during the process of **freezing.** Water near the poles and at high elevations freezes to form snow, glaciers, and sea ice.

There are more than 4,000 different minerals on earth.

Most minerals form in just a few ways. Five common ways that minerals can form are evaporation, crystallization from water, crystallization from magma, metamorphism, and precipitation from hot water.

Ocean water contains many dissolved chemicals. If ocean water evaporates, these chemicals are left behind and may form **crystals** of different minerals. Minerals such as halite form this way.

Minerals and rocks beneath earth's surface may react with each other due to high temperatures and high pressures there. Such **metamorphic reactions** produce new minerals. For example, the garnet and biotite in this schist probably formed when the minerals chlorite, muscovite, and quartz in the original rock reacted during metamorphism.

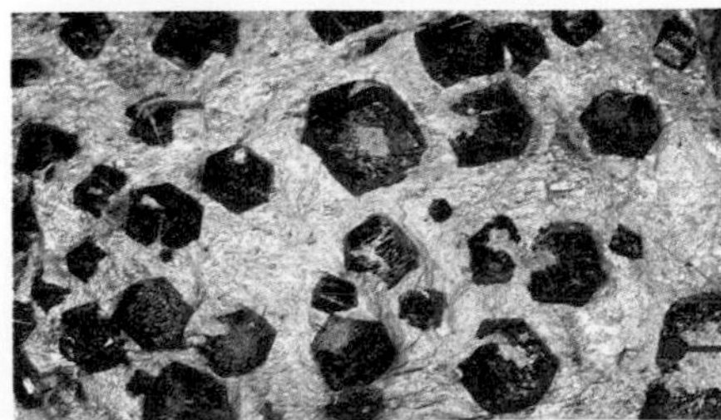

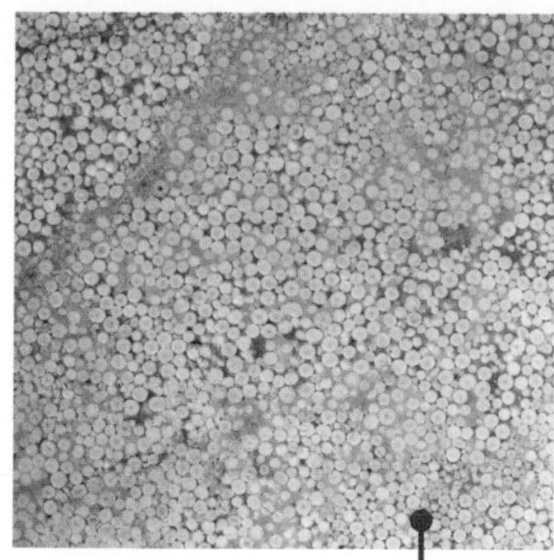

Minerals such as calcite and dolomite dissolve in groundwater and surface water. When water containing these **dissolved minerals** flows into lakes or oceans, the minerals crystallize on the bottom of the water bodies. The calcite and dolomite in some types of limestone form this way.

Rocks deep below earth's surface may be very hot. Groundwater that flows through these rocks becomes heated. The hot water dissolves many different minerals from the rocks that it flows through. When the water rises toward the surface, it cools, and some of these minerals precipitate (form crystals) out of the water solution. This precipitation produces **veins of minerals**, such as the gold shown above.

Magma is melted rock below earth's surface. It contains many minerals. As the magma cools, different minerals begin to crystallize at different temperatures. These minerals can form large crystals if the magma cools slowly enough. Coarse-grained **igneous rocks** such as granite contain large crystals of minerals such as mica, feldspar, and quartz.

Metals and other mineral materials that people use come from ores.

An *ore* is a rock, mineral, or sediment deposit that contains a high concentration of a useful element or material. People mine ores and then refine them to extract the useful materials.

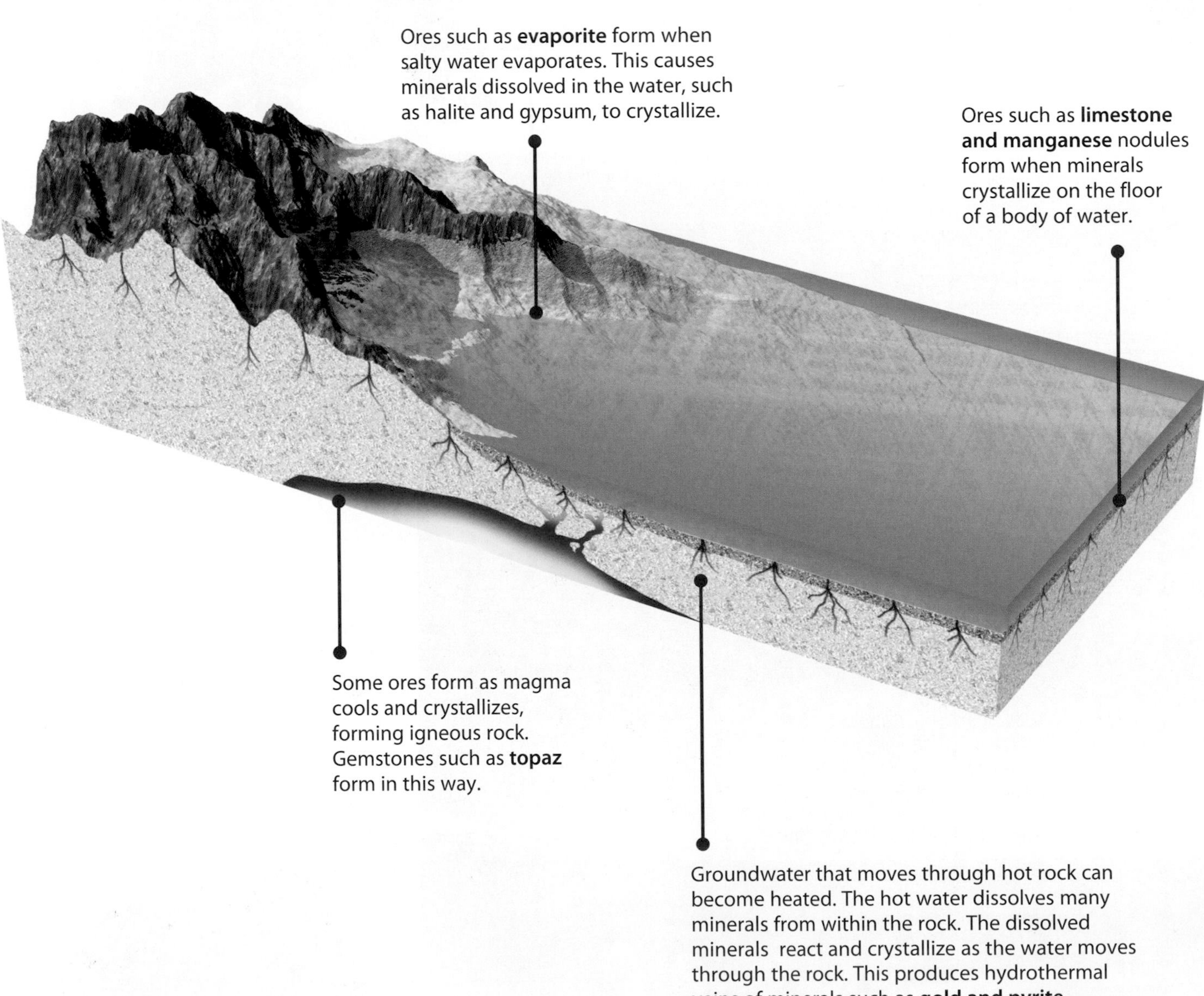

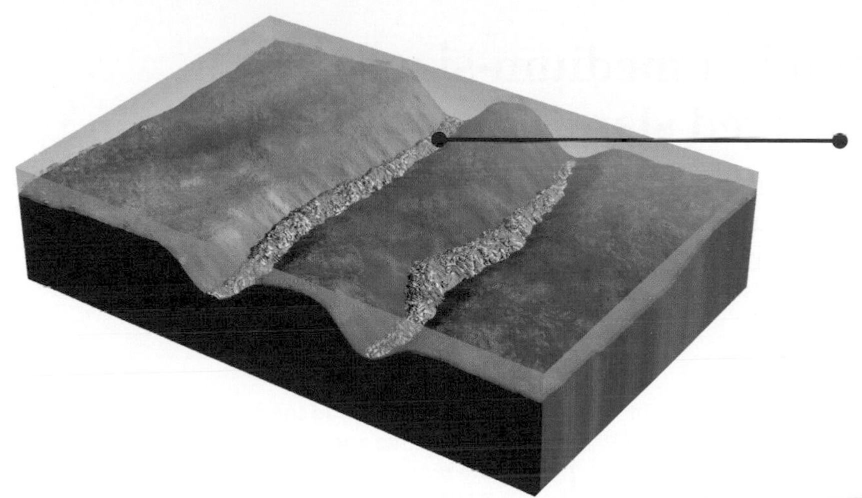

Placer deposits form where fast-moving water slows down. Some placer deposits form in the bottoms of deep spots in rivers and streams, where the water does not move as quickly, enabling heavy minerals to settle out.

Heavy minerals, such as gold, can be carried as **suspended sediment** in fast-moving rivers and streams. When the water in such a stream flows around a bend, the water on the inside of the bend slows down. When that happens, the heavy minerals that are suspended in the water may settle out, forming placer deposits of heavy minerals mixed with other sediment.

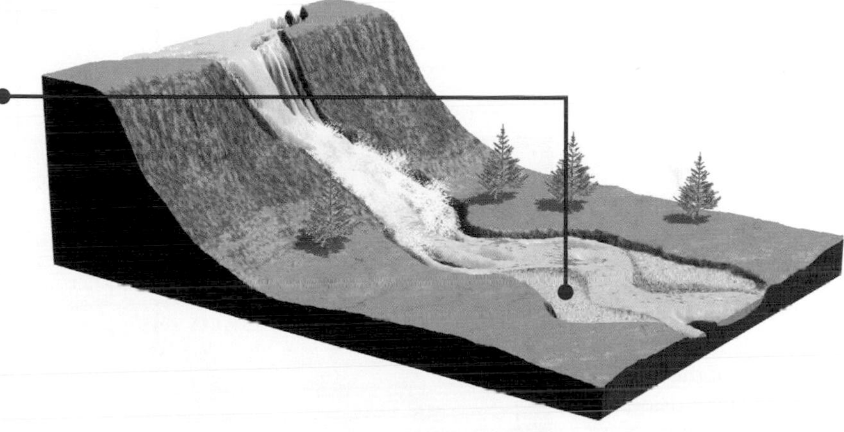

USEFUL MATERIALS OBTAINED FROM ORES

MATERIAL	*SOURCE(S)*	*USES*
calcite	limestone, marble	production of cement; sculpture; building construction
copper	chalcopyrite, bornite, malachite	electronics; coins; jewelry; sculpture
diamond	kimberlite rocks, placer deposits	jewelry; cutting tools; abrasives
gold	placer deposits and hydrothermal veins in rock	electronics; dental work; jewelry; coins
graphite	veins in metamorphic rocks	pencils; paints; lubricants; batteries
gypsum	evaporite deposits, sedimentary rocks	production of plaster; wallboard
halite	evaporite deposits	table salt; road salt; chemical reactions
iron	hematite and magnetite	making steel; tools and cooking utensils; food supplement
kaolinite	clay	cement; ceramics; bricks
lead	galena	batteries; solder; weights; ammunition
quartz	sand, sandstone	glass; construction
silver	ores of lead, zinc, and copper (by-product of refining)	electronics; dental work; jewelry; coins; utensils
sulfur	sulfide minerals (such as galena, pyrite, and sphalerite)	gunpowder; medications; rubber
zinc	sphalerite, zincite, smithsonite	production of brass and galvanized steel

Our sun is a medium-sized, middle-aged star.

Its volume is about 1.3 million times greater than the volume of earth. The sun's energy comes from the fusion of hydrogen nuclei to form helium nuclei. The interior of the sun is divided into layers, each having distinct characteristics.

Fusion within the sun's core produces an enormous amount of energy. Most of this energy is in the form of X rays. The X rays travel outward from the core through the **radiative zone.** Within this zone, the energy travels as radiation. It takes about one million years for radiation to travel through the radiative zone.

The layer outside the radiative zone is the **convective zone.** Within the convective zone, the temperature drops significantly—from 2,000,000°C to about 5,700°C. Energy is transferred through this zone by convection of the matter within it.

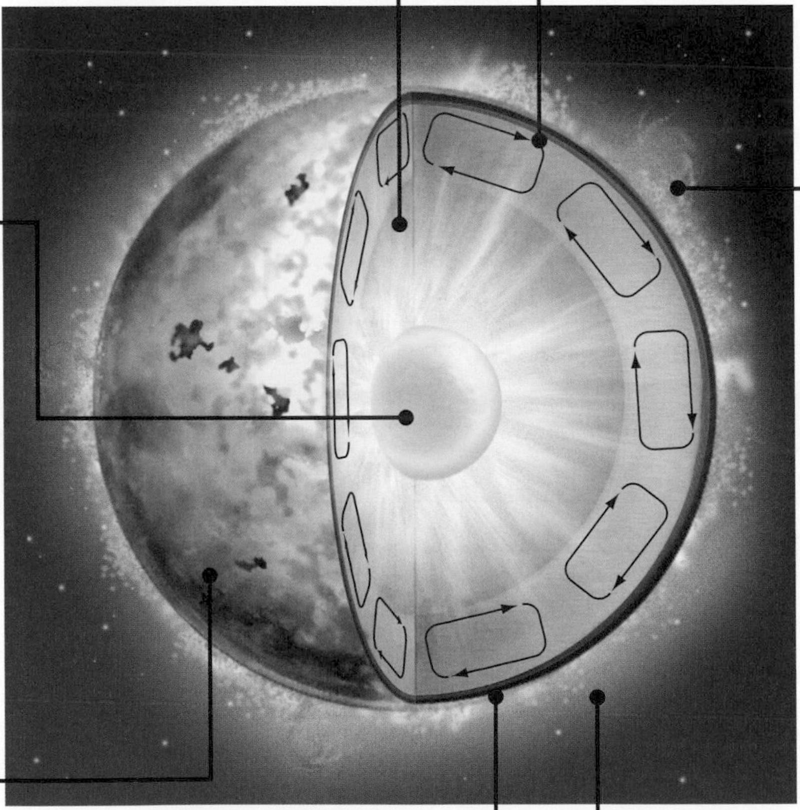

The **core** is the innermost layer of the sun, where nuclear fusion occurs. The radius of the sun's core is about one-quarter of the radius of the sun. Temperatures in the core are estimated to be between 7,000,000 and 15,000,000°C.

The **corona** is the outermost layer of the sun's atmosphere. The temperature in the corona is extremely high—about 1,000,000°C—although scientists are not sure what causes such high temperatures. The solar wind originates in the corona.

The **photosphere** is what we consider to be the surface of the sun. Most of the sunlight that we see on earth is produced in this very thin layer. Light from the sun travels through the sun's atmosphere and then through space to earth.

The **chromosphere** is part of the sun's atmosphere. Within this layer, the temperature increases from 6,000°C to 20,000°C. Most solar flares begin in the chromosphere.

The **solar wind** is a continuous stream of charged particles that moves at extremely high speeds away from the sun in all directions. The interaction of the solar wind with earth's ionosphere produces the aurorae (the Northern and Southern lights).

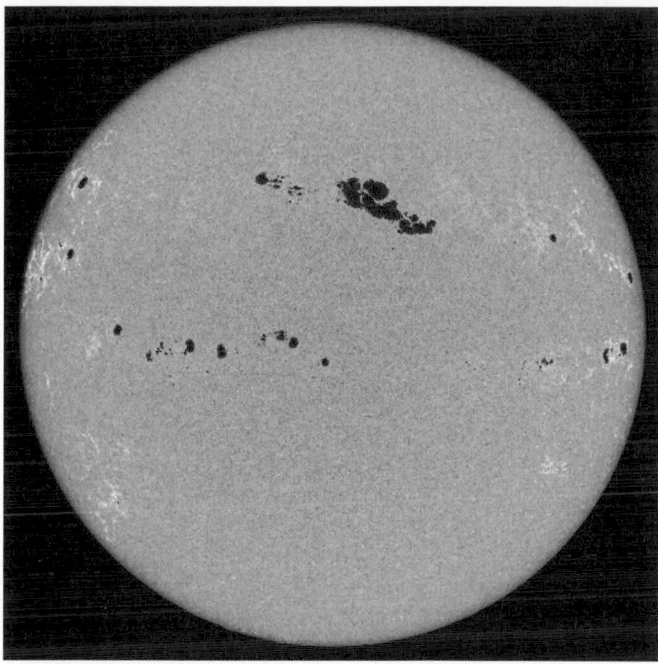

Sunspots are relatively cool areas of the photosphere. They appear darker than the rest of the sun because they are comparatively cooler. Sunspots are the result of unusually strong magnetic fields in the sun. They move across the sun's face in cyclic patterns.

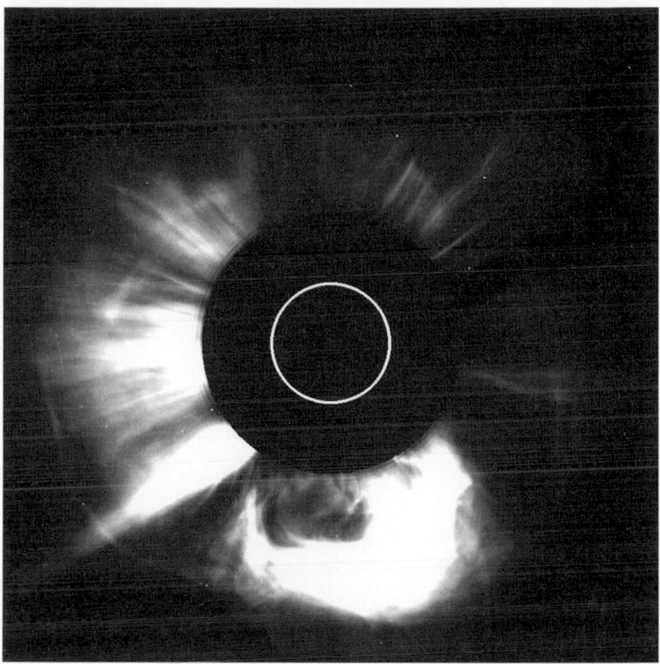

A **coronal mass ejection** is an immense bubble of gas that erupts from the sun's corona. Coronal mass ejections contain highly charged particles and can disrupt the solar wind. The charged particles from these ejections damage satellites and cause problems with radio communications on earth.

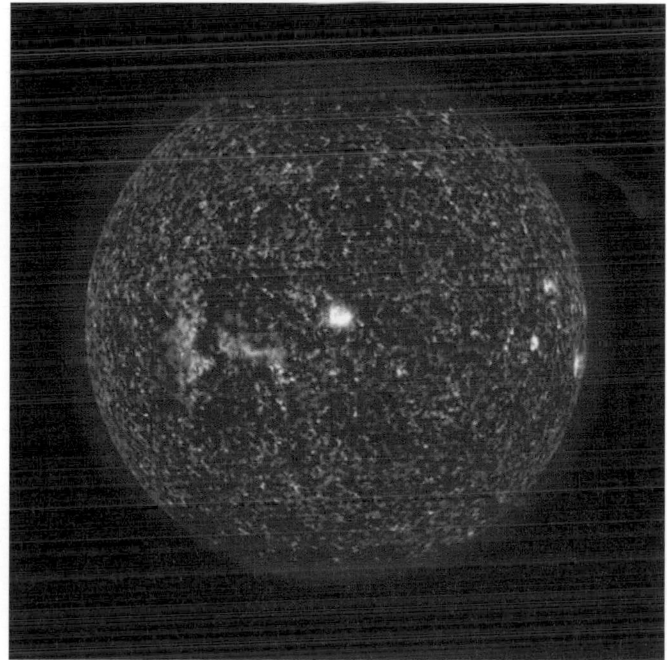

A **solar flare** is a huge, violent ejection of charged particles from the photosphere. Most solar flares form near or above sunspots. They release a huge amount of energy. Like coronal mass ejections, solar flares can disrupt radio communications on earth.

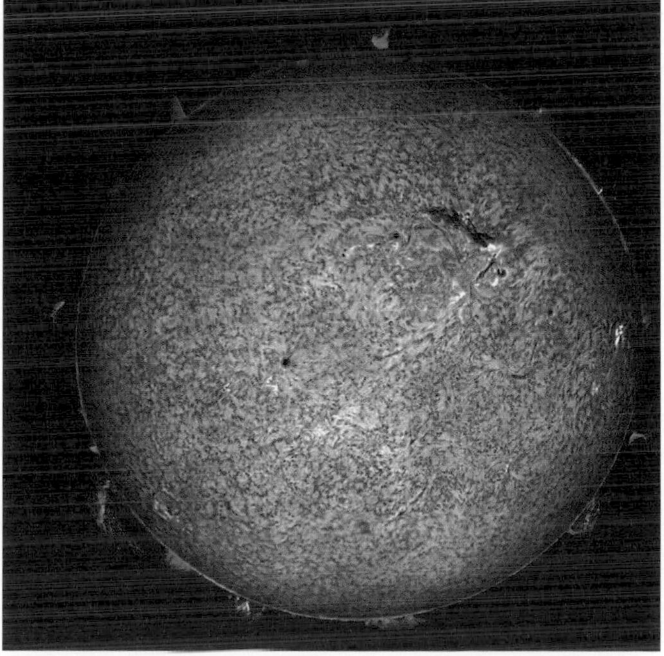

Prominences are gigantic clouds of glowing gas that reach out from the photosphere. The clouds of gas move along the sun's magnetic field in curved paths. Prominences can reach into the corona and can last for weeks.

The moon is earth's closest neighbor in the solar system.

Its appearance seems to change every night. These changes are predictable because they depend on the relative positions and motions of the sun, moon, and earth.

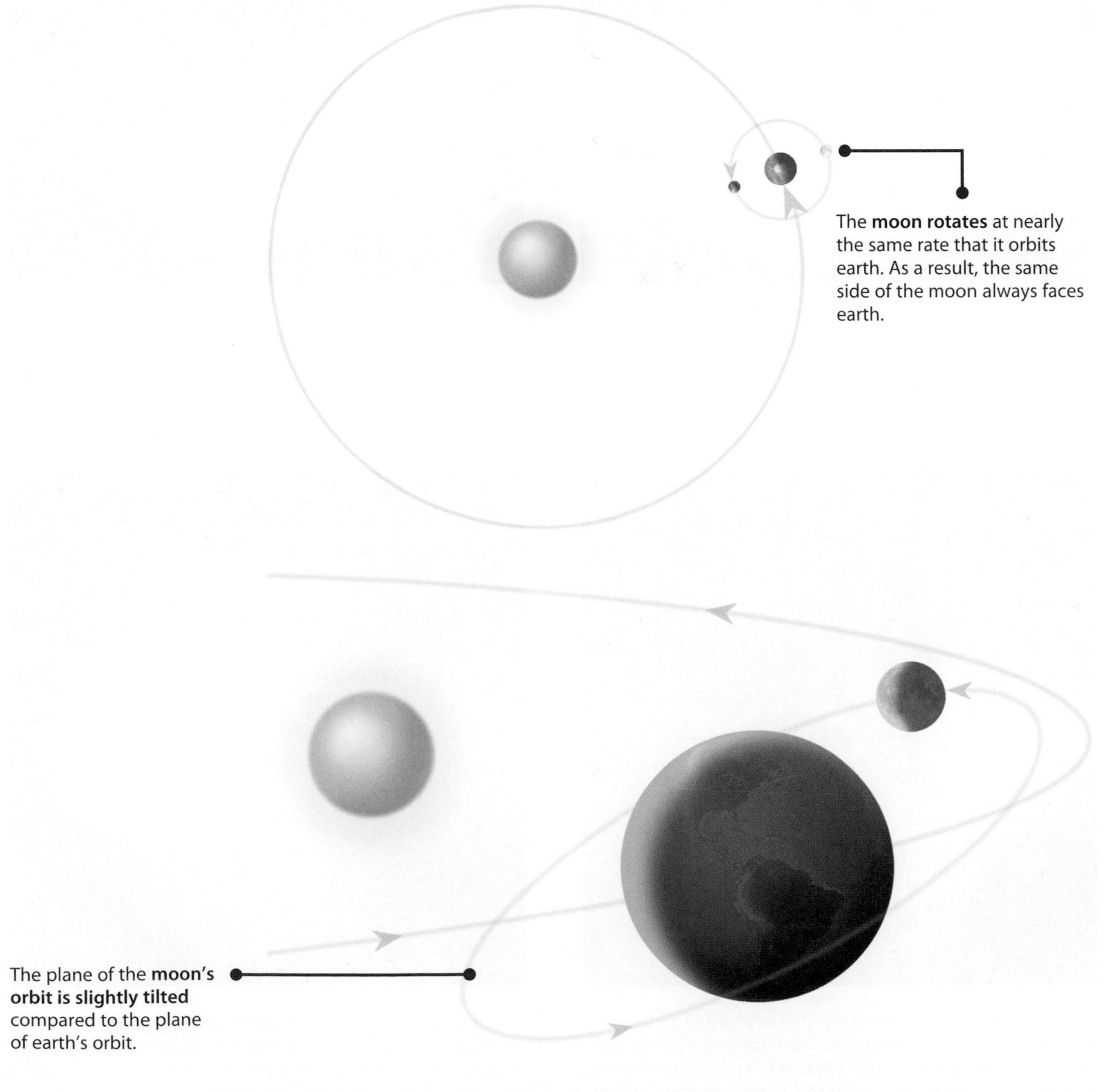

The **full moon** is brightly lit. We see a full moon when the side of the moon facing earth is completely lit.

The **new moon** is dark. We see a new moon when the side of the moon facing earth is not lit.

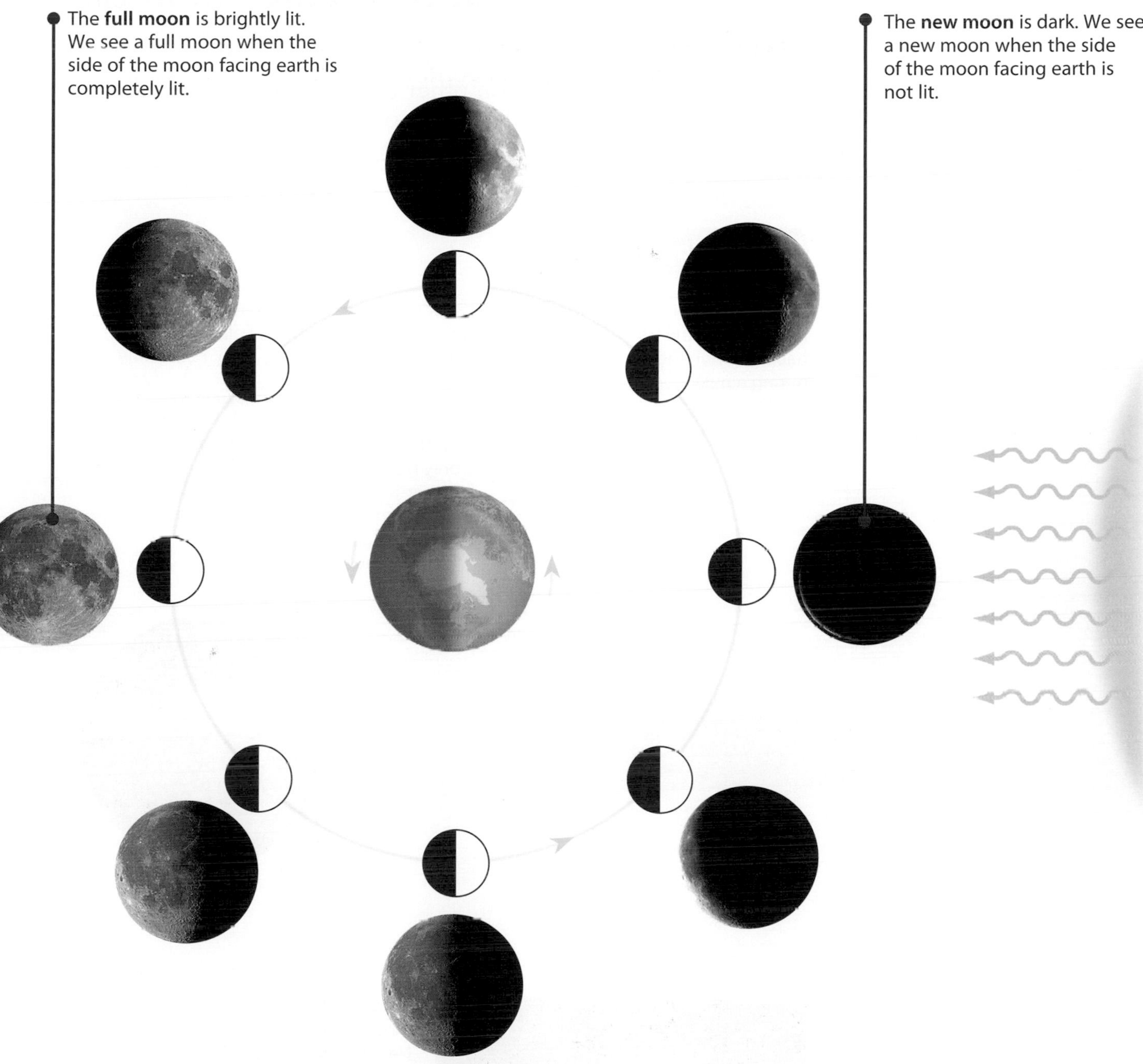

Although there is a "far side" of the moon, there is no such thing as a "dark side." Just like earth, the moon rotates on its axis. As it rotates, it also revolves around earth. Because of this, the side that is lit is not always visible from earth. The appearance of the moon changes slightly every night—we see the same side of the moon every night, but different portions of it are lit. The half that is lit may be only partially visible to us.

The black-and-white circles show how the moon is lit by the sun as it orbits earth. The photographs show what the moon looks like from earth at different points in its cycle.

Just as you cast a shadow on the ground on a sunny day, earth and the moon cast shadows in space.

When these shadows fall on other bodies in the solar system, eclipses occur. Solar and lunar eclipses are visible from earth.

Solar Eclipses

Solar eclipses happen when the moon's shadow falls on earth. In other words, the moon blocks the sun's light from reaching part of earth's surface. Solar eclipses can happen when the moon is between the sun and earth. Therefore, they occur only during a new moon.

Partial solar eclipses happen when only part of the moon's shadow falls on earth. Only part of the sun's light is blocked by the moon's shadow, so the sun appears to be only a "slice" in the sky.

Total solar eclipses happen when the moon's entire shadow falls on earth. Nearly all of the sun's light is blocked by the moon's shadow, and the sun appears to go dark in the sky.

Lunar Eclipses

Lunar eclipses happen when earth is between the sun and the moon, and earth's shadow falls on the moon. This can happen only during a full moon. Because the moon's orbit is tilted relative to earth's orbit, eclipses do not happen every month. They occur only when the sun, moon, and earth line up.

Partial lunar eclipses happen when only part of earth's shadow falls on the moon. Some sunlight reflects off the moon, so the moon appears only slightly darker in the sky.

Total lunar eclipses happen when earth's entire shadow falls on the moon. No direct sunlight reflects off the moon. However, light that is refracted, or bent, by earth's atmosphere does reflect off the moon. Therefore, the moon does not appear totally dark in the sky. The refracted light is red, so the moon looks red-black during a total lunar eclipse.

Earth has seasons because its axis of rotation is tilted. As a result, most places on earth receive different amounts of sunlight during different times of year.

Seasons in the Northern Hemisphere occur opposite those in the Southern Hemisphere—that is, when it's winter in one, it's summer in the other.

Between about March 21 and September 21, the Northern Hemisphere is tilted toward the sun. As a result, the weather is warmer in the Northern Hemisphere. Around **June 21**, sunlight hits the Northern Hemisphere most directly. This is the summer solstice in the Northern Hemisphere. It corresponds to the winter solstice in the Southern Hemisphere.

About **September 21**, day and night are the same length. After that, days in the Northern Hemisphere begin to get shorter. This is the autumnal equinox in the Northern Hemisphere.

About **March 21**, day and night are the same length. After this time, days in the Northern Hemisphere begin to get longer. This is the vernal (spring) equinox in the Northern Hemisphere.

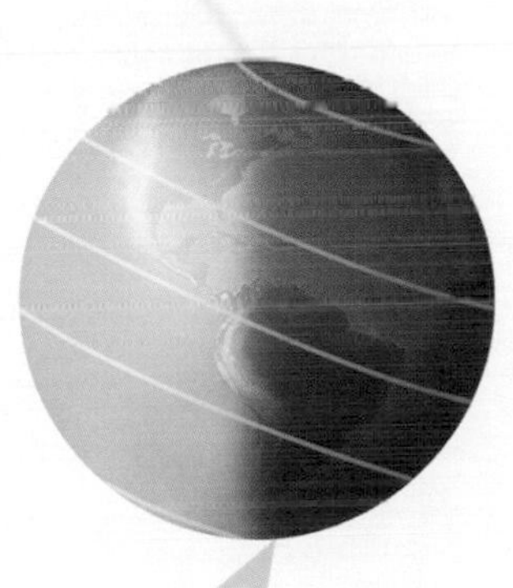

Between about September 21 and about March 21, the Northern Hemisphere is tilted away from the sun. As a result, the weather is cooler in the Northern Hemisphere during this time. About **December 21**, sunlight hits the Northern Hemisphere least directly. This is the winter solstice, the shortest day of the year, in the Northern Hemisphere. It corresponds to the summer solstice, the longest day of the year, in the Southern Hemisphere.

Our solar system consists of our sun, eight planets and their moons, comets, asteroids, and many other objects.

Each object in our solar system has unique features. Although we know of many solar systems in the universe, our solar system is the only one we know to contain life.

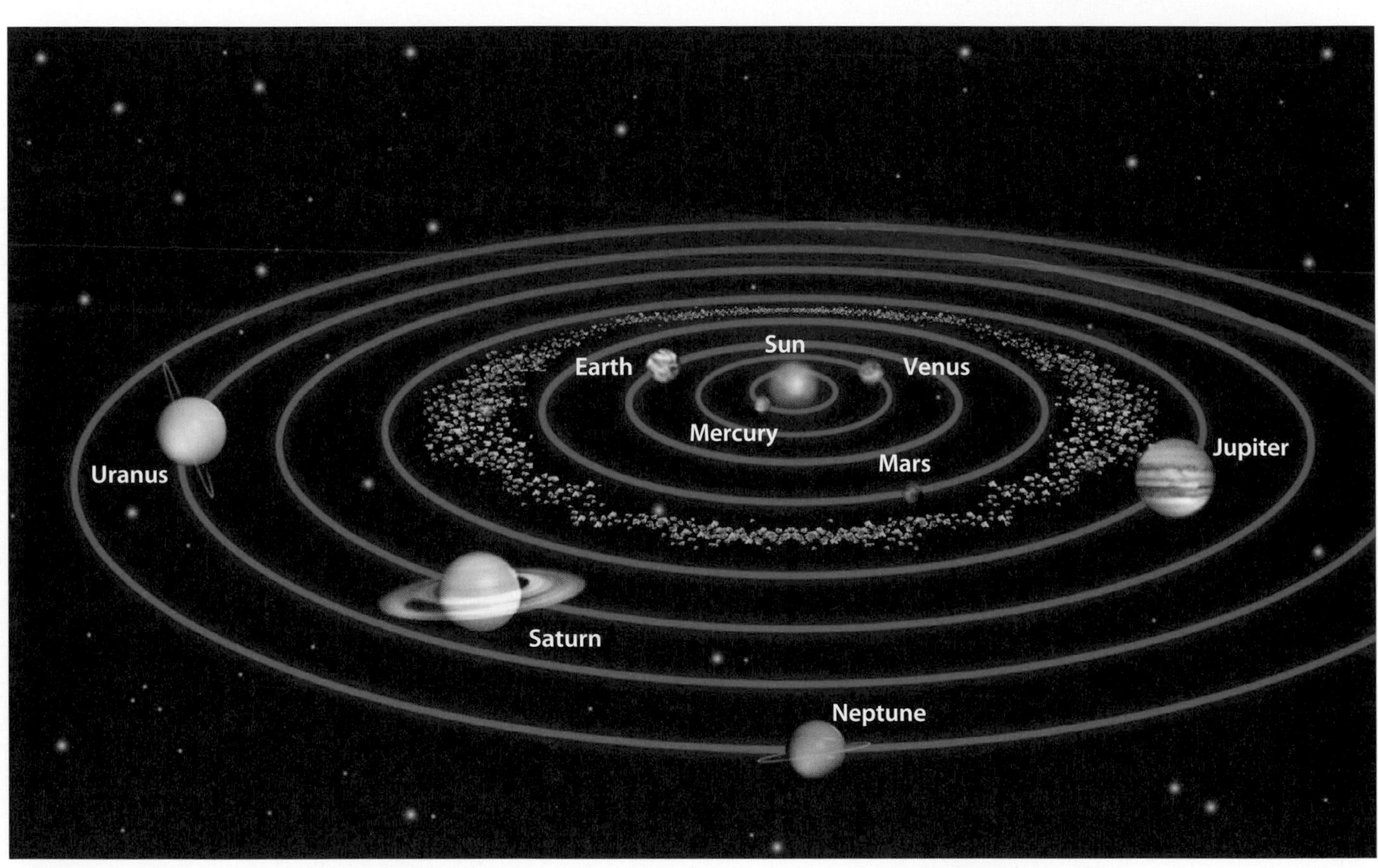

CHARACTERISTICS OF THE PLANETS IN OUR SOLAR SYSTEM

PLANET	*AVERAGE DISTANCE FROM THE SUN*	*PERIOD OF ROTATION*	*PERIOD OF REVOLUTION*	*AVERAGE DENSITY (g/cm^3)*
Mercury	36 million mi; 58 million km; 0.39 AU	59 Earth days	88 Earth days	5.4
Venus	67 million mi; 108 million km; 0.72 AU	244 Earth days	225 Earth days	5.2
Earth	93 million mi; 150 million km; 1.00 AU	23 hours, 56 minutes	365.25 Earth days	5.5
Mars	142 million mi; 228 million km; 1.52 AU	24 hours, 37 minutes	687 Earth days	3.9
Jupiter	483 million mi; 778 million km; 5.20 AU	9 hours, 50 minutes	12 Earth years	1.3
Saturn	886 million mi; 1.43 billion km; 9.54 AU	10 hours, 14 minutes	29.5 Earth years	0.7
Uranus	1.78 billion mi; 2.87 billion km; 19.2 AU	17 hours, 14 minutes	84 Earth years	1.2
Neptune	2.79 billion mi; 4.50 billion km; 30.1 AU	16 hours, 3 minutes	165 Earth years	1.7

PLANET	*AVERAGE SURFACE TEMPERATURE (°C)*	*AVERAGE SURFACE GRAVITY (COMPARED TO EARTH'S)*	*DIAMETER*	*NUMBER OF KNOWN MOONS*
Mercury	–173 to 427	38%	3,015 mi; 4,878 km	0
Venus	464	91%	7,526 mi; 12,104 km	0
Earth	–13 to 37	100%	7,920 mi; 12,756 km	1
Mars	–123 to 37	38%	4,216 mi; 6,794 km	2
Jupiter	–110	236%	88,700 mi; 143,884 km	at least 60
Saturn	–140	92%	75,000 mi; 120,536 km	at least 30
Uranus	–195	89%	29,000 mi; 50,530 km	at least 20
Neptune	–200	112%	28,900 mi; 51,118 km	at least 14

The inner, terrestrial planets are small and dense, with rocky crusts and interiors.

Mercury, Venus, Earth, and Mars are the inner, terrestrial planets in our solar system. Of all the planets in our solar system, Venus is closest to Earth in both size and distance.

CHARACTERISTICS OF VENUS

AVERAGE DISTANCE FROM THE SUN	PERIOD OF ROTATION	PERIOD OF REVOLUTION	AVERAGE DENSITY (g/cm^3)
67 million mi; 108 million km; 0.72 AU	244 Earth days	225 Earth days	5.2

AVERAGE SURFACE TEMPERATURE (°C)	AVERAGE SURFACE GRAVITY (COMPARED TO EARTH'S)	DIAMETER	NUMBER OF KNOWN MOONS
464	91%	7,526 mi; 12,104 km	0

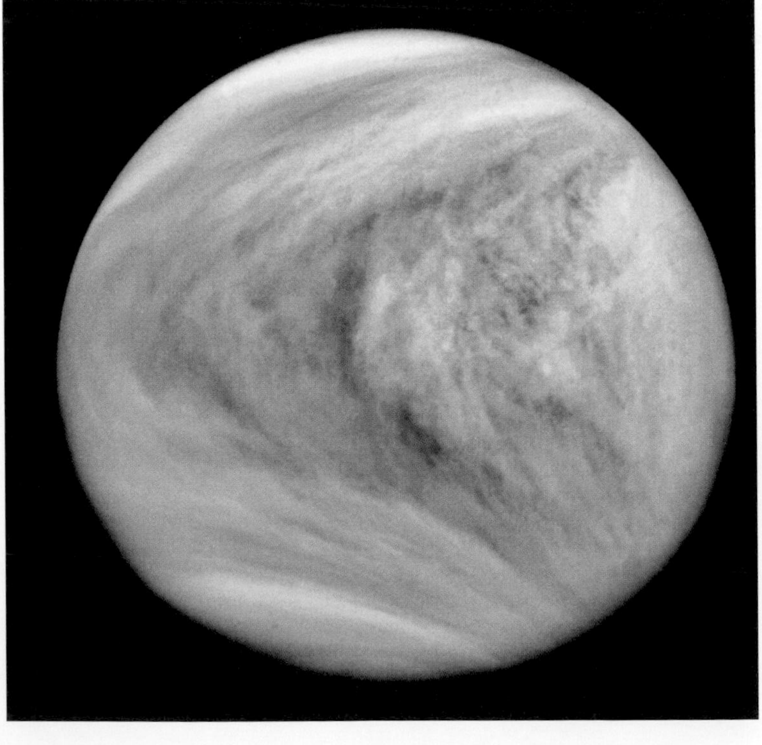

Venus's **atmosphere** is much thicker than Earth's. It contains about 96 percent carbon dioxide. Because the atmosphere is so thick, atmospheric pressure on the surface of Venus is 90 times greater than atmospheric pressure on Earth's surface. Very high concentrations of greenhouse gases, combined with this high atmospheric pressure, make Venus's surface temperatures very high—greater than 400°C.

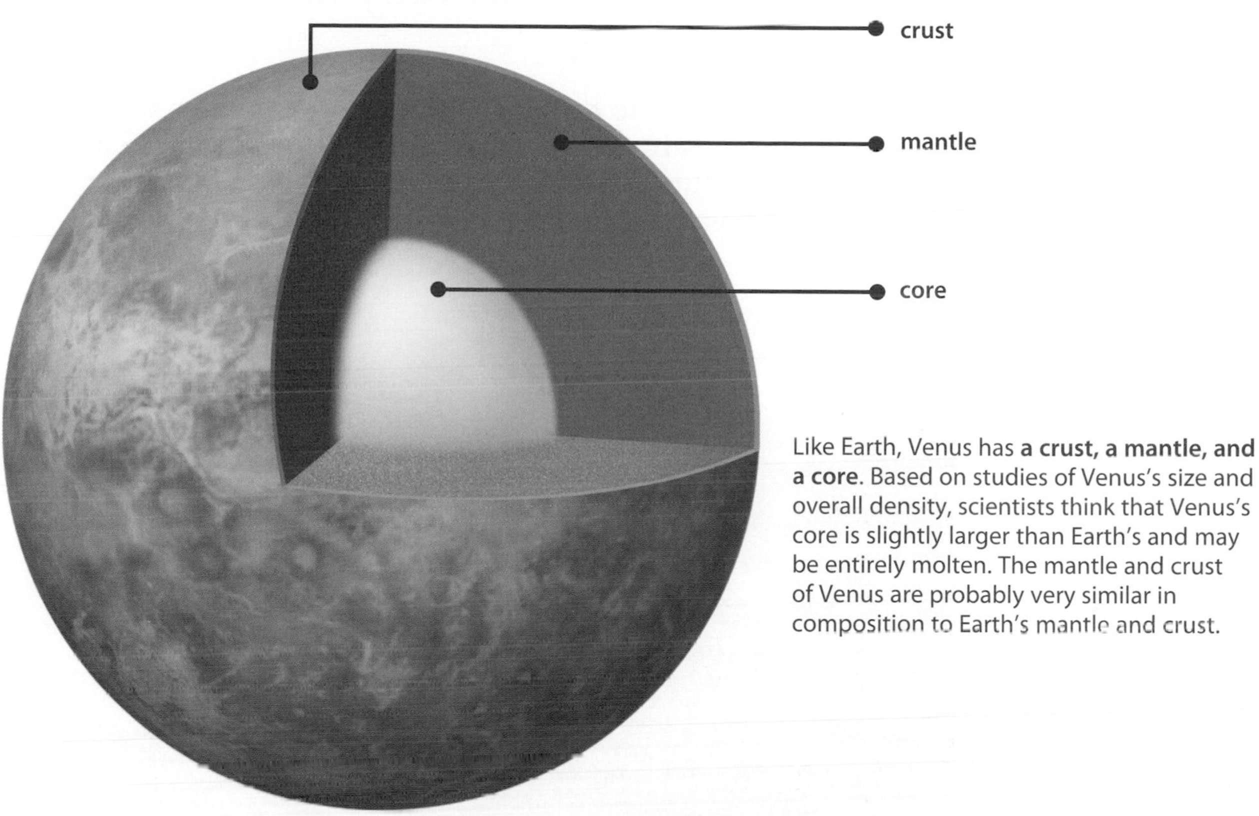

Like Earth, Venus has **a crust, a mantle, and a core**. Based on studies of Venus's size and overall density, scientists think that Venus's core is slightly larger than Earth's and may be entirely molten. The mantle and crust of Venus are probably very similar in composition to Earth's mantle and crust.

Venus is similar to Earth in many ways. For example, Venus's radius is only about 5 percent smaller than Earth's, and Venus's gravitational pull is only about 10 percent less than Earth's. For these reasons, Venus is sometimes called "Earth's twin."

The **similarities between Earth and Venus** extend to surface features as well as size and gravity. Venus's surface has mountain ranges, rifts, and volcanoes, like Earth's surface. However, scientists do not think these features on Venus were produced by plate tectonics.

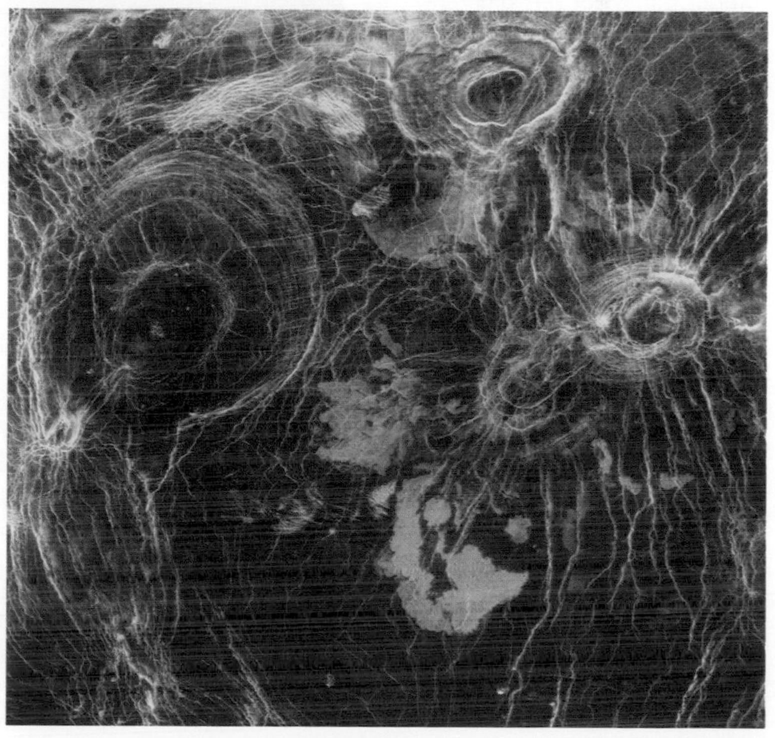

The gas giant planets are all much larger than Earth, but they are also much less dense.

This is primarily due to their very thick atmospheres, which are composed mostly of hydrogen and helium. Jupiter, Saturn, Uranus, and Neptune are the gas giant planets in our solar system. Jupiter is the largest of all the planets.

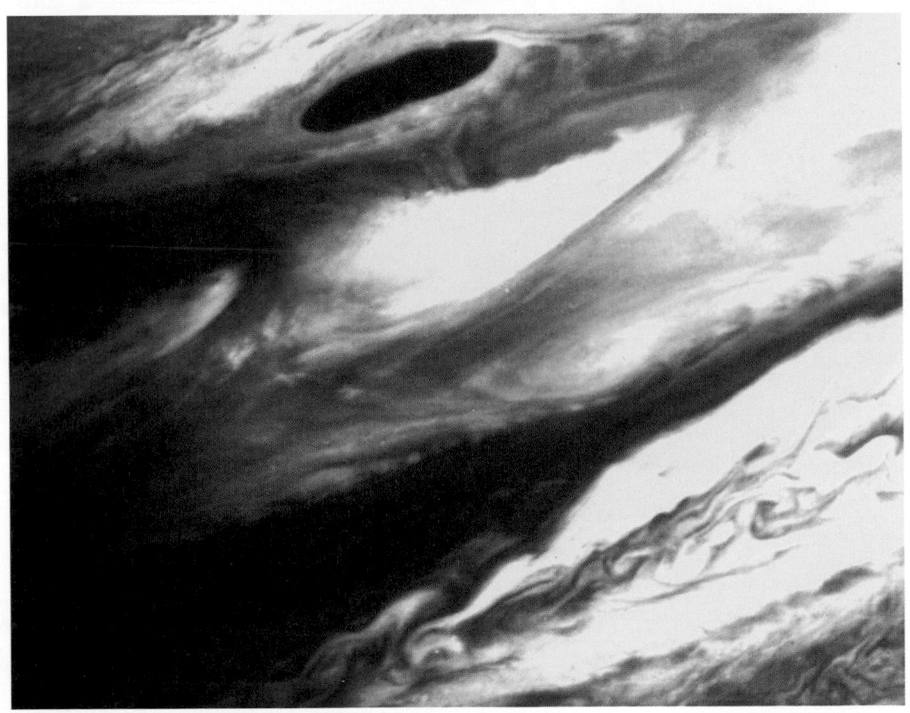

Jupiter's **atmosphere** is composed almost entirely of hydrogen and helium. In fact, Jupiter's overall composition is not very different from that of our sun. If Jupiter's mass were greater, nuclear fusion could begin in its core, forming a small star.

CHARACTERISTICS OF JUPITER

AVERAGE DISTANCE FROM THE SUN	*PERIOD OF ROTATION*	*PERIOD OF REVOLUTION*	*AVERAGE DENSITY (g/cm³)*
483 million mi; 778 million km; 5.20 AU	9 hours, 50 minutes	12 Earth days	1.3

AVERAGE SURFACE TEMPERATURE (°C)	*AVERAGE SURFACE GRAVITY (COMPARED TO EARTH'S)*	*DIAMETER*	*NUMBER OF KNOWN MOONS*
−110	236%	88,700 mi; 143,884 km	at least 60

Scientists think that Jupiter may have a tiny, rocky core at its center. Much of the rest of Jupiter's **interior** is made of hot, liquid, metallic hydrogen. The hydrogen exists in this state because of the immensely high temperatures and pressures in the interior of the planet.

Jupiter's atmosphere is marked by many huge **storms**, which appear as round or oval-shaped spots in its atmosphere. The Great Red Spot is one of the largest of these storms. This giant rotating storm in Jupiter's southern hemisphere has existed for at least several hundred years. The diameter of the Great Red Spot is about twice that of Earth.

Visible light, radio waves, microwaves, and gamma rays are all examples of electromagnetic radiation.

Each type of electromagnetic radiation has a specific range of frequencies. Together, all of these types of radiation make up the *electromagnetic spectrum*.

Gamma rays are very high-energy electromagnetic waves. They have the shortest wavelength and the highest frequency of all electromagnetic waves. When people talk about harmful radiation, such as that produced by radioactive elements like radium, they are often referring to gamma rays. Marie Curie, a Polish scientist, was one of the first to describe a radioactive element that produces gamma rays. Astronomers use gamma rays that travel through space to study the history and evolution of the universe.

gamma rays

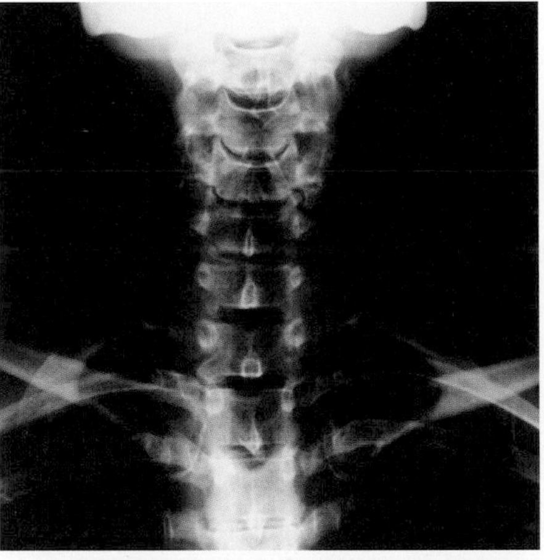

X rays are high-energy electromagnetic waves. They easily travel through skin and muscle. However, bones and teeth absorb X rays. Therefore, doctors and dentists use X rays to take pictures of bones and teeth. Many stars and other objects in space emit X rays. Astronomers use these X rays to study such objects.

X rays

Ultraviolet radiation has wavelengths slightly shorter than violet visible light. Ultraviolet radiation contains more energy than visible light. When our skin or eyes absorb this energy, it can damage cells and lead to sunburn or eye damage. It can also cause genetic damage, leading to skin cancer.

ultraviolet

Infrared light is electromagnetic radiation with wavelengths slightly longer than red visible light. People detect infrared radiation as heat. People also use infrared radiation in remote controls for televisions and other devices.

Radio waves have the longest wavelengths and the lowest frequencies of all electromagnetic waves. People use radio waves to carry information, such as music, television programs, and communications, from place to place. Objects in space, such as stars and galaxies, also give off radio waves. These waves can be detected here on earth using special telescopes.

Microwaves have slightly shorter wavelengths than radio waves. Microwave ovens produce this form of radiation when they are turned on. As food in a microwave oven absorbs this radiation, its temperature increases. This is what allows people to cook food in a microwave oven. Many cellular phone systems use microwaves to transmit information from phone to phone.

Visible light is light that people can see. The wavelengths of visible light correspond to colors. Red visible light has the longest wavelength, and violet visible light has the shortest wavelength.

visible light | infrared | microwaves | radio waves

How bright a star looks to us depends on its age, its size, and its distance from earth.

Stars that are farther away may appear dimmer than closer stars of the same age and size. Astronomers use the brightness of a star to learn its age and approximate distance.

The **Big Dipper** is part of the constellation Ursa Major. Each star in the Big Dipper has a different magnitude, or brightness.

MAGNITUDES AND DISTANCES OF STARS IN THE BIG DIPPER			
STAR NAME	***APPARENT MAGNITUDE***	***ABSOLUTE MAGNITUDE***	***APPROXIMATE DISTANCE FROM EARTH (LIGHT-YEARS)***
Alkaid	1.85	−0.60	100
Mizar	2.23	0.33	78
Alioth	1.76	−0.22	81
Megrez	3.32	1.33	81
Dubhe	1.81	−1.09	124
Merak	2.34	0.41	79
Phecda	2.41	0.36	84

Astronomers use two scales to measure the magnitude, or brightness, of stars. The *apparent magnitude* of a star measures how bright the star is compared to other stars. The *absolute magnitude* of a star is a measure of how bright the star would be if it were at a standard distance of about 33 light-years from earth. In both scales, larger numbers indicate dimmer stars. For example, a star with an apparent magnitude of −0.3 appears brighter to us on earth than a star with an apparent magnitude of 4.5. A star with an absolute magnitude of 2.3 is brighter than a star with an absolute magnitude of 6.7.

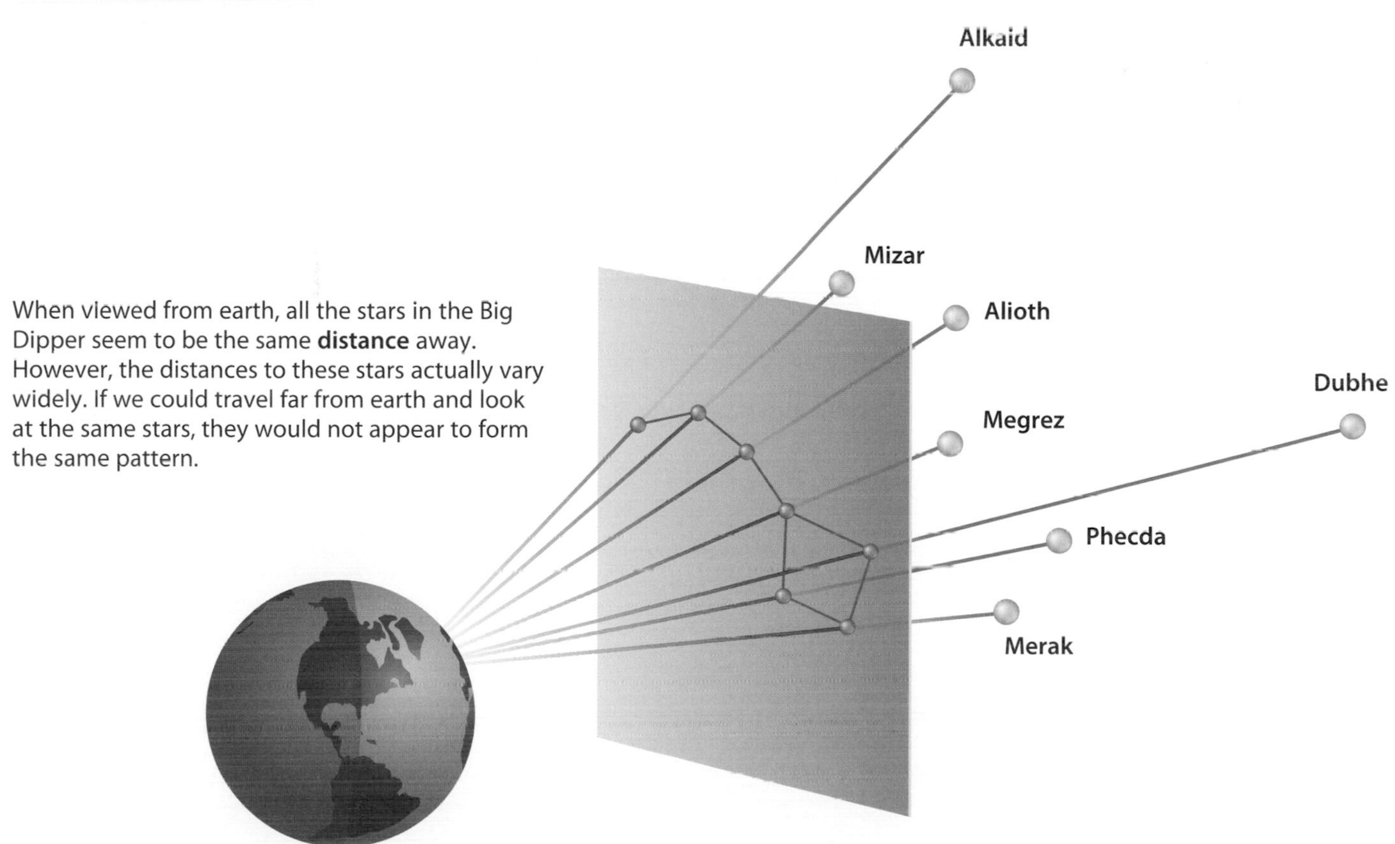

When viewed from earth, all the stars in the Big Dipper seem to be the same **distance** away. However, the distances to these stars actually vary widely. If we could travel far from earth and look at the same stars, they would not appear to form the same pattern.

Although they may seem constant, stars change slowly over time.

Stars form from large clouds of gas that condense to form a protostar. The protostar further condenses to form a star. The end of a star's existence may be marked by an immense explosion or a quiet collapse. How a star changes and how it dies depend mainly on its mass.

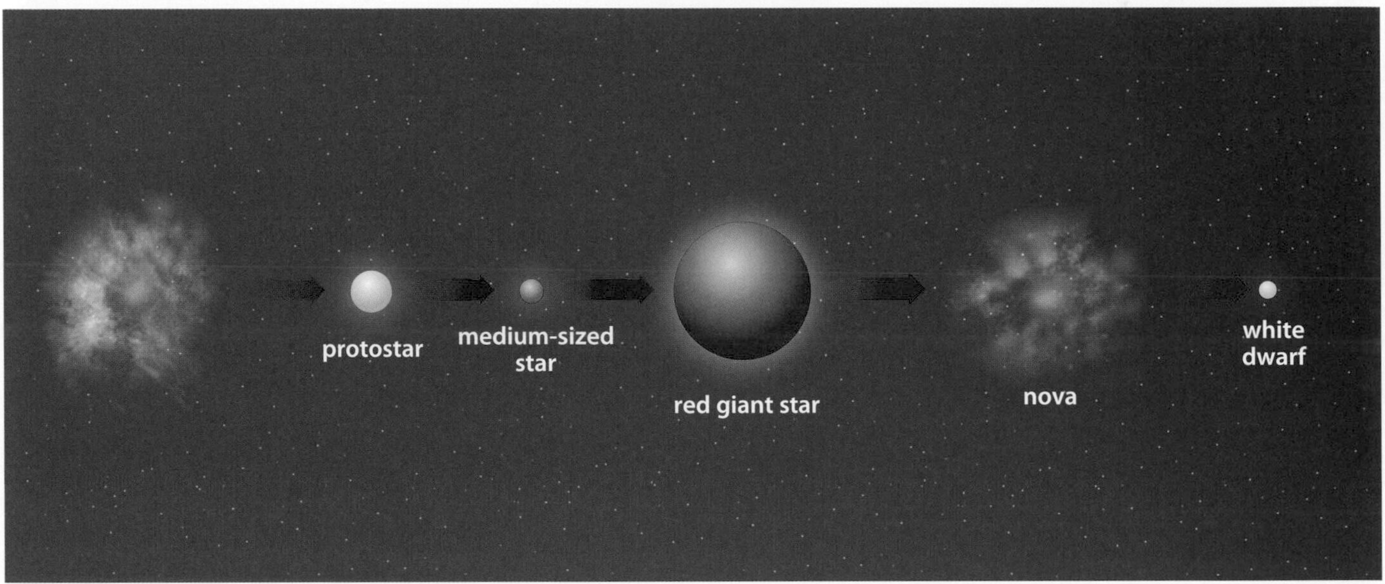

Stars with masses of one-half that of our sun and three times larger than our sun are considered **medium-sized stars**. These stars may spend billions of years on the main sequence (see Hertzsprung-Russell diagram) before exhausting their supply of hydrogen. When all their hydrogen is used up, they begin to burn other gases and expand to become red giants. Then, they collapse to form white dwarf stars. This is what scientists think will happen to our sun in about 5 billion years.

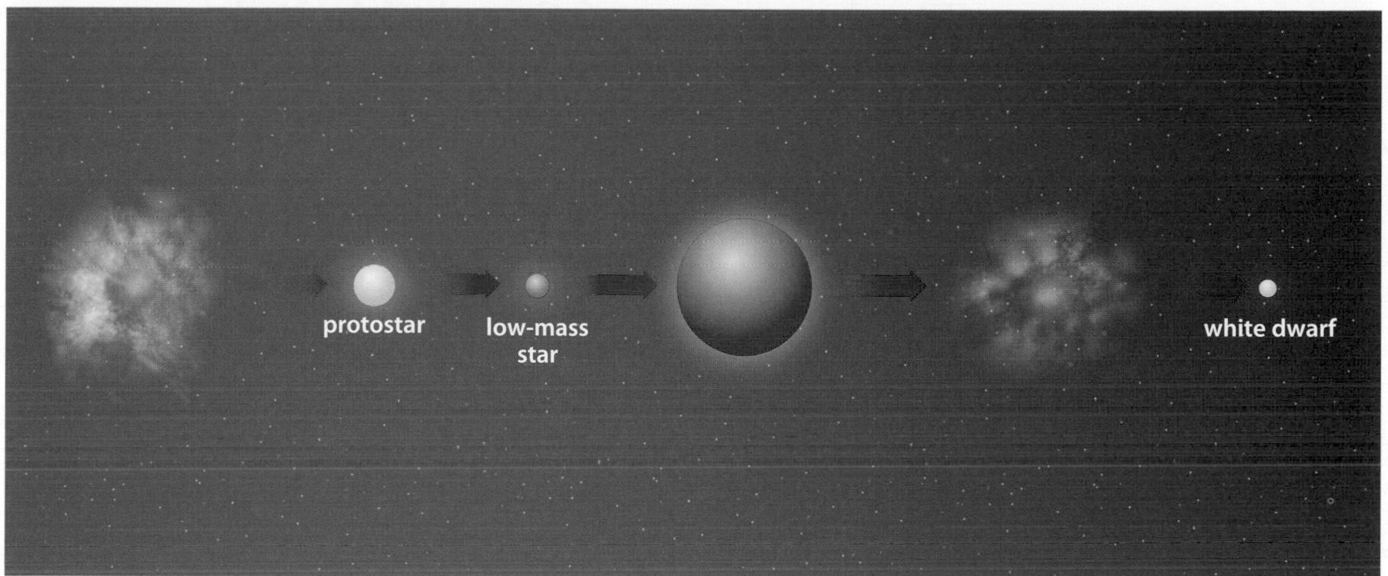

Stars with masses of less than one-half that of our sun are considered **low-mass stars.** These stars burn their hydrogen fuel very slowly. When this fuel is used up, low-mass stars collapse to form tiny, hot, white dwarf stars.

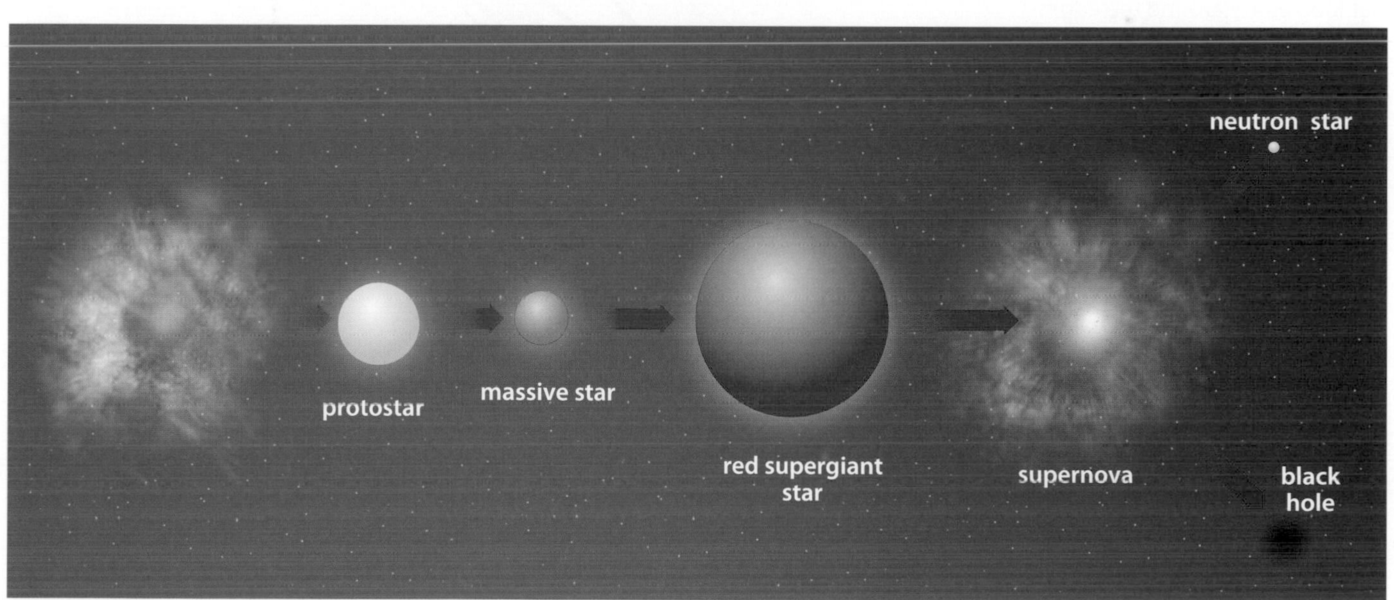

Stars with masses of more than three times that of our sun are considered **massive stars.** These stars burn fuel very quickly. As their fuel is used up, the stars expand to become red supergiant stars. These stars eventually collapse and then explode in supernovae. Their remains may become dense neutron stars or even denser black holes.

The invention of the visible light telescope forever changed how we perceive the solar system.

Telescopes allowed early astronomers to study and describe the features of our moon and to recognize that the planets are different from the stars. Today's visible light telescopes can reveal even more details about our universe.

Refracting telescopes use lenses to collect and focus light from distant objects. In most refracting telescopes, light passes through an objective lens, which focuses the light, and an eyepiece lens, which magnifies the image. Because the lenses focus different colors of light at different locations, refracting telescopes produce images that look blurry around the edges.

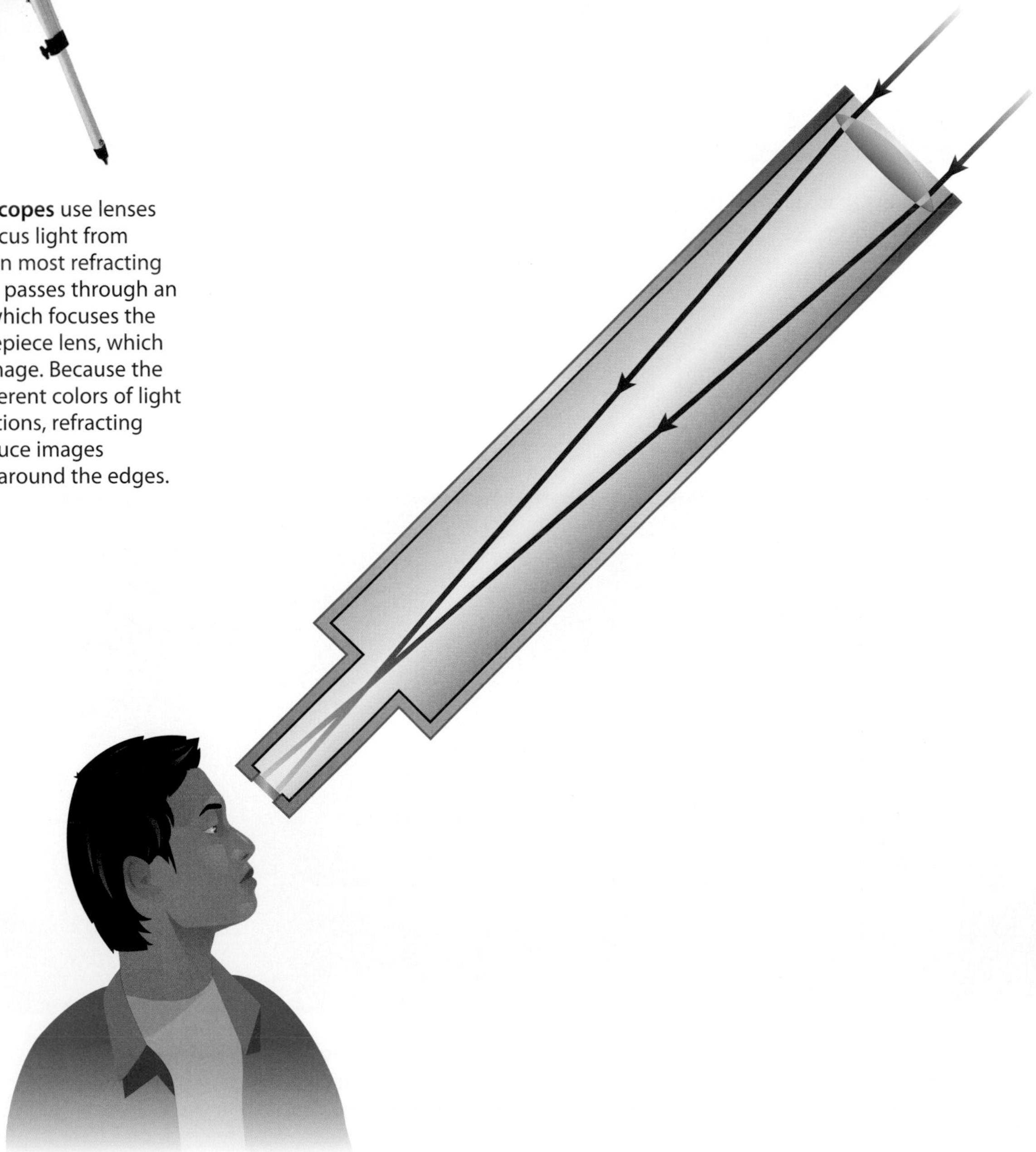

Reflecting telescopes use mirrors to collect and focus light. First, light enters the telescope and strikes a concave mirror. This focuses the light onto a flat mirror, which reflects the light into an eyepiece lens. The eyepiece lens magnifies the image for viewing. Reflecting telescopes can be much larger than refracting telescopes and still produce clear images.

The color of a star is related to its temperature.

The Hertzsprung-Russell (HR) diagram shows how the color, temperature, and brightness of a star are related. Cool red stars are on the right-hand side of the diagram. Hot blue stars are on the left-hand side of the diagram. As stars grow older, their locations on the HR diagram change.

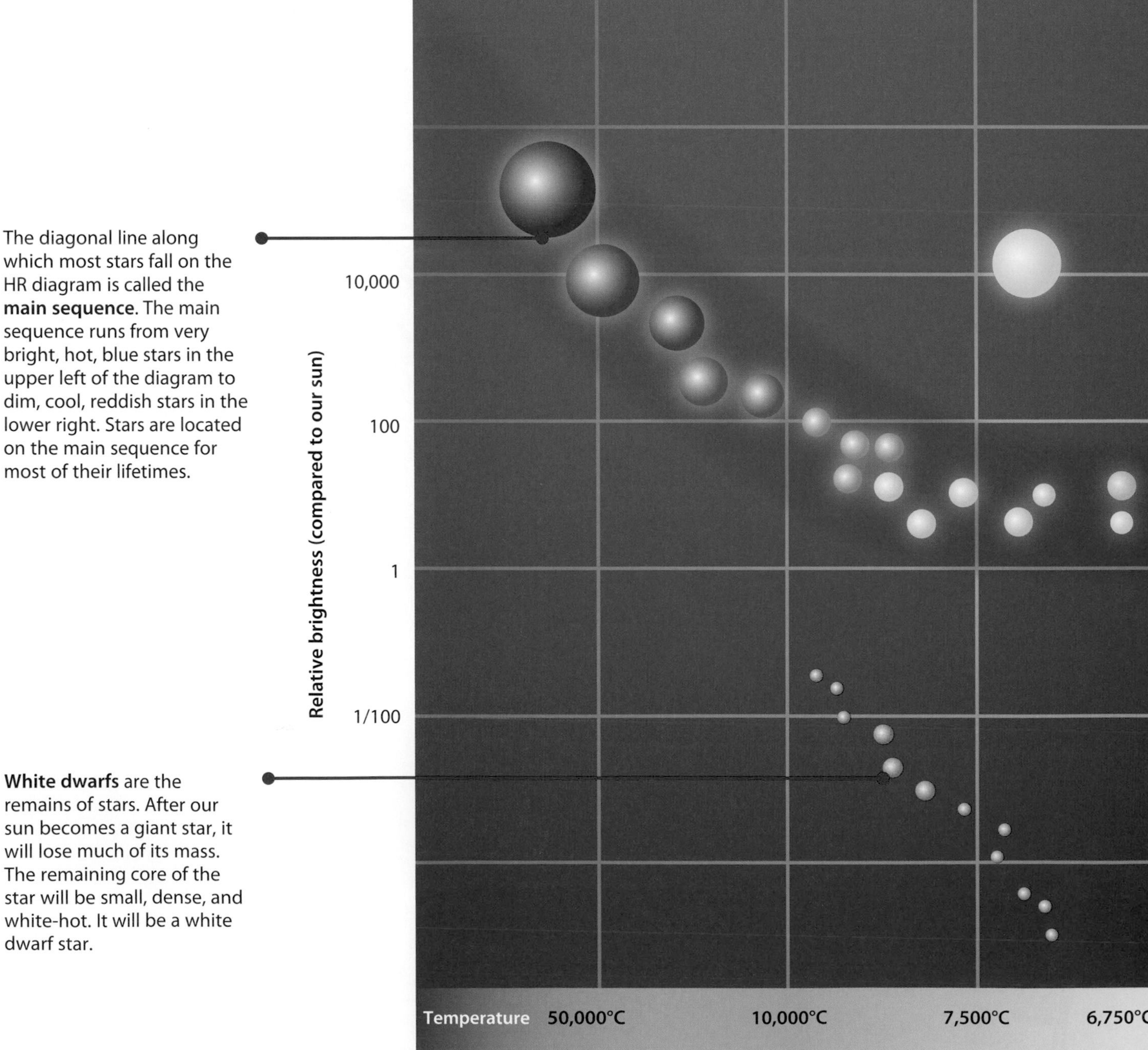

When a star uses up all its hydrogen fuel, it expands and cools. It can then become a large, **reddish giant or supergiant star**. Most scientists agree that our sun will become a giant star in a few billion years. Giant stars visible from earth today include Betelgeuse, Aldebaran, and Antares.

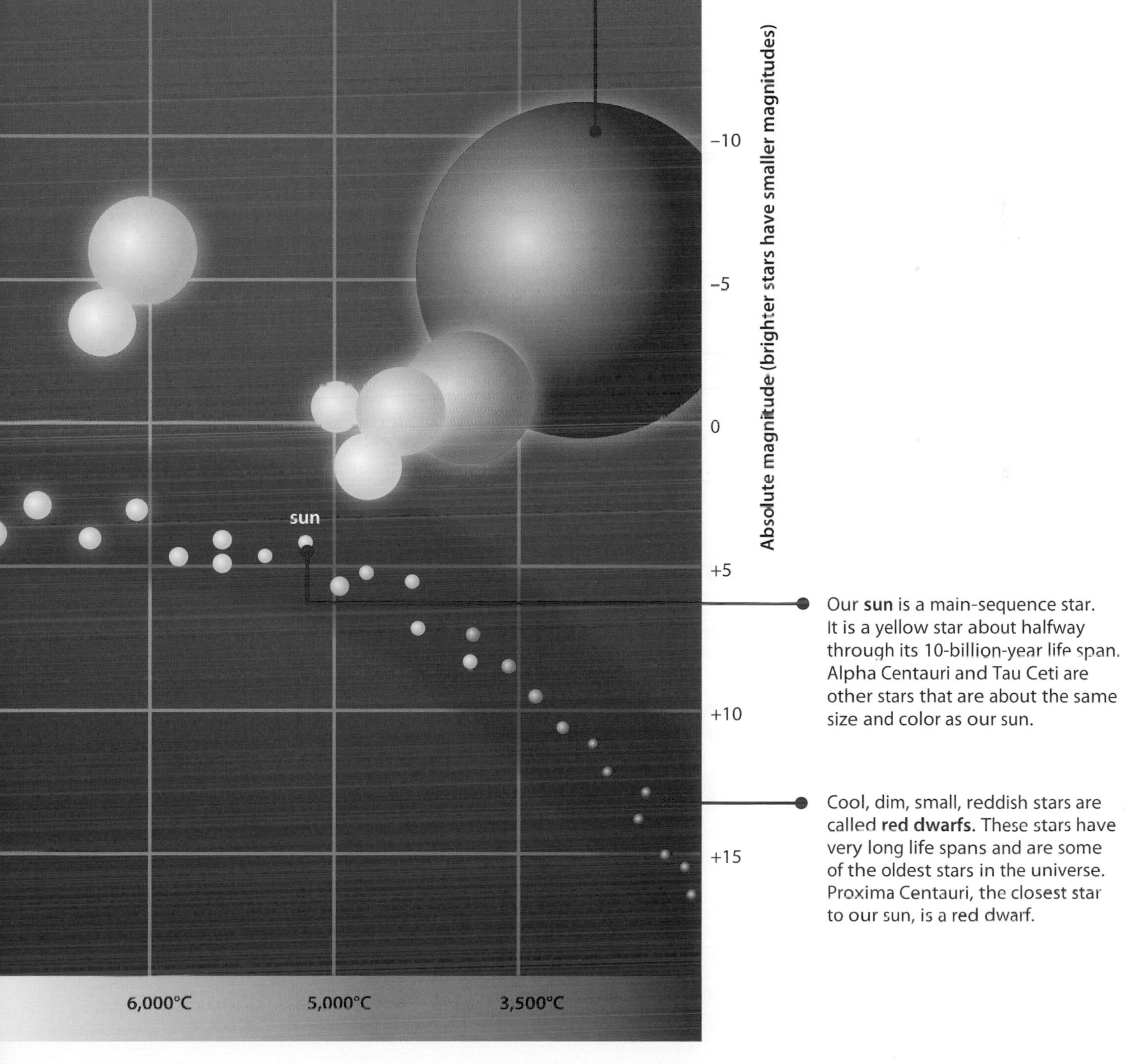

Our **sun** is a main-sequence star. It is a yellow star about halfway through its 10-billion-year life span. Alpha Centauri and Tau Ceti are other stars that are about the same size and color as our sun.

Cool, dim, small, reddish stars are called **red dwarfs.** These stars have very long life spans and are some of the oldest stars in the universe. Proxima Centauri, the closest star to our sun, is a red dwarf.

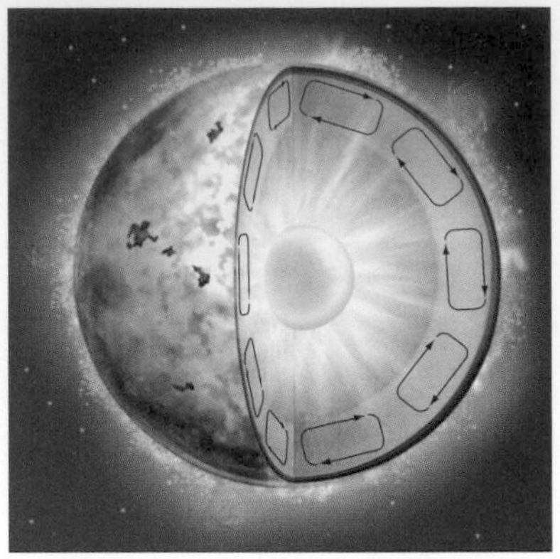

Nuclear reactions within stars produce their tremendous energy.

In stars like our sun, hydrogen nuclei fuse to form helium. Within older stars, elements heavier than helium are produced by fusion of different nuclei.

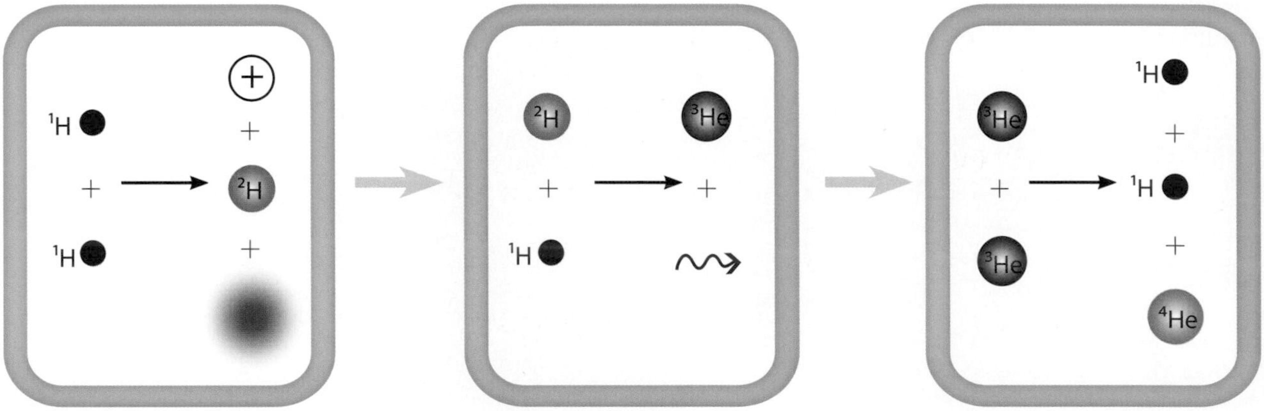

Within our sun, 2 hydrogen nuclei (^1H), or single protons, can combine to form a deuterium nucleus (^2H), a positron, and a neutrino.

The ^2H and another ^1H can combine to form a nucleus of helium-3 (^3He). This process releases energy in the form of photons.

Two ^3He can combine to produce a nucleus of helium-4 (^4He) and 2 protons.

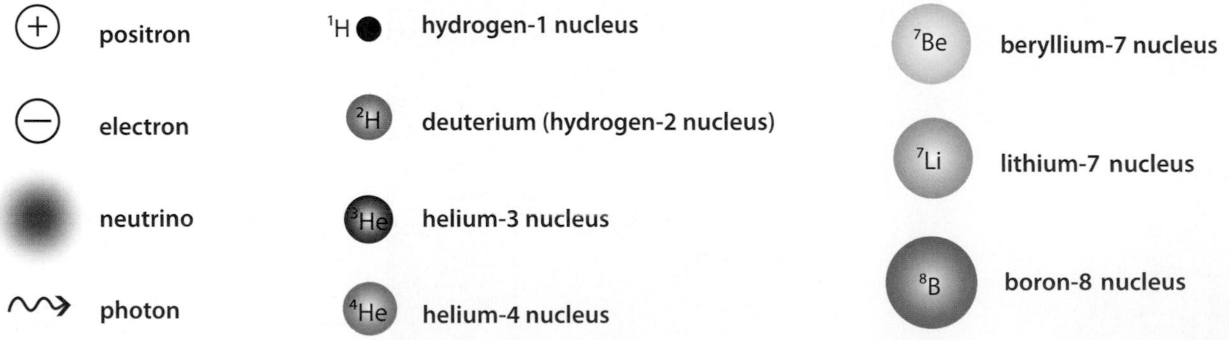

positron

electron

neutrino

photon

^1H **hydrogen-1 nucleus**

^2H **deuterium (hydrogen-2 nucleus)**

^3He **helium-3 nucleus**

^4He **helium-4 nucleus**

^7Be **beryllium-7 nucleus**

^7Li **lithium-7 nucleus**

^8B **boron-8 nucleus**

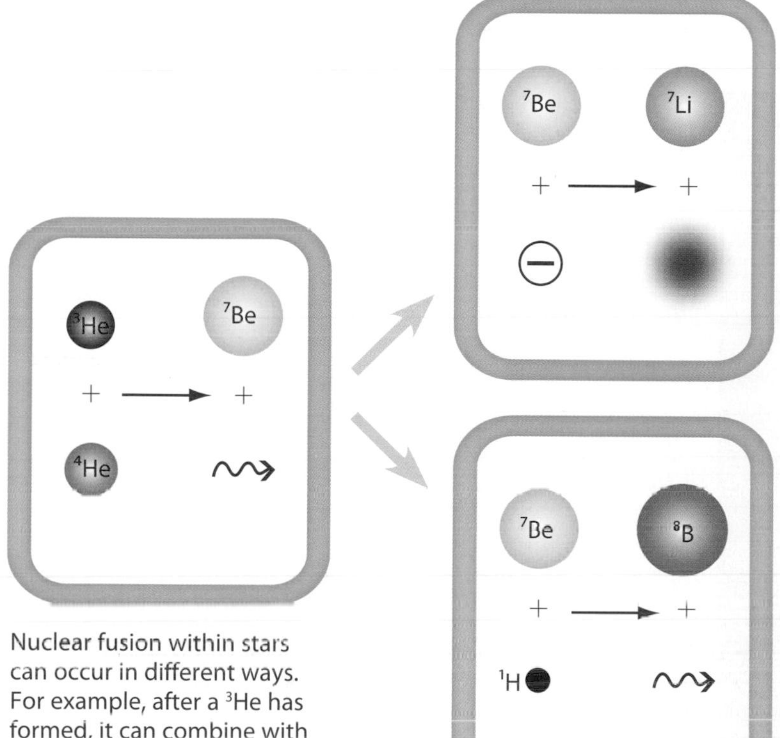

Nuclear fusion within stars can occur in different ways. For example, after a ^{3}He has formed, it can combine with a ^{4}He to produce a nucleus of beryllium-7 (^{7}Be) and energy.

A ^{7}Be can combine with an electron to form a nucleus of lithium-7 (^{7}Li) and a neutrino.

A ^{7}Be can combine with a proton to produce a nucleus of boron-8 (^{8}B) and energy.

Abundance of Elements in Our Solar System

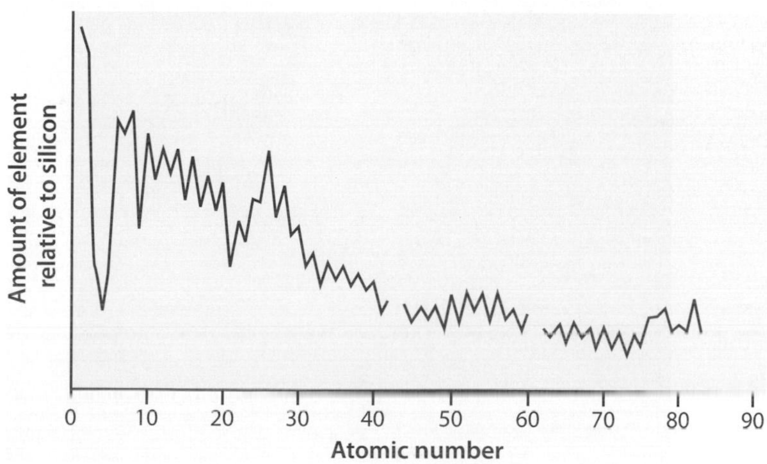

The reactions shown here occur in stars like our sun. As the star uses up its supply of hydrogen, other reactions begin to occur. These reactions involve fusing ^{4}He to heavier nuclei to produce even heavier elements. For example, ^{28}Si combines with ^{4}He to produce ^{32}S. These reactions are the main cause of the zigzag pattern of element abundances in our solar system and the rest of the universe. Elements with even atomic masses are more common than other elements because many even-numbered atomic masses can be produced through helium fusion.

Galaxies come in many shapes and sizes and are in constant motion.

If two galaxies get close enough to each other, their gravitational forces may cause them to merge, forming a single larger galaxy. In addition, galaxies spin in different directions and at different rates.

Spiral galaxies have thin, curved "arms" surrounding a central region. Some spiral galaxies have several sets of these arms. The rotation and gravitational forces in a spiral galaxy cause the arms to curve around the central region, forming the characteristic spiral shape.

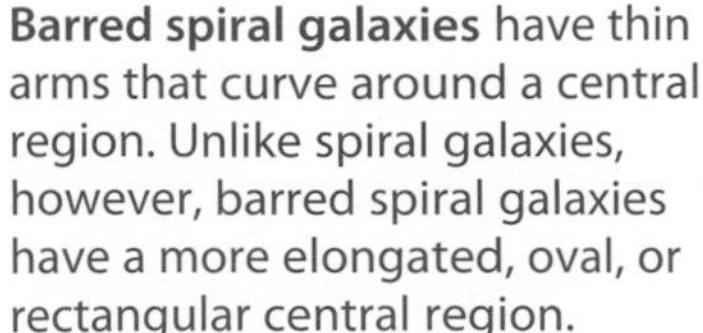

Barred spiral galaxies have thin arms that curve around a central region. Unlike spiral galaxies, however, barred spiral galaxies have a more elongated, oval, or rectangular central region.

Elliptical galaxies do not have outer arms. They consist of a large, oval mass of stars. Some scientists think that elliptical galaxies form when two spiral galaxies collide or merge.

Irregular galaxies can have a variety of different shapes. The shapes of some irregular galaxies may be caused by the collision or near-collision of two galaxies.

Scientists think that **dwarf galaxies** may be the most common kind of galaxy in the universe. Many dwarf galaxies consist mainly of gas and dust, with fewer stars than larger galaxies.

The universe contains objects of many sizes, from tiny atoms of hydrogen to huge clusters of galaxies.

The distance from which we observe an object affects what the object looks like.

From about **10 m** above earth's surface (for example, on the third floor of a building), you can see the many people on this beach.

From about **1,000 m** (1 km) above earth's surface (for example, in an airplane), you can see the whole shoreline. However, you can no longer see individual people.

From about **100 km** above earth's surface, satellites can record the shape of the coastline. However, it is difficult to find a particular beach from this distance.

If you could travel about **100,000 km** from earth's surface, you would be able to see earth and the moon. However, it would be difficult to make out the details of the coastlines from this distance.

10 m | 1,000 m | 100 km | 100,000 km

If you could travel **83 light-minutes** (about 1.5 billion km) away from earth, you would be able to see the sun and the inner planets of the solar system. However, you would not be able to see any details of the surfaces of the planets.

If you could look back at the solar system from **150 light-days** (about 30 billion km) away, you would be able to see the whole solar system, but you would probably not be able to see all the planets and moons.

From **10 light-years** (about 9.5 trillion km) away, our sun would look like just another bright star in the sky. It would not be possible to see any of the planets in our solar system.

From **1 million light-years** away, you would be able to see our galaxy, the Milky Way. However, you would not be able to see our sun.

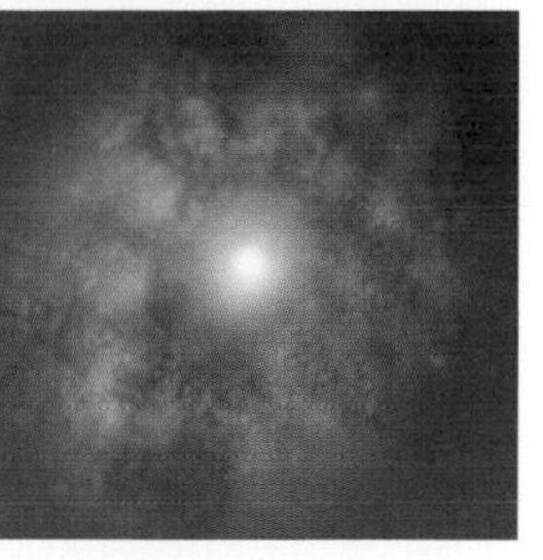

If you could travel **10 million light-years** away from earth, you would be able to see that the Milky Way is one of several galaxies in a cluster called the *Local Group*.

83 light-minutes	150 light-days	10 light-years	1 million light-years	10 million light-years

The universe is an immensely vast system made up of many smaller systems.

Each system has its own unique scale and characteristics.

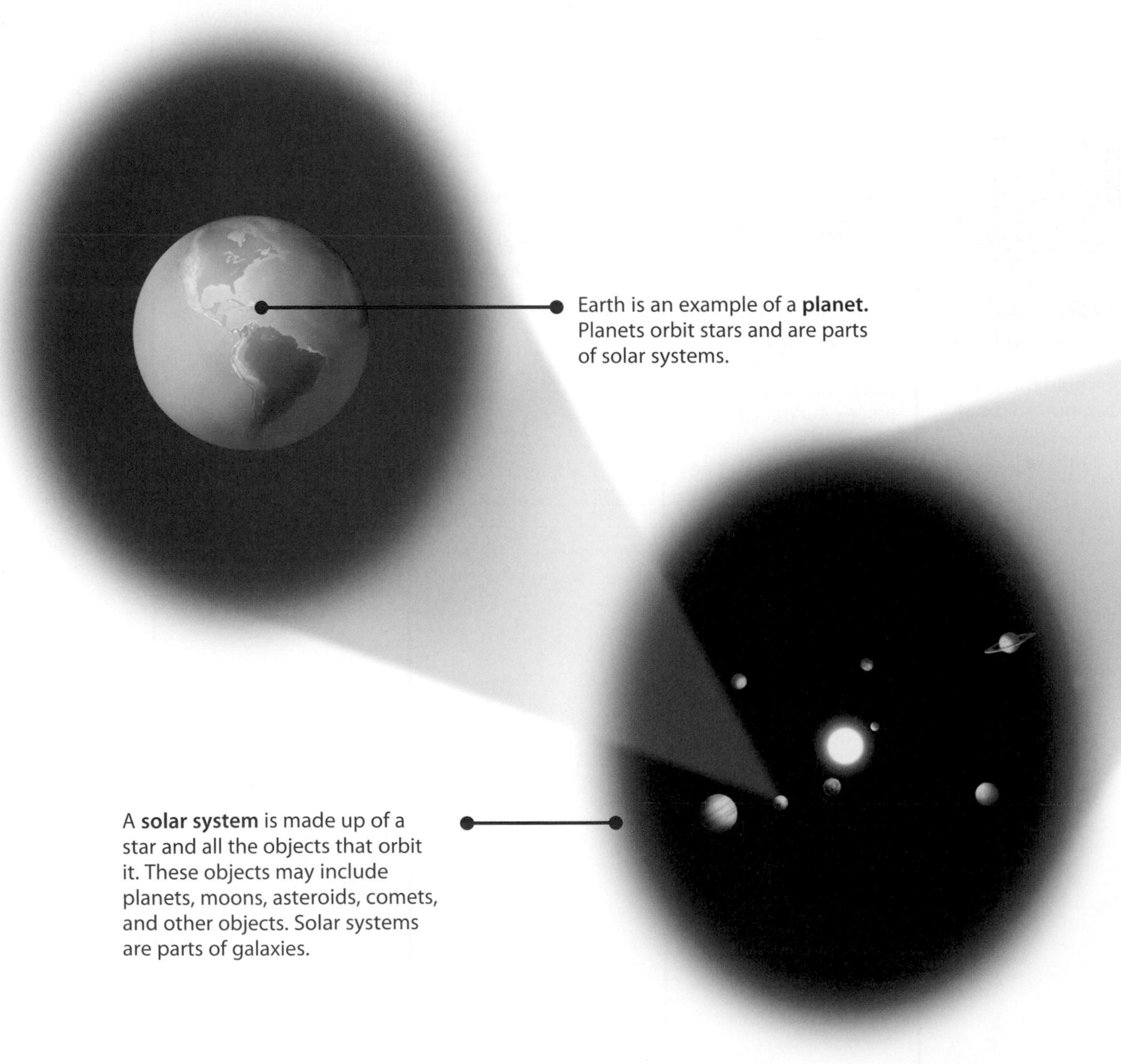

A **galaxy** is made up of hundreds of millions to billions of stars and their solar systems. Galaxies may have different shapes and sizes. There are millions of galaxies in the universe.

All the galaxy clusters, along with all the other matter and space in between, make up the **universe.**

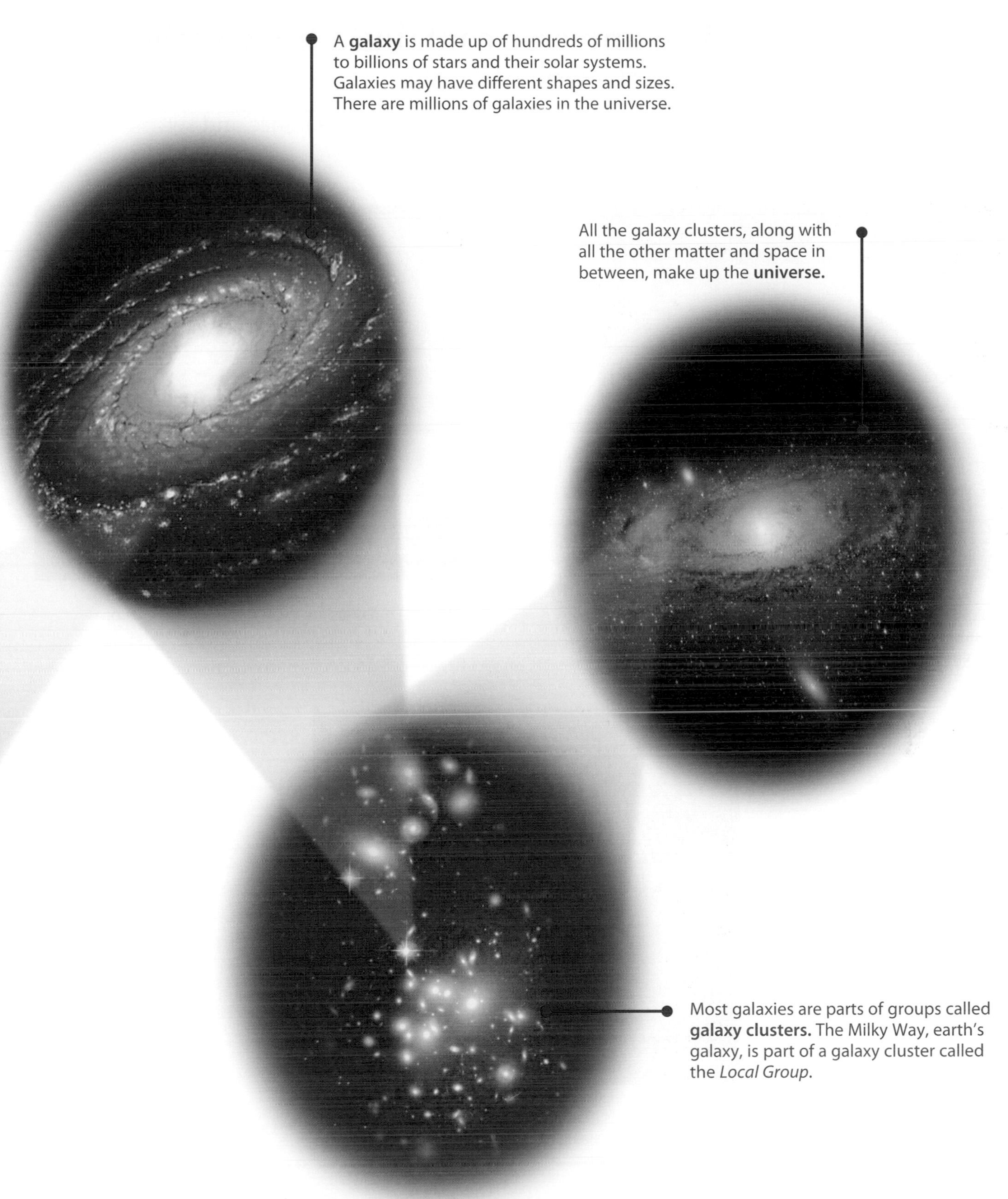

Most galaxies are parts of groups called **galaxy clusters.** The Milky Way, earth's galaxy, is part of a galaxy cluster called the *Local Group*.

The universe contains a huge variety of different objects.

Each type of object has a distinct appearance and set of properties.

Spiral galaxies have bright, central masses of stars and long, thin "arms" that extend away from the center. Our galaxy, the Milky Way, is a spiral galaxy.

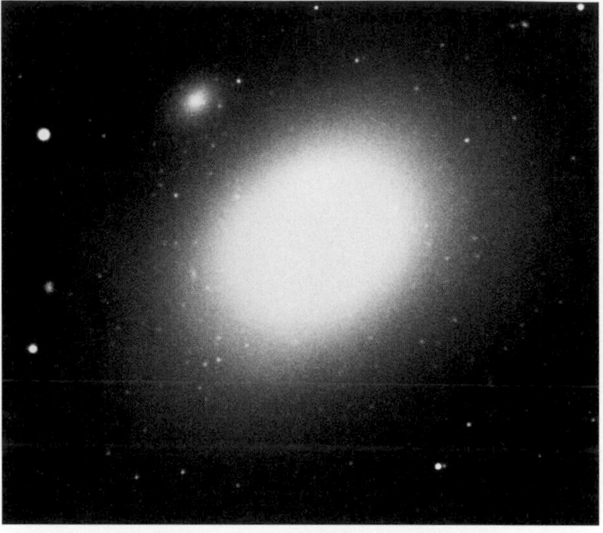

Elliptical galaxies are oval groups of millions of stars. Unlike spiral galaxies, elliptical galaxies do not have arms.

Irregular galaxies are neither spiral shaped nor elliptical. Like all galaxies, they contain a mixture of stars, gas, and dust.

The **Crab Nebula** formed as a result of a supernova. Chinese astronomers observed this supernova in 1054.

The **Horsehead Nebula** is located in the constellation Orion. It consists of a huge cloud of gas and dust and is approximately 150 light years from earth.

A pair of stars that orbit a common center are known as a **binary star system.** The two stars in a binary star system may be so close together that they appear to be one star when observed from earth. Astronomers think that as many as half of the stars visible from earth are members of binary star systems.

Comets are bodies of ice, rock, and dust that orbit our sun. As a comet moves through its orbit, it gets closer to the sun. This can cause some of the ice in the comet to vaporize. The gases form the comet's tail.

The big bang theory is a scientific theory that describes the formation and history of the universe.

According to this theory, before the universe existed, all matter and energy was condensed into an infinitely small point. Then, it began to expand extremely rapidly, forming the universe.

The early expansion of the universe is called *inflation*. The **inflation of the universe** began about 10^{-35} seconds—about 1 hundred-millionth of 1 trillionth of 1 trillionth of 1 second—after the big bang. At this time, the universe was still very tiny and was extremely hot—more than 1 quadrillion trillion degrees Celsius. The temperature and pressure at this time were so high that even subatomic particles such as protons and neutrons could not exist.

One second after the big bang, the universe had cooled and expanded enough for the **first hydrogen nuclei** to form. Nuclei of some other light elements, such as lithium, also began to form at this time.

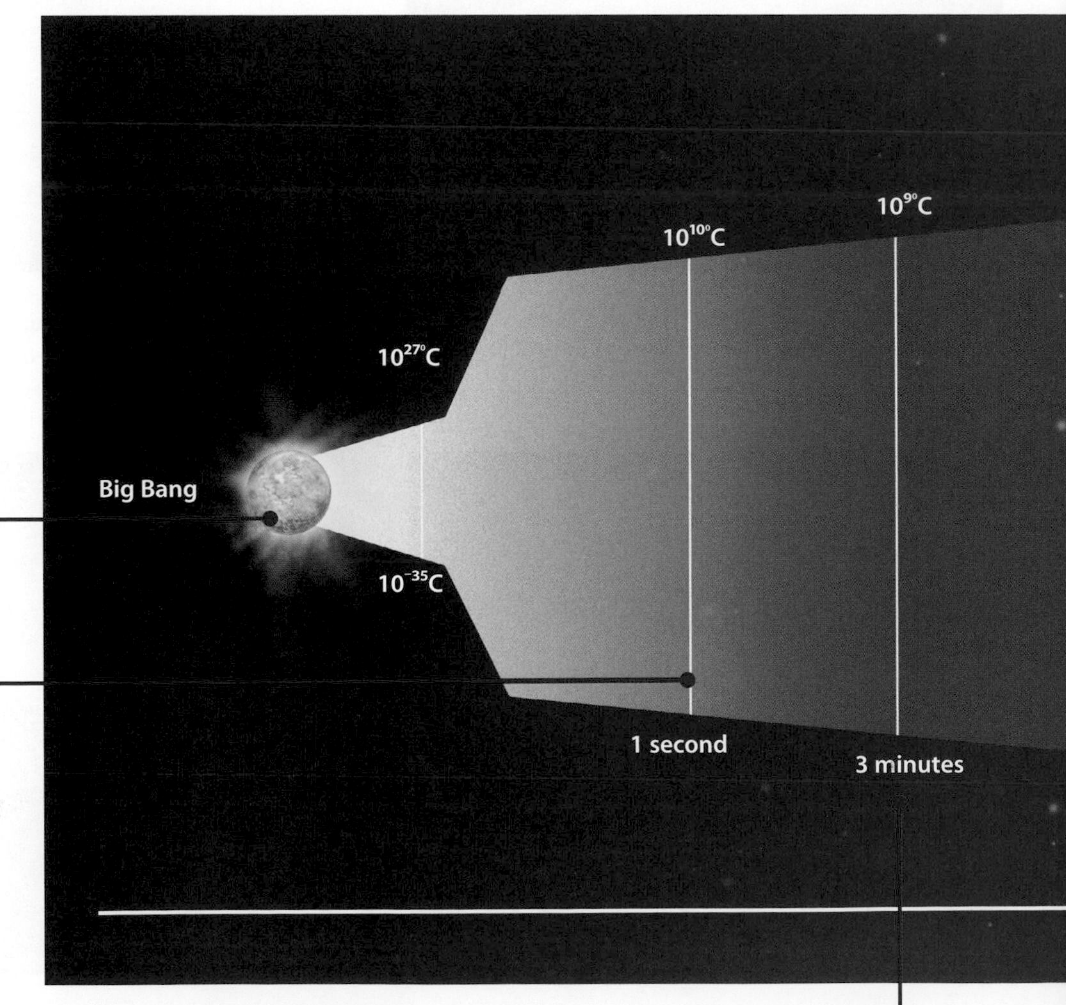

After about 3 minutes, the temperature of the universe had cooled to only 1 billion degrees Celsius. By this time, **nuclei of many light elements** had formed. About 75 percent of them were hydrogen nuclei, and about 25 percent were helium nuclei. There were also a tiny number of other nuclei present.

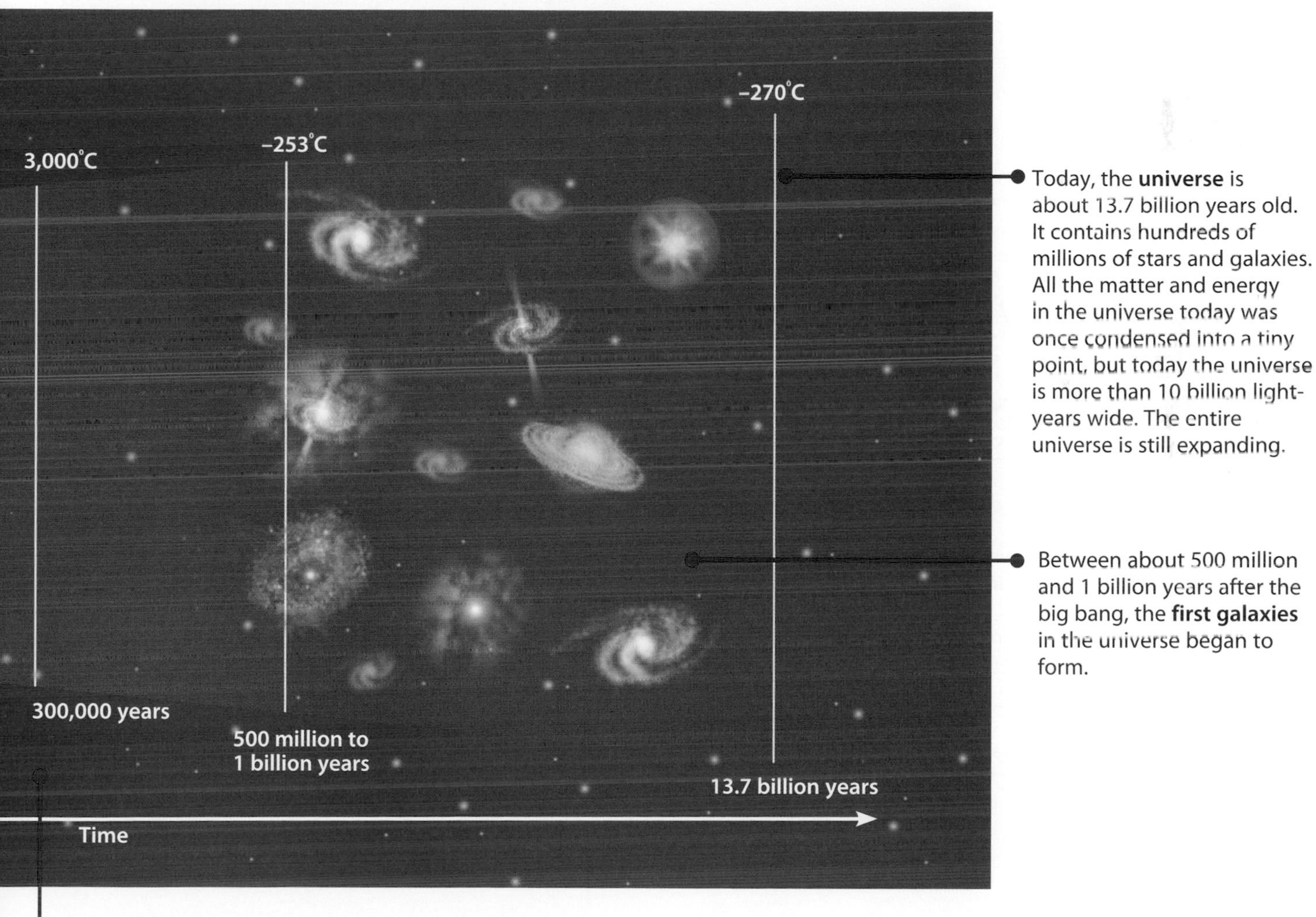

Today, the **universe** is about 13.7 billion years old. It contains hundreds of millions of stars and galaxies. All the matter and energy in the universe today was once condensed into a tiny point, but today the universe is more than 10 billion light-years wide. The entire universe is still expanding.

Between about 500 million and 1 billion years after the big bang, the **first galaxies** in the universe began to form.

Stable atoms began to form as electrons combined with nuclei about 300,000 years after the big bang. The cosmic background radiation that we observe from earth today was generated at about this time.

Important evidence that supports the big bang theory is the observation that light from galaxies shows evidence of red shift.

This observation suggests that these galaxies are all moving away from each other and from earth, which, in turn, indicates that the universe is expanding in all directions.

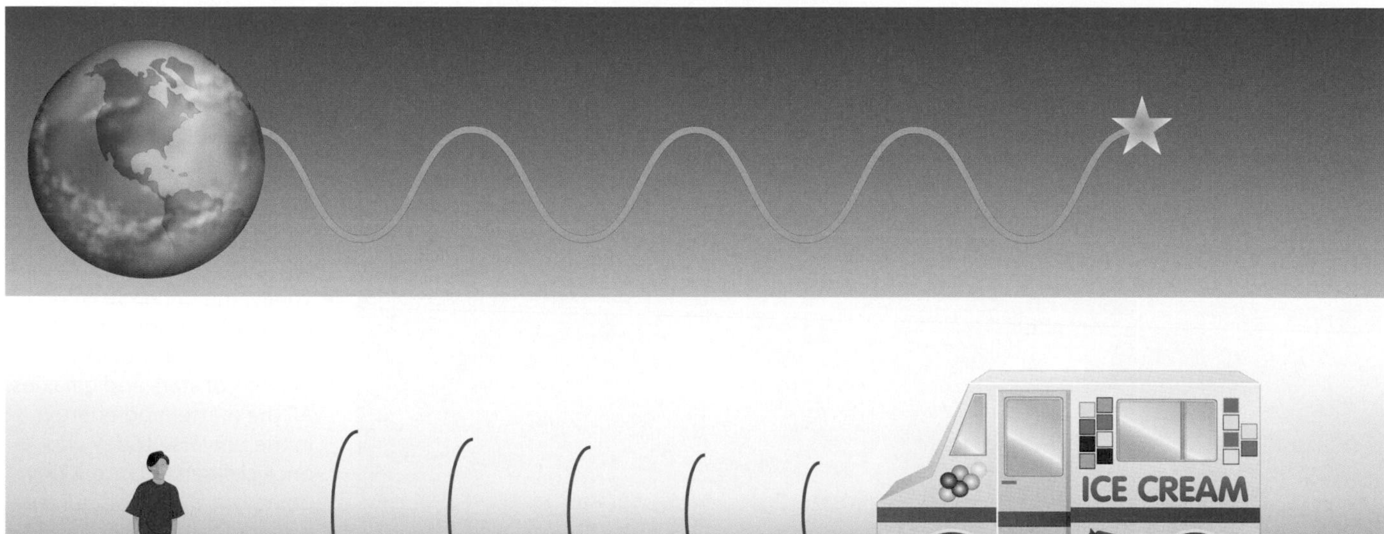

When a light source is not moving relative to an observer, the observer sees the light as exactly the same color as the source is emitting. For example, if the light source is a star emitting yellow light, an observer who is stationary relative to the star will see the star's light as yellow. This is similar to a situation in which a person hears the sound from a vehicle that is not in relative motion. The pitch of sound the person hears is exactly the same as the pitch of the sound emitted by the vehicle.

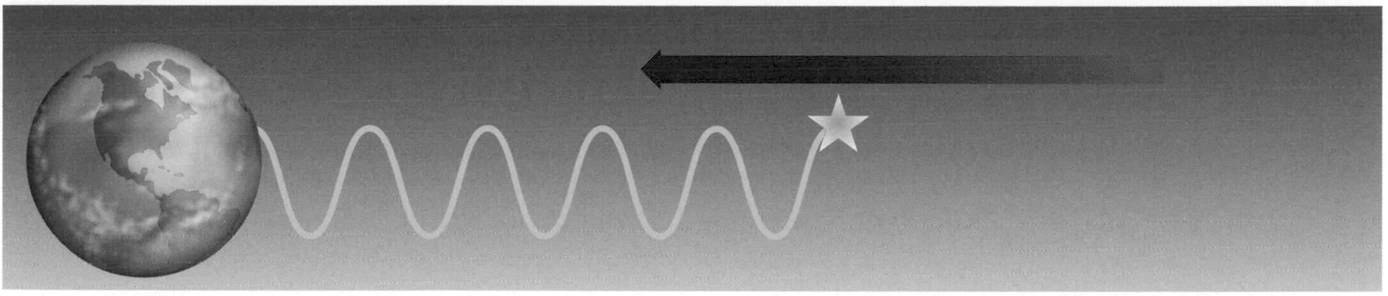

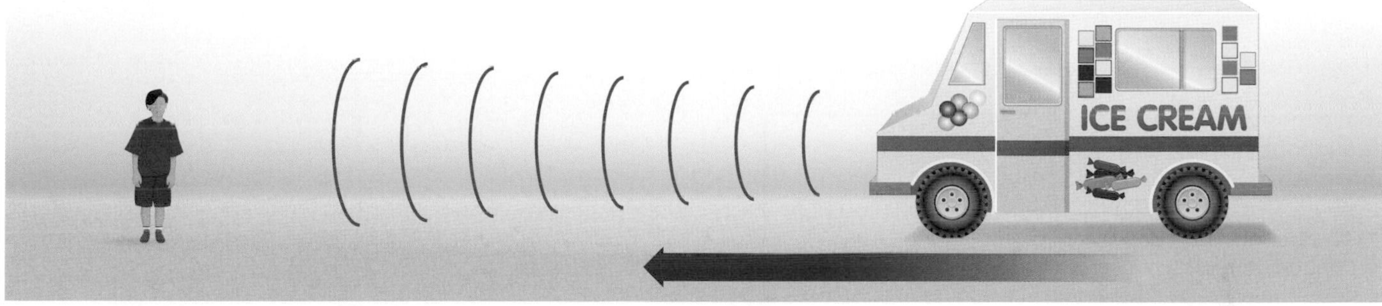

When a light source and an observer are moving toward each other, the light from the source appears bluer to the observer than the light that is emitted by the object. For example, if a star emitting yellow light is moving toward an observer, the observer may see the light as green or blue. The light is blue shifted. This is similar to the situation in which a vehicle is moving toward a person. The sound of the vehicle seems to have a higher pitch than it actually has.

When a light source and an observer are moving apart, the light from the source appears redder than it actually is. The light is red shifted, which is similar to the situation in which a vehicle is moving away from a person. The person hears a lower pitch of sound than the vehicle is actually emitting.

People use earth materials in almost every aspect of their lives.

We use coal to heat our homes and run power plants. We process oil into gasoline. From diamond-tipped drill bits to steel bars made from iron ore, earth materials are used in a wide variety of products.

Coal deposits can be found throughout most of North America and in other parts of the world. People burn coal and other fossil fuels to release energy. In the United States, most coal is used to power electric generators.

Iron ore can be found across the globe. The major deposits in the United States lie in the eastern mountain ranges. The majority of iron ore mined in the world is used to make steel.

Large deposits of **petroleum** (crude oil) are found in many parts of the world, including much of the Middle East and parts of North and South America. Crude oil is a mixture of many different chemicals. It can be refined into a variety of products, such as gasoline, kerosene, and tar. Most petroleum products are used for transportation.

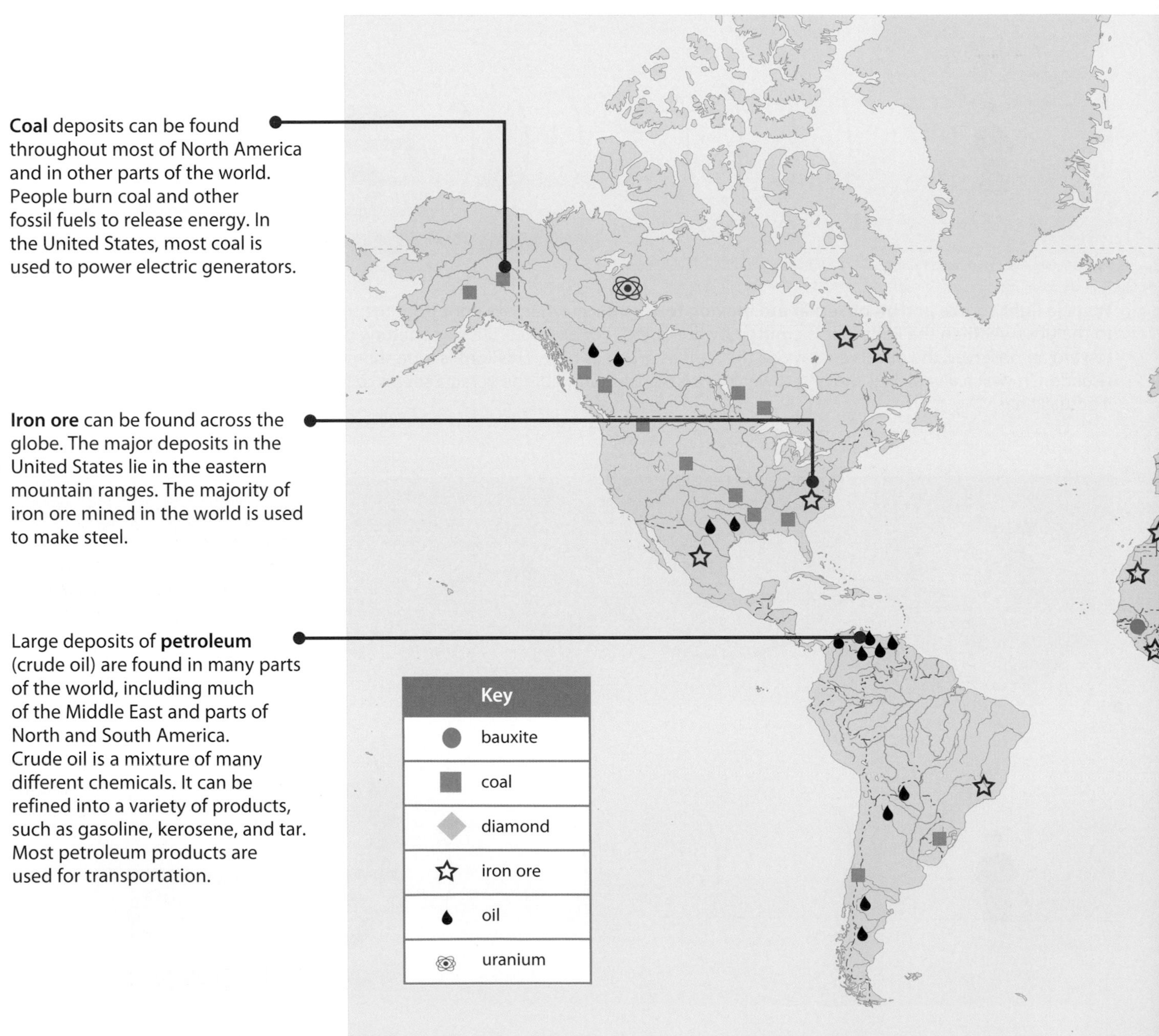

The largest known deposits of **diamonds** are found in South Africa. Diamond is the hardest known mineral. Therefore, in addition to its value as a gemstone, diamond is very useful in industry. Lower-quality diamonds can be crushed into a powder that is used as an abrasive.

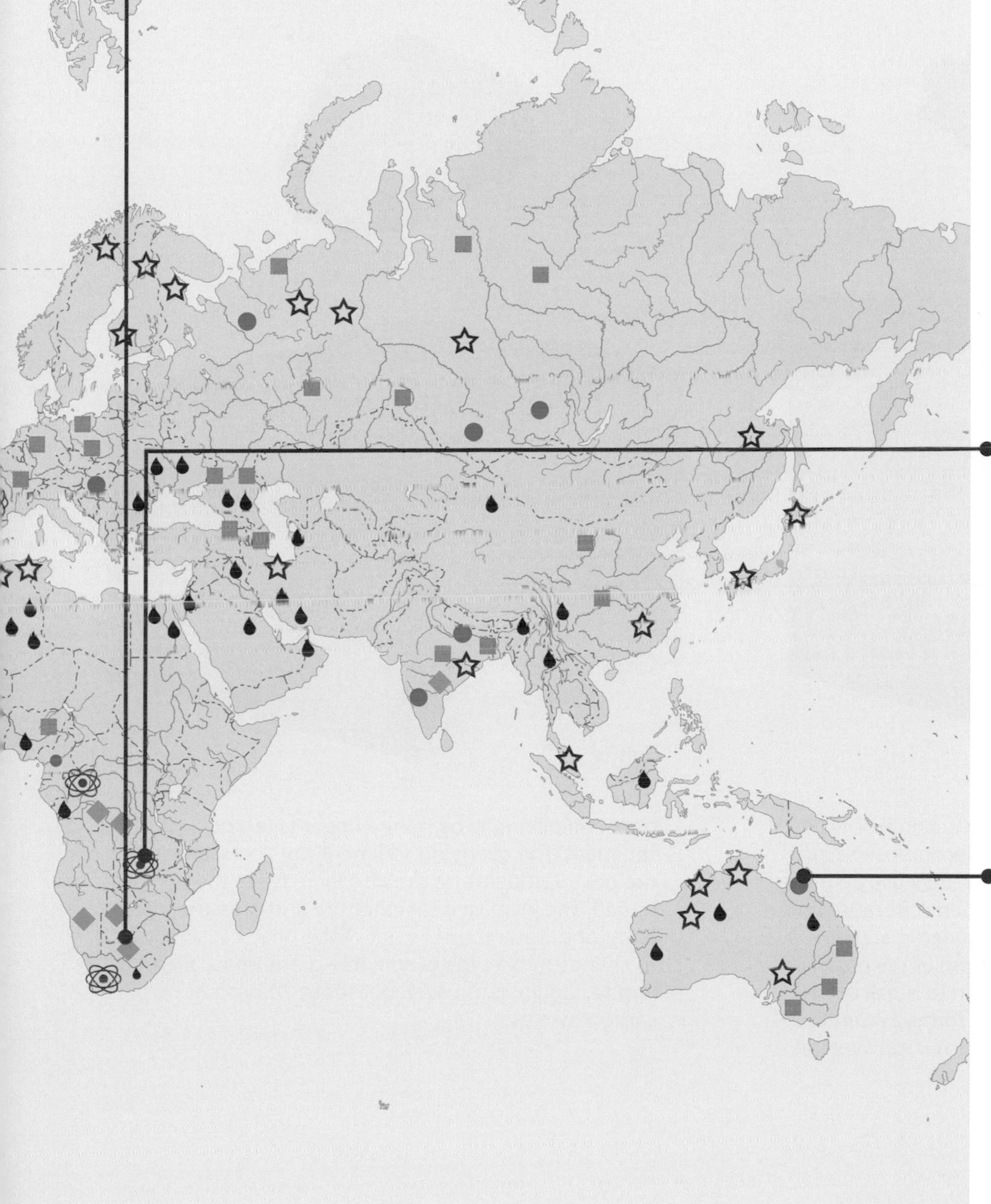

Uranium is a very dense metallic element. Its density makes it useful in armor for military vehicles and in missiles and bullets. Some uranium contains high concentrations of radioactive isotopes. This type of uranium, known as *enriched uranium,* is used in nuclear power plants to produce electricity. Major uranium deposits are found in Africa and Europe.

Bauxite is the main ore of aluminum. Major deposits of bauxite are found in Australia, with smaller deposits in Africa, Russia, and India. Bauxite is processed into aluminum to make everything from soda cans to aircraft.

Coal is a fossil fuel that forms from the remains of land plants.

Much of the coal on earth today formed from plants that lived hundreds of millions of years ago, when huge swamps that were full of plant life covered parts of the land.

The first step in the formation of coal is the burial of dead plant materials, usually underwater. As decomposers (such as bacteria) break down the plant matter, they use up most of the oxygen that is buried with the material. Because little oxygen can move from the air into the buried material, the decomposers cannot break down more of the organic matter. Instead, other decomposers begin to break down the organic matter without oxygen. This forms a crumbly, dark brown material called **peat.** Dried peat is, on average, about 60 percent carbon by mass.

Next, the peat may be buried under layers of sediment. The overlying sediment puts pressure on the peat. Temperature can also increase as the peat is buried. The increased temperature and pressure force gases and water to move out of the peat, causing it to change into a type of soft, low-grade coal called **lignite.** Lignite is, on average, about 70 percent carbon by mass.

Next, more sediment may continue to cover the lignite. This causes even greater temperatures and pressure, which can drive more water and gas out of the lignite. This produces **bituminous coal,** which is about 80 percent carbon by mass on average.

If temperature and pressure continue to increase due to more sediment accumulation, bituminous coal can become **anthracite.** Anthracite contains about 90 percent carbon by mass on average. Anthracite is the highest grade of coal.

Coal is an important source of energy for people. Most coal is located below earth's surface and can be difficult to locate and obtain.

Scientists, engineers, and miners use tools and technology to find and extract coal.

Distribution of Coal Resources in the United States

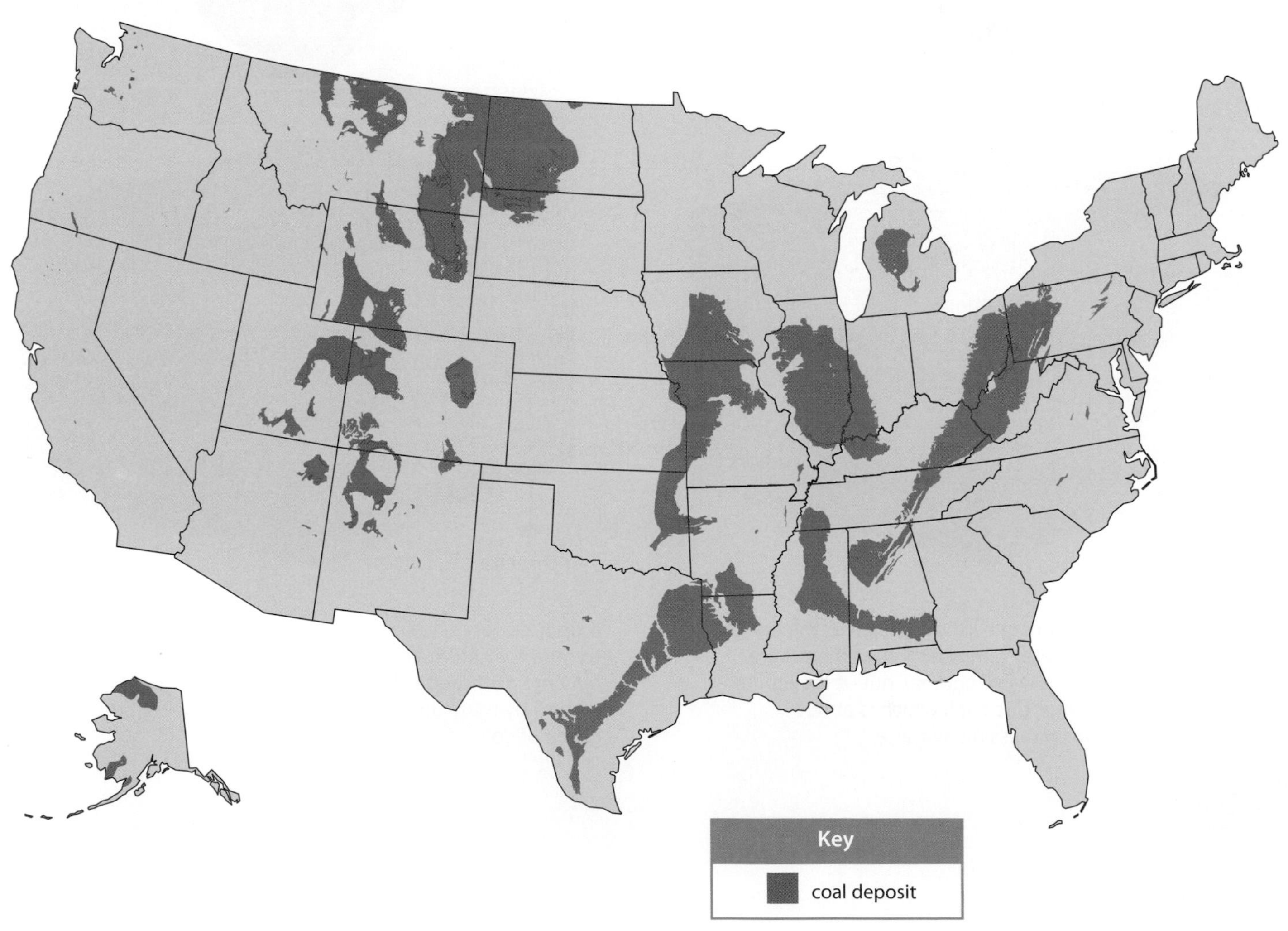

Geologists and mining engineers use geologic maps, subsurface imaging, and other tools to infer the **location of a coal deposit**. To determine whether it will be profitable to mine the coal, they first estimate how much coal can be obtained from the deposit. Then, they calculate how much that amount of coal is worth. Next, they estimate the costs associated with mining the coal: the cost of equipment, workers, and repairing the environmental damage that mining causes. Finally, they determine whether the value of the coal is enough to offset the costs of mining. If it is, they begin to plan how to extract the coal.

Most coal is taken from the ground by **strip mining**. During strip mining, people use explosives to remove the layers of rock and soil above the coal deposit. Then, they remove the coal in large pieces. Some coal is also extracted using subsurface mining. During subsurface mining, miners dig tunnels into coal deposits. As they dig the tunnels, more coal is removed.

After the coal is removed from the ground, it must be **transported** to where it will be used. People use trains, trucks, and ships to transport coal from place to place.

Coal mining causes significant **environmental damage**. Strip mines destroy huge areas of soil, as well as the organisms that live on the land. In addition, most coal contains acids that cause water pollution when rainwater percolates through the coal and mine waste. Therefore, many countries have laws requiring that coal mining companies protect and restore the land that they mine. In many cases, this means reclaiming the land by putting back the rock and soil layers that were removed.

Iron is an important natural resource.

People use iron in buildings, cookware, automobiles, and many other objects. Most of the iron that people use is found in earth's crust as iron ores. These ores must be refined before people can use the iron.

Key
☆ major iron ore deposit

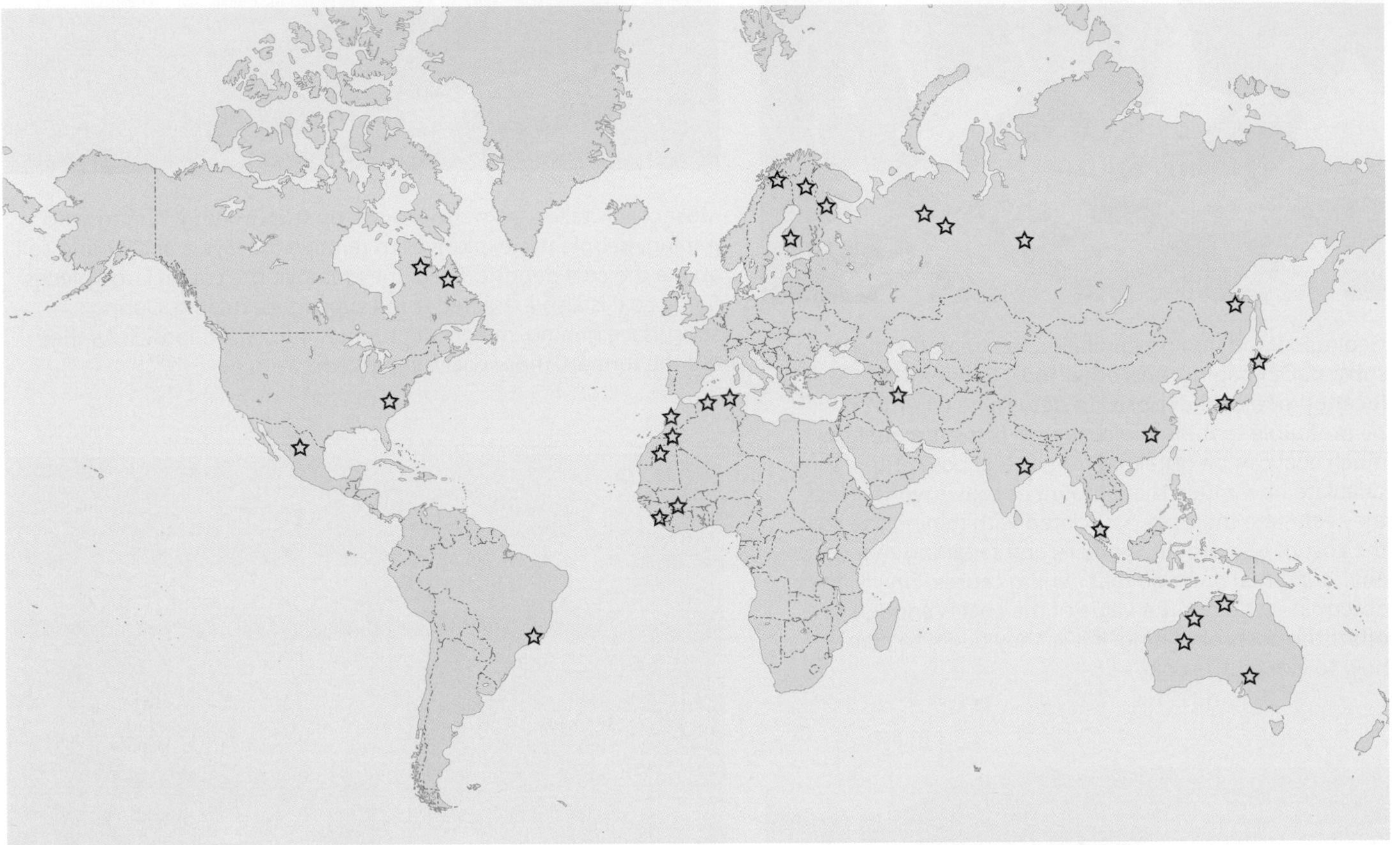

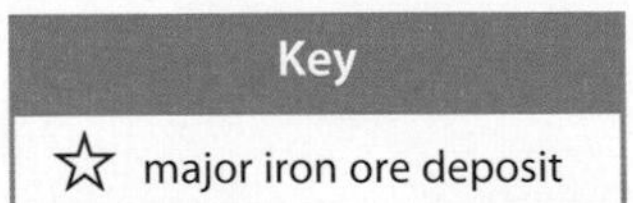

Iron oxides, such as hematite and magnetite, are the most common **iron ores**. Scientists test samples of rocks to determine how much iron is contained in the rocks. Engineers and miners use geologic maps, subsurface imaging, and other tools to predict where iron-rich rocks may be located.

People mainly use **open-pit mining** to obtain iron ore. During open-pit mining, explosives and machinery remove layers of soil and rock above the iron ore. Then, the iron ore is removed and taken to a refinery.

Iron ore and metal must be **transported** from the mines to the refineries to places where it will be used. People use trains, trucks, and ships to transport iron from place to place.

During the process of **smelting**, iron ore is melted. Chemicals such as calcium carbonate and carbon react with the oxygen in the iron ore. The iron metal is left behind. In most cases, the metal still contains some impurities, so it may be further refined through a process called *electrolysis*.

As the population of the world continues to increase, so will the demand for electricity.

Earth's future population will need to develop new methods for producing electricity, as well as more efficient ways of using it. The figure below shows the average amount of electricity used in different countries.

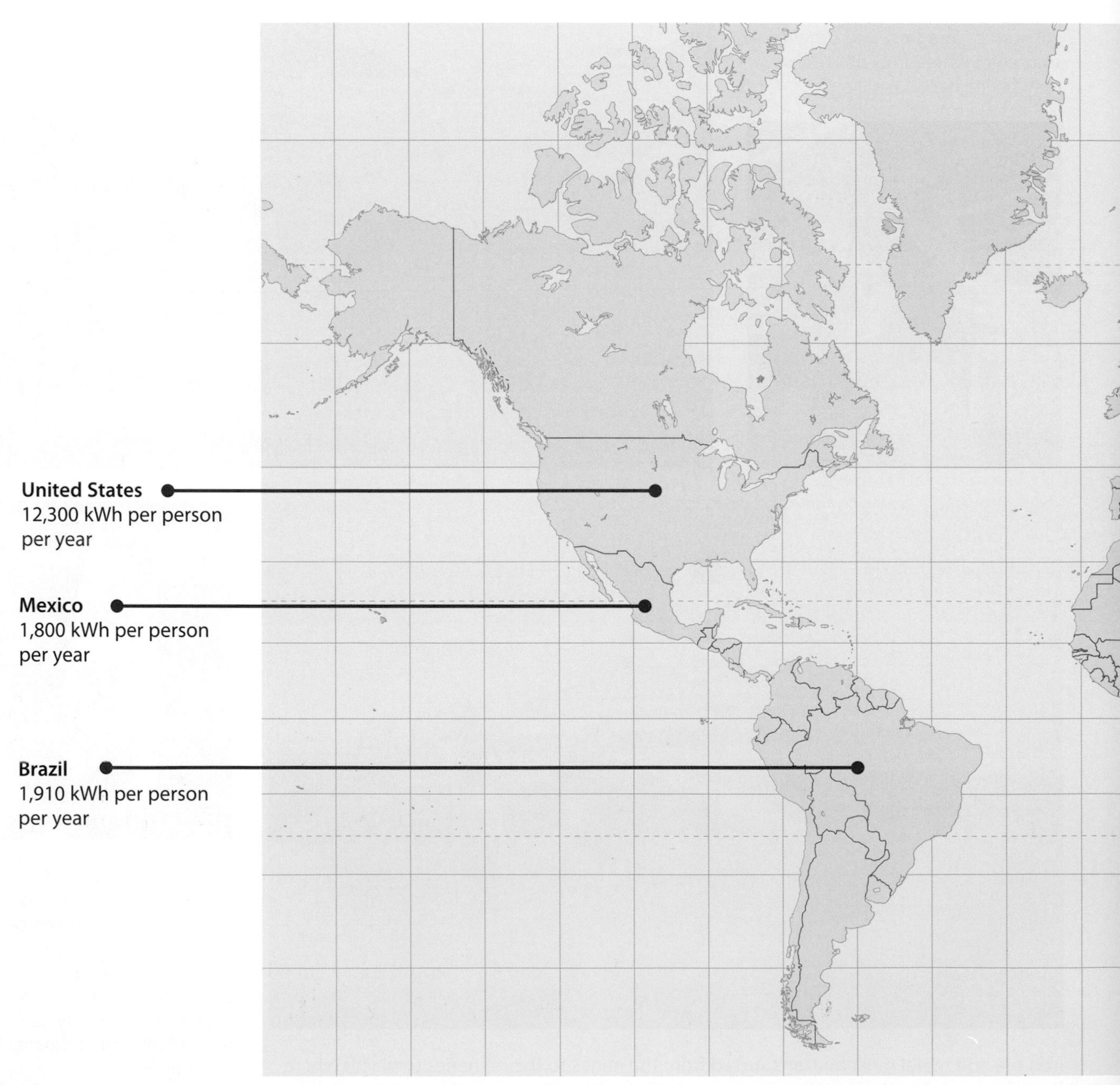

kWh = 5 kilowatt-hour

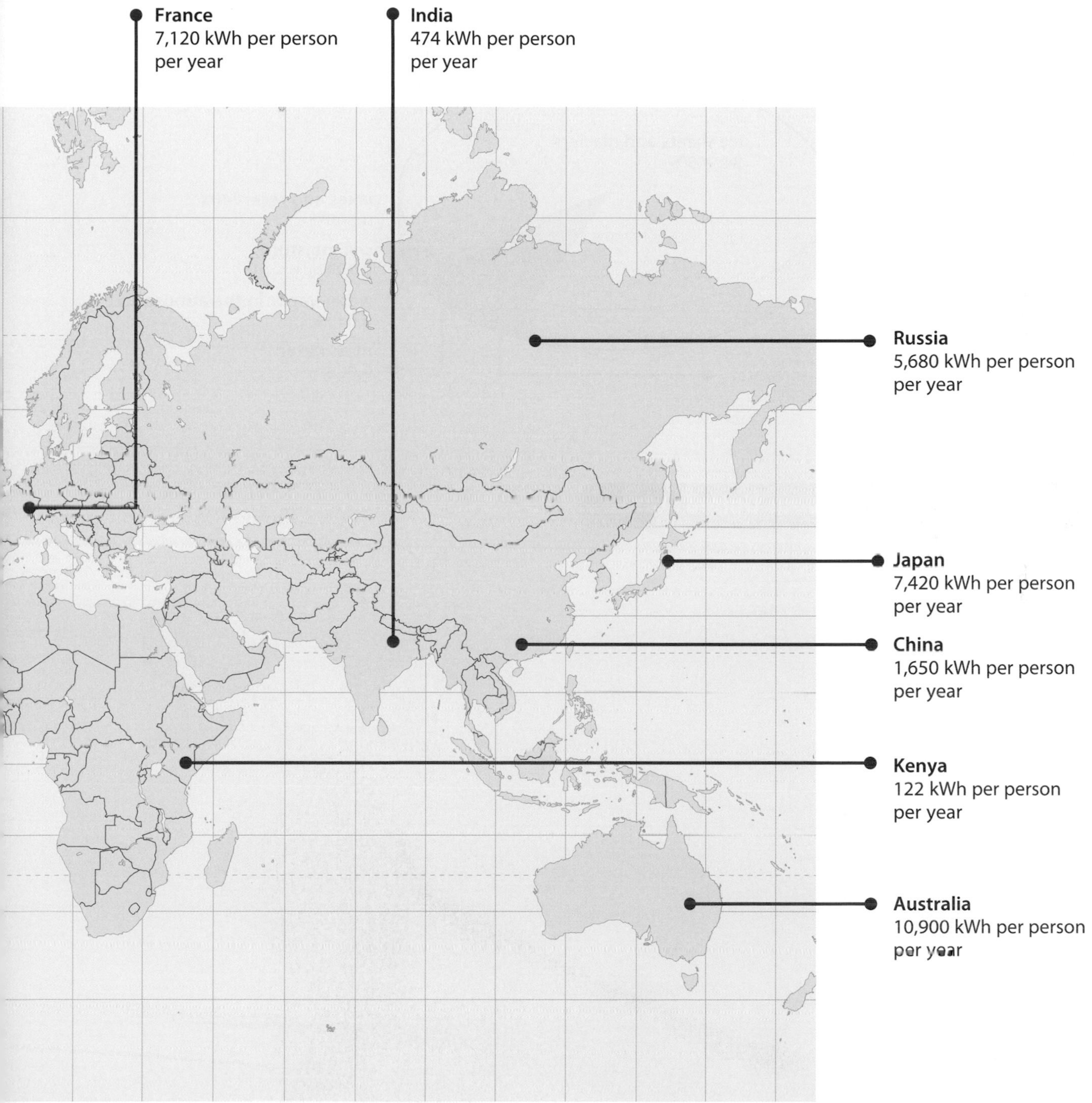

France
7,120 kWh per person per year
India
474 kWh per person per year
Russia
5,680 kWh per person per year
Japan
7,420 kWh per person per year
China
1,650 kWh per person per year
Kenya
122 kWh per person per year
Australia
10,900 kWh per person per year

Water is one of our most important natural resources.

Most of the water on earth—about 97 percent—is in the oceans. People can't drink this water because it is too salty. Instead, people must use the small fraction of earth's water that is fresh.

Most of earth's **fresh water** is stored in ice. The rest is found either underground or on the surface in lakes, rivers, ponds, or other surface waters.

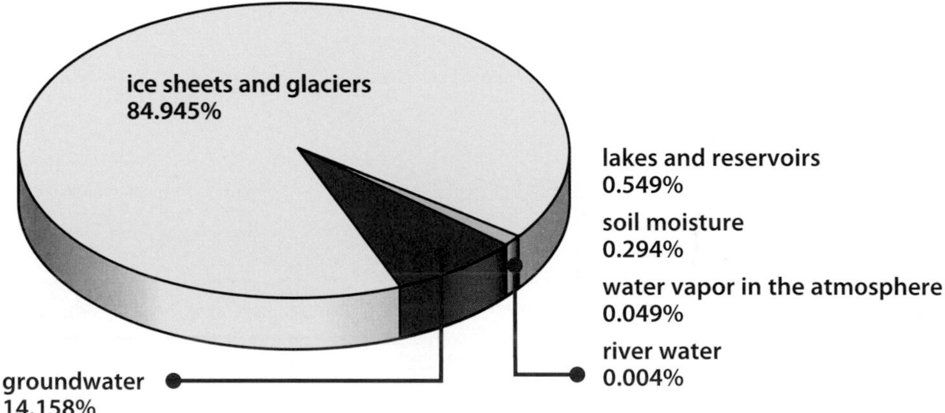

Major Aquifers in the Contiguous United States

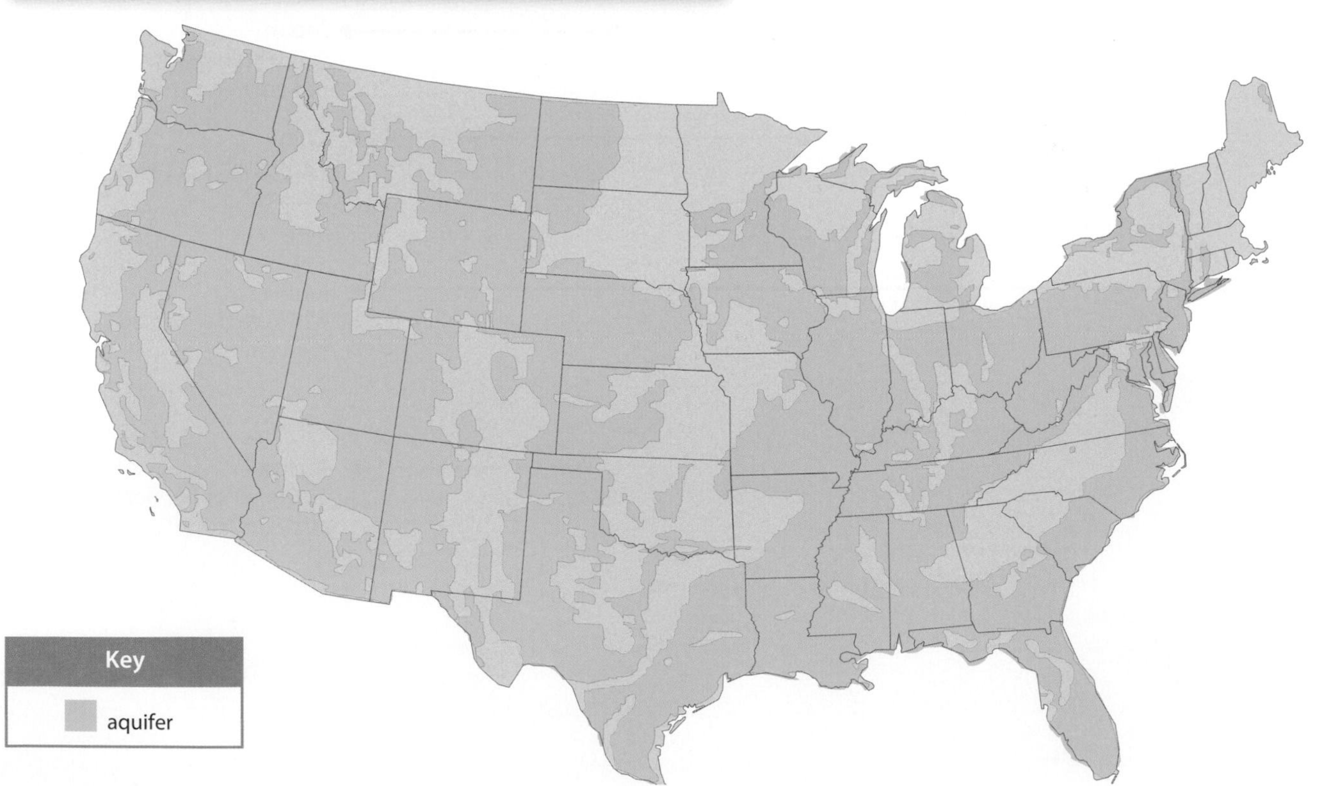

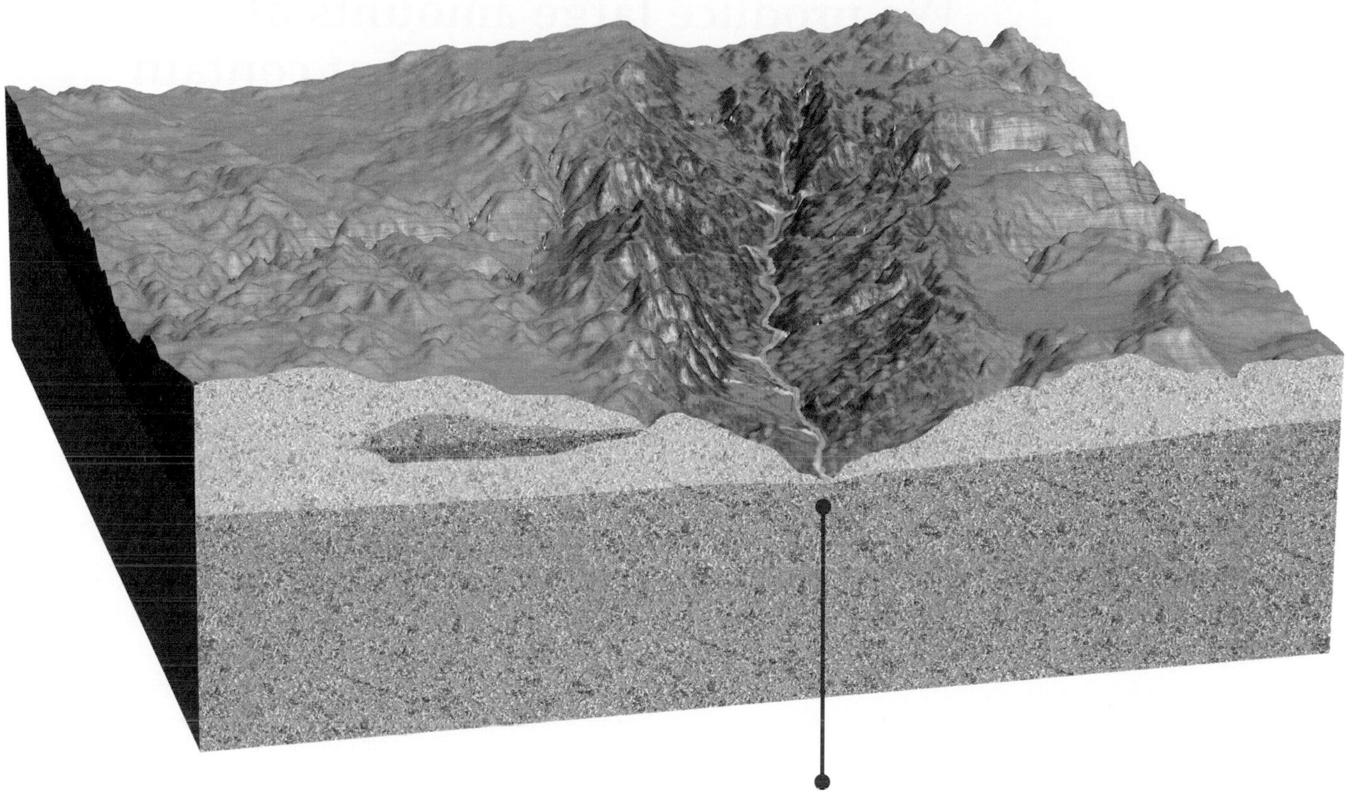

An **aquifer** is an underground layer of rock that holds (or stores) water. Most aquifers consist of sandstone or other porous rock.

A **well** is a hole that is drilled into an aquifer. Water from the aquifer can flow up through the well to the surface. If water is removed from the aquifer faster than it can be recharged by precipitation, the water level of the aquifer can drop and the well can go dry.

We produce large amounts of wastewater every day that contain materials toxic to many organisms.

Before wastewater can be returned to the environment, it must be treated to remove harmful materials.

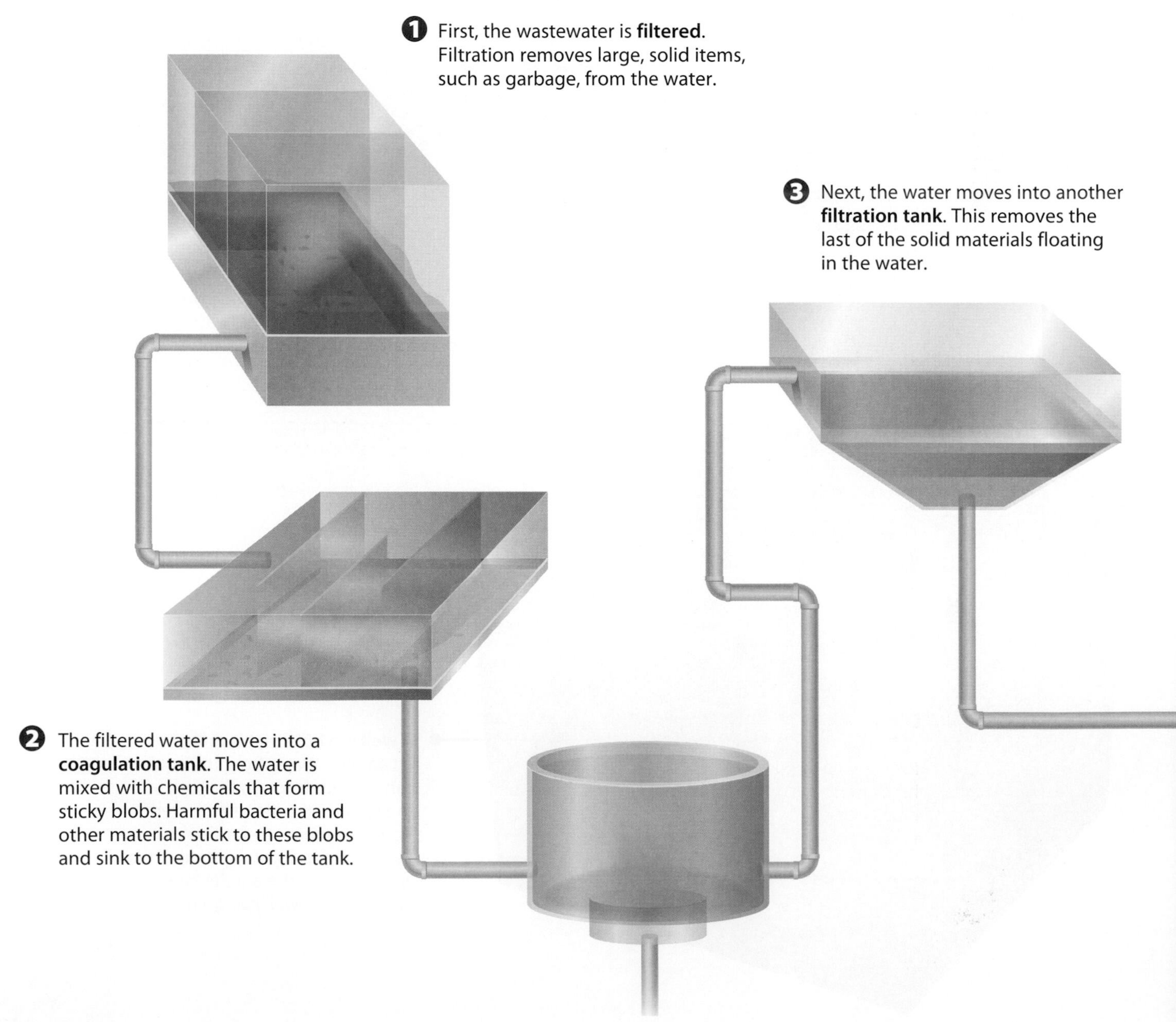

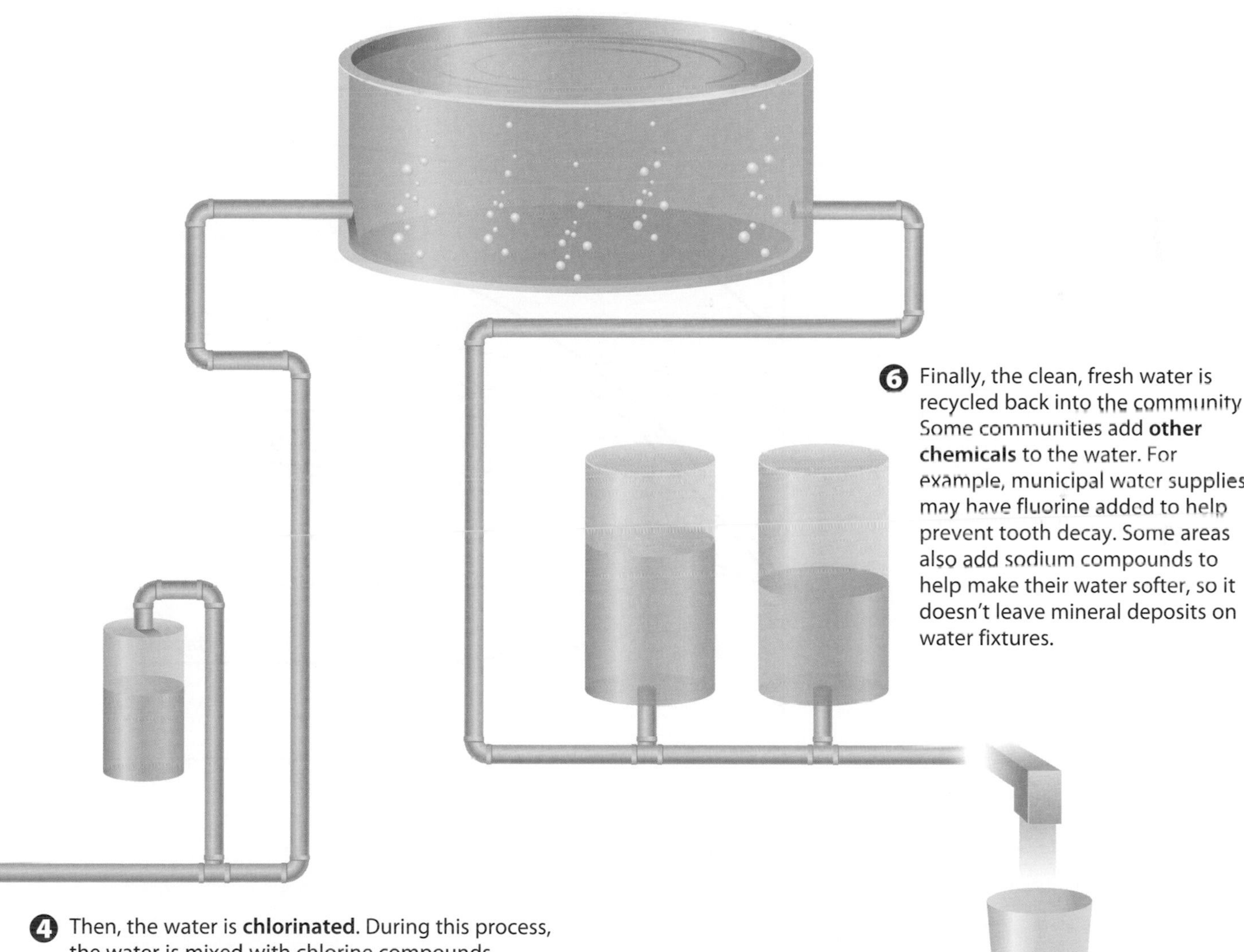

5 The water moves into an **aeration tank**. There, air moves through the water. This helps drive off gases that are dissolved in the water. It improves the water's taste and odor.

6 Finally, the clean, fresh water is recycled back into the community Some communities add **other chemicals** to the water. For example, municipal water supplies may have fluorine added to help prevent tooth decay. Some areas also add sodium compounds to help make their water softer, so it doesn't leave mineral deposits on water fixtures.

4 Then, the water is **chlorinated**. During this process, the water is mixed with chlorine compounds that kill harmful bacteria and prevent other microorganisms from growing in the water.

The number of people on earth has increased through history and the population continues to grow.

Earth's human population passed the 6 billion mark in the late 1990s. If current population growth rates continue, there could be more than 9 billion people living on earth by the year 2050.

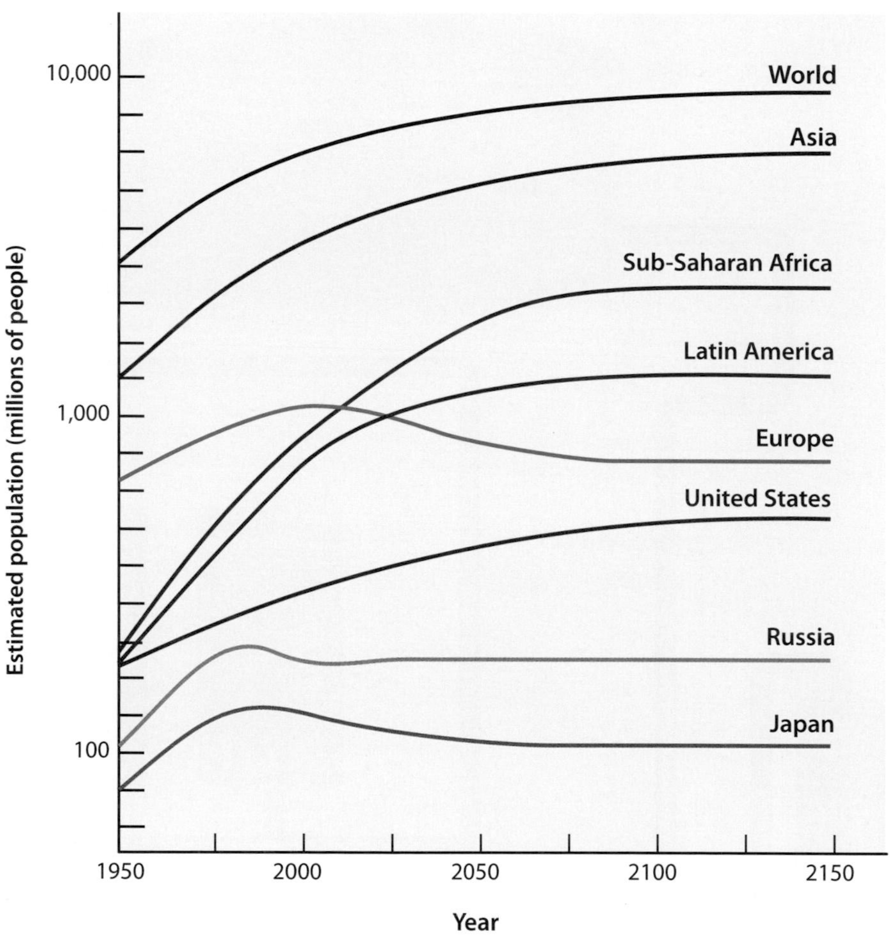

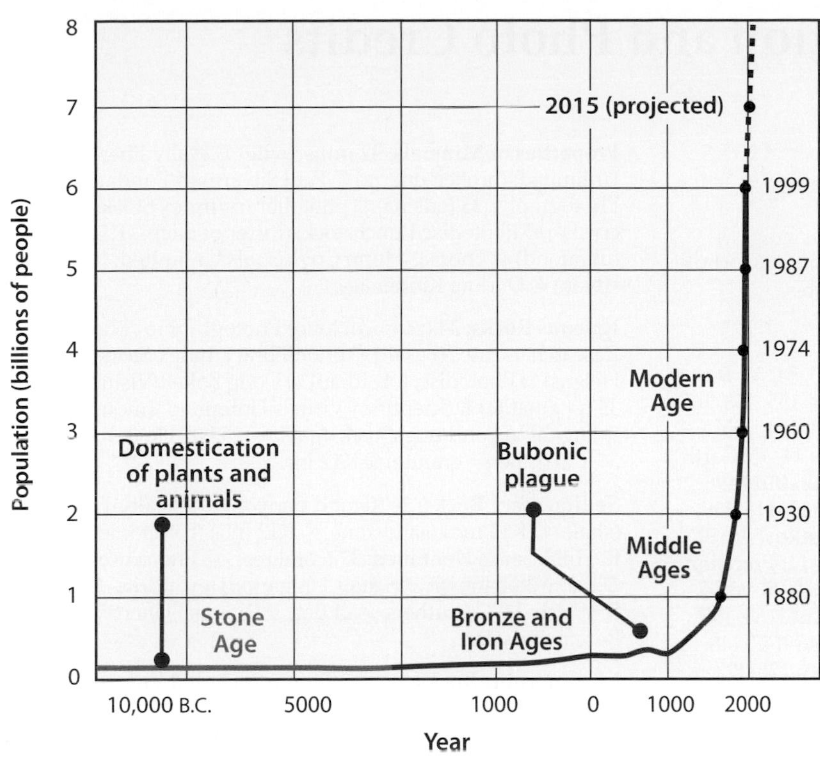

POPULATIONS AND POPULATION DENSITIES OF SELECTED COUNTRIES		
COUNTRY	***POPULATION (MILLIONS OF PEOPLE) JULY 2006 ESTIMATE***	***AVERAGE POPULATION DENSITY (PEOPLE PER SQUARE KILOMETER OF LAND AREA)***
Australia	20.26	2.659
Brazil	188.1	22.24
China	1,313	140.9
France	60.88	111.6
India	1,095	368.3
Japan	127.5	340.3
Kenya	34.71	60.97
Mexico	107.4	55.85
Russia	142.9	8.406
United States	298.4	32.57

Illustration and Photo Credits

All illustrations © K12 Inc. unless otherwise noted.

Front cover: (Arches National Park) © Linde Waidhofer/Taxi/Getty Images

Back cover: (Jupiter) © Corbis; (Aurora borealis) © Roman Krochuk/Alamy; (fluorite) © José Manuel Sanchis Calvete/Corbis

Illustration Credits

Argosy Publishing: 24, 25, 26, 27, 30, 42, 43, 48, 52, 53, 54, 55, 60, 61, 64, 65, 68, 70, 71, 76, 80, 82, 83, 84, 88, 115, 140, 154–155, 158, 166, 175, 176, 177, 178, 179, 180–181, 182, 186, 187, 192–193, 194, 195, 198, 199, 200, 208–209; **Kim Battista:** 27; **James Bravo:** 14–15; Garth Glazier / AA Reps, Inc.: 134 135, 138, 139; **Jennifer Horsburgh:** 18–19, 44–45, 86–87, 102–103, 106–107, 110, 111, 112–113, 118–119, 122, 123, 130, 132, 136–137, 141, 196–197, 202; **Julie Jankowski:** 91, 92–93, 182, 183; **Pennsylvania Geolo:ical Survey:** 10–11; **Precision Graphics:** 2–3, 50–51, 56, 57, 58–59, 60–61, 62–63, 67, 94–95, 96, 97, 98, 99, 100, 101, 116, 117, 120–121, 160, 162, 163, 164, 164–165, 165, 169, 171, 188–189; **Tony Randazzo / AA Reps, Inc.:** 146–147, 148–149, 150–151, 152–153; **Space Channel / AA Reps, Inc.:** 126–127, 142–143, 144–145, 156, 157, 206, 207; **US Geological Survey:** 16–17.

Photo Credits

Our Place in the Universe: **4** (solar eclipse) © blickwinkel/Alamy; (Earth view) NASA Headquarters: Greatest Images of NASA (NASA-HQ-GRIN); (sun) NASA Jet Propulsion Laboratory (NASA-JPL); (meteor shower) © John Chumack/Photo Researchers, Inc.; (Copernican solar system) The Granger Collection, New York. **5** (Galileo) Mary Evans Picture Library/Photo Researchers, Inc.; (Newton) The Granger Collection, New York; (Hubble and telescope) Margaret Bourke-White/Time Life Pictures/Getty Images; (big bang) © Markowitz Jeffrey/Corbis Sygma; (cosmic radiation) NASA/Photo Researchers, Inc.

Topographic Maps: **8** (Berkshires) © Robert Landau/Corbis.

Geologic Maps: **11** (Fallingwater) © Richard A. Cooke/Corbis.

Seismographs: **12** (seismograph) Zephyr/Photo Researchers, Inc. **13** (damaged building) © Joseph Sohm; ChromoSohm Inc./Corbis; (seismogram) © Science VU/Visuals Unlimited.

The Spheres of Earth: **20** (ocean) © Photodisc/Punchstock; (Grand Canyon) © Photodisc/Punchstock. **20–21** (Earth) NASA/Science Source. **21** (glacier) © Ralph A. Clevenger/Corbis; (coral reef) © Photodisc; (aurora) © Roman Krochuk/Alamy.

Groups of Minerals: **28** (dolomite) © Scientifica/Visuals Unlimited; (calcite) © K12 Inc.; (halite) © photolibrary/Index Stock; (fluorite) © José Manuel Sanchis Calvete/Corbis; (gold) © Ken Lucas/Visuals Unlimited; (copper) © Scientifica/Visuals Unlimited. **29** (corundum) © ephotocorp/Alamy; (hematite) © Roberto de Gugliemo/Photo Researchers, Inc.; (gypsum) © ephotocorp/Alamy; (anhydrite) © Mark A. Schneider/Visuals Unlimited; (galena) © photolibrary/Index Stock; (pyrite) © John Lens/Alamy; (olivine) © Mark A. Schneider/Photo Researchers, Inc.; (plagioclase) © Marli Miller/Visuals Unlimited.

Silicate Minerals: **30** (olivine) © Mark A. Schneider/Photo Researchers, Inc.; (pyroxene) © Marli Miller/Visuals Unlimited. **31** (amphibole) © Marli Miller/Visuals Unlimited; (muscovite) © Wally Eberhart/Visuals Unlimited; (quartz) © Photodisc/Punchstock.

Properties of Minerals: **32** (muscovite) © Wally Eberhart/Visuals Unlimited; (broken quartz) © Paul Silverman/Fundamental Photographs. **33** (galena) © photolibrary/Index Stock; (quartz crystal) © Photodisc/Punchstock; (luster of talc) © K12 Inc.; (diamond) © Thomas Hunn Co./Visuals Unlimited; (hematite streak) © Dorling Kindersley.

Igneous Rocks: **34** (pumice) Joyce Photographics/Photo Researchers, Inc.; (basalt) © photolibrary/Index Stock; (Mount Saint Helens) © Photodisc; (obsidian) © Doug Sokell/Visuals Unlimited. **35** (pegmatite) © Scientifica/Visuals Unlimited; (mountain) © Digital Vision/PunchStock; (gabbro) © J. Beckett, B. Walsh, M. Carruthers; (granite) © K12 Inc.

Sedimentary Rocks: **36** (Grand Canyon) © Royalty-Free/Corbis; (shale) © K12 Inc.; (sandstone) © K12 Inc.; (conglomerate) © Arthur R. Hill/Visuals Unlimited. **37** (coral reef) © Photodisc; (coquina) © Larry Stepanowicz/Visuals Unlimited; (evaporite) © J. Beckett, B. Walsh, M. Carruthers; (salt flat) © Dennis Flaherty/Photo Researchers, Inc.

Metamorphic Rocks: **38** (mountain) © Digital Vision/PunchStock; (slate) © Hot Ideas/Index Stock; (gneiss) © K12 Inc.; (schist) © K12 Inc. **39** (marble quarry) © Marli Miller/Visuals Unlimited; (marble) © Scientifica/Visuals Unlimited.

The Rock Cycle: **40** (El Capitan) © Photodisc/Punchstock. **41** (folded rocks) © Marli Miller/Visuals Unlimited; (Dover cliffs) © david martyn hughes/Alamy.

Changes in Earth's Surface: **46** (mudslide) © Gene Blevins/LA Daily News/Corbis; (acid rain lion) © Adam Hart-Davis/Photo Researchers, Inc.; (cracked rock) © Walt Anderson/Visuals Unlimited; (mountain range) © Digital Vision/PunchStock. **47** (glacier) © Morley Read/Photo Researchers, Inc.; (White Cliffs of Dover) © david martyn hughes/Alamy; (ventifact) © Peter Arnold, Inc./Alamy.

Plate Tectonics: Development of a Theory: **66** (seismogram) © Science VU/Visuals Unlimited; (David Griggs) Courtesy of UCLA Institute of Geophysics and Planetary Physics. **67** (Atlantis) © Dean Conger/Corbis; (Harry Hess) Courtesy of Department of Geological and Geophysical Sciences, Princeton University; (Glomar Challenger) © Corbis.

Volcanic Processes: **69** (Paricutin, 1943) © Bettmann/Corbis; (Paracutin, 1952) © Corbis.

Kinds of Volcanoes: **70** (Paricutin) © Photodisc/Punchstock. **71** (Mount Saint Helens) © Gary Braasch/Corbis; (Mauna Kea) © Ted Streshinsky/Corbis.

History of Life on Earth: **75** (Neanderthal) The Bridgeman Art Library/Getty Images; (*Eohippus*) © Tierbild Okapia/Photo Researchers, Inc.; (magnolia) © Sinclair Stammers/Photo Researchers, Inc.; (*Tyrannosaurus rex*) © Tom McHugh/Photo Researchers, Inc.; (*Archaeopteryx* fossil) © James L. Amos/Photo Researchers, Inc.; (fossil fish) © Gary Braasch/Corbis; (trilobite) © Ken Lucas/Visuals Unlimited; (*Tiktaalik*) Zina Deretsky/ National Science Foundation.

Radiometric Dating: **78** (chondrite meteorite) © Scott Camazine/Alamy.

The Grand Canyon: 80 (Grand Canyon) © Royalty-Free/Corbis.

History of Earth's Atmosphere: 84 (volcanic eruption) © Gary Braasch/Corbis; (torrential rain) © Bruce Peebles/Corbis. **85** (stromatolite) © Ken Lucas/Visuals Unlimited; (African trilobite fossil) © Eurelios/PhototakeUSA.com; (banded iron formation) © Dirk Wiersma/Photo Researchers, Inc.

Precambrian Life and Rocks: 86 (Burgess Shale) © Albert J. Copley/Visuals Unlimited; (Gunflint Chert) © Sinclair Stammers/ Photo Researchers, Inc.; (stromatolite) © Ken Lucas/Visuals Unlimited. **87** (African microfossil) © Eurelios/PhototakeUSA.com; (Bitter Springs fossils) © Science VU/Visuals Unlimited; (Ediacaran fossil) © Ken Lucas/Visuals Unlimited.

Local Influences on the Weather: 104 (forest) © Dave Porter/ Alamy; (skiing) © Galen Rowell/Corbis; (Nevada desert) © Royalty-Free/Corbis. **105** (snow) © Buffalo News/J.P.Mc Coy/Corbis Sygma; (blizzard) © Alaska Stock LLC/Alamy; (beach) © Neil Rabinowitz/ Corbis; (tornado) © Eric Nguyen/Jim Reed Photography/Photo Researchers, Inc.

Weather Instruments: 115 (rain gauge) © Wm. Baker/GhostWorx Images/Alamy.

Marine Organisms: 144 (ghost crab) © Millard H. Sharp/Photo Researchers, Inc.; (coral reef) © Photodisc; (octopus) © Brandon Cole Marine Photography/Alamy. **145** (manatee) © Royalty-Free/Corbis; (plankton) © Peter Arnold, Inc./Alamy; (whale) © Photodisc/Punchstock; (anglerfish) © Bruce Robison/ Corbis; (tube worms) © Ralph White/Corbis.

Mineral Formation: 154 (halite) © photolibrary/Index Stock; (garnet schist) © Biophoto Associates/Photo Researchers, Inc. **155** (limestone) © Albert Copley/Visuals Unlimited; (gold) © Ken Lucas/Visuals Unlimited; (granite) Courtesy of Margaret Carruthers.

Our Sun: 159 (sunspots) SOHO/ESA/NASA/Photo Researchers, Inc.; (solar flare) ESA/Photo Researchers, Inc.; (coronal mass ejection) NASA/Photo Researchers, Inc.; (prominences) © Phototake Inc./Alamy.

Phases of the Moon: 161 (first quarter) © Jason Ware/Photo Researchers, Inc.; (waxing gibbous) © Eckhard Slawik/Photo Researchers, Inc.; (full) © Eckhard Slawik/Photo Researchers, Inc.; (waning gibbous) © Eckhard Slawik/Photo Researchers, Inc.; (last quarter) © Eckhard Slawik/Photo Researchers, Inc.; (waning crescent) © John Chumack/Photo Researchers, Inc.; (new) © Larry Landolfi/Photo Researchers, Inc.; (waxing crescent) © John Chumack/Photo Researchers, Inc.

Eclipses: 162 (solar eclipse) © Fred Espenak/Photo Researchers, Inc. **163** (lunar eclipse) © Eckhard Slawik/Photo Researchers, Inc.

Venus: A Terrestrial Planet: 168 (planet Venus) © Bettmann/Corbis. **169** (surface of Venus) NASA/Photo Researchers, Inc.

Jupiter: A Gas Giant: 170 (Jupiter's atmosphere) Time Life Pictures/NASA/Time Life Pictures/Getty Images. **171** (Great Red Spot) © Corbis.

The Electromagnetic Spectrum: 172 (Marie Curie) © Hulton-Deutsch Collection/Corbis; (X ray) © Hot Ideas/Index Stock; (sunglasses) © Photos.com Select/Index Stock. **173** (remote control) © Photo Objects.net; (cell phone) © Photos.com Select/Index Stock; (television) © Photos.com Select/Index Stock.

Star Magnitudes: 174–175 (Big Dipper) © Roger Ressmeyer/Corbis.

Telescopes: 178 (refracting telescope) © photolibrary/Index Stock. **179** (reflecting telescope) © photolibrary/Index Stock.

Types of Galaxies: 184 (spiral) NASA Jet Propulsion Laboratory (NASA-JPL); (barred spiral) © Celestial Image Co./Photo Researchers, Inc. **185** (elliptical) © NOAO/Photo Researchers, Inc.; (irregular) © John Chumack/Photo Researchers, Inc.; (dwarf) NASA Jet Propulsion Laboratory (NASA-JPL).

The Scale of the Universe: 186 (beach with people) © Neil Rabinowitz/Corbis; (shoreline) Photodisc/Getty Images; (satellite photo of Florida) Orbimage/Photo Researchers, Inc.

Unique Objects in the Universe: 190 (spiral) NASA Jet Propulsion Laboratory (NASA-JPL); (elliptical) NOAO/Photo Researchers, Inc.; (irregular) © John Chumack/Photo Researchers, Inc. **191** (Crab Nebula) NASA Jet Propulsion Laboratory (NASA-JPL); (Horsehead Nebula) European Southern Observatory/Photo Researchers, Inc.; (binary stars) © SPL/Photo Researchers, Inc.; (comet) Walter Pacholka, Astropics/Photo Researchers, Inc.

The Formation of Coal: 198 (peat) © Andrew J. Martinez/Photo Researchers, Inc.; (lignite) © Scientifica/Visuals Unlimited. **199** (bituminous) © Scientifica/Visuals Unlimited; (anthracite) © Scientifica/Visuals Unlimited.

Coal Resources: 201 (geologists) © Steve Chenn/Corbis; (coal mine) © Jonathan Blair/Corbis; (coal cars) © Royalty-Free/Corbis; (reclaimed mine) © Charles E. Rotkin/Corbis.

Iron Resources: 203 (chemist) © ImageDJ/Index Stock; (open pit mining) © photolibrary/Index Stock; (smelting iron ore) © C. Voigt/ zefa/Corbis; (iron ore ship) © James L. Amos/Corbis.

Pronunciation Guide

The table below provides sample words to explain the sounds associated with specific letters and letter combinations used in the respellings in this book. For example, *a* represents the short "a" sound in *cat,* while *ay* represents the long "a" sound in *day.*

Letter combinations are used to approximate certain more complex sounds. For example, in the respelling of *Celsius*—SEL-see-uhs—the letters *uhs* represent the vowel sound you hear in *shut* and *other.*

Vowels

a	short a: apple, cat
ay	long a: cane, day
e, eh	short e: hen, bed
ee	long e: feed, team
i, ih	short i: lip, active
iy	long i: try, might
ah	short o: hot, father
oh	long o: home, throw
uh	short u: shut, other
yoo	long u: union, cute

Letter Combinations

ch	chin, ancient
sh	show, mission
zh	vision, azure
th	thin, health
th	then, heather
ur	bird, further, word
us	bus, crust
or	court, formal
ehr	error, care
oo	cool, true, rule
ow	now, out
ou	look, pull, would
oy	coin, toy
aw	saw, maul, fall
ng	song, finger
air	Aristotle, barrister
ahr	cart, martyr

Consonants

b	butter, baby
d	dog, cradle
f	fun, phone
g	grade, angle
h	hat, ahead
j	judge, gorge
k	kite, car, black
l	lily, mile
m	mom, camel
n	next, candid
p	price, copper
r	rubber, free
s	small, circle, hassle
t	ton, pottery
v	vase, vivid
w	wall, away
y	yellow, kayak
z	zebra, haze

Glossary

absolute magnitude a measure of the brightness of a star when viewed from a standard distance

abyssal (uh-BIH-suhl) plain the flat plain of the deep ocean floor

abyssal (uh-BIH-suhl) zone a region of the deep ocean floor, where the water depth is between about 2,000 m and 6,000 m

aeration a process in wastewater treatment in which air is passed through the water to drive off dissolved gases

aftershock an earthquake that follows a more powerful earthquake after a short time

air mass a body of air with a specific temperature and humidity throughout

air pollution anything in the atmosphere that is harmful to living things

alfisol a clay- and nutrient-rich soil that forms in humid, tropical, subtropical, and mid-latitude forest regions

alpha emission a type of radioactive decay in which a parent isotope breaks down into a daughter isotope and an alpha particle

alpha particle a particle that consists of two protons and two neutrons and is emitted during alpha decay; compositionally equivalent to a ^4He nucleus

altitude a measure of an object's height above sea level

ampere (AM-pir) the SI unit used to describe the rate of electrical current flow

amplitude the vertical distance between the crest or trough of a wave and the middle of the wave

analog seismograph an instrument that records earth movement on a revolving paper roll

andisol a generally fertile soil formed mainly from volcanic ash

anthracite (AN-thruh-siyt) the highest grade of coal; contains about 90 percent carbon by mass

anticline a fold in rock in which the oldest rock layers are located in the interior of the fold; most anticlines are ∩-shaped

apparent magnitude the measure of the brightness of a star relative to other stars

aquifer an underground body of rock that contains water

aridisol a soil that is found in dry regions and that contains little water or organic matter

asthenosphere the zone of solid, but weak, mantle rock below the lithosphere

atmosphere a layer of gases that surrounds a planet

atmospheric pressure the force exerted by the weight of the gases above the surface of a planet

atom the smallest unit of an element that has the properties of that element

aurora colored lights in the sky caused by the interaction of the solar wind with earth's ionosphere

autumnal equinox the day on which the lengths of day and night are equal and after which the number of daylight hours in a hemisphere begin to decrease; also called the fall equinox

axis an imaginary line through the center of a planet, about which the planet rotates

barometer an instrument used to measure atmospheric pressure

barred spiral galaxy a spiral galaxy with an elongated, oval, or rectangular central region

bathyal (BA-thee-uhl) zone a region of the ocean floor that extends from the sublittoral zone to the abyssal zone

benthic zone a region of the ocean that includes the ocean floor and the water just above it

beta emission a type of radioactive decay in which a parent isotope breaks down into a daughter isotope and a beta particle

beta particle an electron; usually used to refer to an electron emitted during radioactive decay

big bang theory the theory that states that the universe formed by rapid expansion of matter and energy from an initial infinitely small, dense point

binary star system a pair of stars orbiting a common center of mass

biome an area of earth defined by its climate and the organisms that live there

biosphere the zone of earth that contains all living things

bituminous coal coal that is about 80 percent carbon by mass

boiling point the temperature at which a liquid changes into a gas

candela (kan-DEE-luh) the SI base unit used to measure luminous intensity

carbon (C) the element with an atomic number of 12; carbon forms the basis of most biologically important compounds on earth

Celsius the temperature scale on which water freezes at 0° and boils at 100°

chemical weathering a type of weathering in which chemical reactions change the chemical composition of a rock or mineral

chlorination a process during wastewater treatment in which chlorine is mixed with the water to kill harmful microorganisms

chromosphere the part of the sun's atmosphere located just above the photosphere

cinder cone volcano a relatively small, steep-sided volcano made mainly of pyroclastic material

cirrus (SIHR-uhs) cloud a thin, wispy cloud, composed primarily of ice crystals, that forms at high altitudes

cleavage the tendency of a mineral to break along one or more distinct planes

climate the average weather conditions in an area

cloud a visible mass of tiny water droplets, ice crystals, and solid particles in the atmosphere

coagulation a process used in wastewater treatment in which a chemical is mixed with the water to form sticky blobs, to which harmful microorganisms and other small particles stick

coal a fossil fuel formed from the remains of plants that have been buried, heated, and compressed over millions of years

cold desert a region with low average temperatures that receives less than 25 cm of precipitation per year

cold front a front that forms where a cold air mass moves in and replaces a warm air mass, generally causing heavy precipitation

comet a body of ice, rock, and dust that orbits a star

composite volcano a volcano that consists of alternating layers of pyroclastic material and lava flows

concave mirror a mirror that curves inward and reflects light to a single point; used in reflecting telescopes

concentration the amount of a substance dissolved in a specific volume of liquid

conifer a type of tree that does not produce flowers; most have seeds in cones and are evergreen

continental air mass an air mass that forms over a continent

continental drift the theory that states that the continents were once connected in one large landmass and have since moved apart

continental margin the region of the ocean floor extending from the coast of a continent to the continental rise and abyssal plain

continental rise the area of the ocean floor located at the bottom of the continental slope

continental shelf the part of the ocean floor that starts at the coastline and slopes gently toward the open ocean

continental slope the part of the ocean floor that slopes steeply from the edge of the continental shelf to the continental rise

contour line a line on a topographic map that connects points of the same elevation

convection the process in which heat energy is transferred through the movement of matter

convection cell a region in which hot, less-dense matter rises and is replaced by cold, more-dense matter; convection cells help transport heat in earth's atmosphere and oceans

convective zone the zone in the sun through which energy travels by convection

convergent plate boundary a boundary between two tectonic plates that are moving toward each other

core the innermost compositional layer of earth, consisting primarily of iron and nickel and having a radius of approximately 3,500 km

Coriolis effect the tendency of material traveling along a north-south path in earth's atmosphere or oceans to curve due to earth's rotation

corona the outermost layer of the sun's atmosphere

coronal mass ejection a bubble of gas that erupts from the sun's corona

covalent bond a chemical bond in which two atoms share electrons

crest the highest point in a transverse wave

crust the rocky, solid, outermost compositional layer of earth

cryosphere all of the frozen water on earth

cumulus (KYOO-myuh-luhs) cloud a large, puffy cloud composed primarily of liquid water droplets

current a mass of water or air that is moving in a consistent direction

daughter isotope an isotope that forms through the radioactive decay of another isotope

deciduous tree a tree that loses all of its leaves during one part of the year

denitrifying (dee-NIY-truh-fi-ying) bacteria bacteria that convert nitrate to gaseous nitrogen

density a measure of how closely packed the particles in a substance are; the mass of an object divided by its volume

depositional basin a place in which eroded materials from mountains are deposited

desert a region that receives less than 25 cm of precipitation per year

deuterium (dou-TIR-ee-uhm) an isotope of hydrogen with an atomic number of 1 and an atomic mass number of 2

diamond the hardest known mineral, composed of covalently bound carbon atoms

digital seismograph instrument that records ground movement as electronic data on a computer

dike a sheetlike body of igneous rock that cuts across other rock layers

divergent plate boundary a boundary between two tectonic plates that are moving away from each other

dwarf galaxy a relatively small galaxy that contains fewer stars than most other galaxies

earthquake the ground motion produced when rock below earth's surface breaks because of stress

El Niño–Southern Oscillation (ENSO) a short-term, cyclic climate variation caused by the weakening of trade winds in the equatorial Pacific Ocean

electrolysis (ih-lek-TRAH-luh-suhs) in mining, the process in which an electric current is passed through a metal to remove impurities

electromagnetic radiation energy that moves as a wave at the speed of light

electromagnetic spectrum all types of electromagnetic radiation, including gamma rays, X rays, ultraviolet rays, visible light, infrared light, microwaves, and radio waves

electron a subatomic particle with a negative charge

electron capture a type of radioactive decay in which a proton and an electron in a parent isotope combine to form a neutron in a daughter isotope

element a substance that cannot be broken down into simpler substances

elliptical galaxy a galaxy that is oval shaped

enriched uranium uranium with high concentrations of radioactive isotopes

entisol a relatively undeveloped soil with a similar composition to the bedrock beneath it

equator an imaginary line around earth that divides it into Northern and Southern hemispheres; the line of 0° latitude

erosion the process in which soil and sediment are transported over earth's surface

evaporation the process in which a liquid changes into a gas

evaporation rate a measure of how quickly a liquid becomes a gas

extinction the disappearance of all members of a species

extrusive igneous rock rock that forms when lava solidifies on Earth's surface

Fahrenheit the temperature scale on which water freezes at 32° and boils at 212°

fault a break in the lithosphere along which bodies of rock can move

fault block a body of rock that is separated from other bodies of rock by one or more faults

filtration a process in which solid materials are separated from liquids

fold a bend in a layer of rock

foliated a texture in a metamorphic rock, characterized by layers or sheets of minerals

footwall the fault block that lies below the fault plane of a normal or reverse fault

foreshock a small earthquake that precedes a more powerful earthquake

fossil the trace or remains of an organism preserved in rock

fracture the tendency of a mineral to break unevenly

freezing point the temperature at which a liquid changes into a solid

frequency the number of waves that pass through a point during a specific amount of time

freshwater water with a very low (less than 0.05%) concentration of dissolved salts

front the region where two air masses meet

gamma ray electromagnetic radiation with a wavelength between about 10^{-16} m and 10^{-11} m

gas giants large planets composed mostly of gas, such as Jupiter, Saturn, Uranus, and Neptune

gelisol a soil found in cold regions that contains a layer of permanently frozen soil, or permafrost, beneath the surface

geologic map a map that shows rock formations at or near earth's surface

geologist a scientist who specializes in the study of earth and its geologic history

geosphere all the rock on and just below earth's surface

glacier a large body of slowly moving ice

glucose a simple sugar used as an energy source by most living things

greenhouse gas a gas in the atmosphere that reradiates energy

groundwater water stored in layers of rock or soil below earth's surface

half-life the time required for half the atoms of a parent isotope in a sample to decay

hanging wall the fault block that lies above a fault plane in a normal or reverse fault

hardness the resistance of a mineral to being scratched

Hertzsprung-Russell diagram a diagram that shows the relationship among color, temperature, and brightness of stars

high cloud a cloud at an altitude of 6,000 m or more; generally composed of ice crystals

histosol a soil that is rich in organic matter; most histosols are also very low in oxygen

hurricane a severe tropical storm with winds of more than 120 km/h

hydrogen bond an electrostatic attraction between an oxygen, nitrogen, or fluorine atom and a hydrogen atom that is covalently bonded to another oxygen, nitrogen, or fluorine atom

hydrosphere all the liquid water on earth's surface

hypothesis a testable explanation of a phenomenon based on scientific research and observations

igneous rock a rock that forms through the solidification of liquid rock

inceptisol a young soil common in floodplains and deltas

inflation in astronomy, the early expansion of the universe

infrared light electromagnetic radiation with wavelengths between about 10^{-6} m and 10^{-3} m

inner core the solid, innermost physical layer of earth, with a radius of about 1,200 km

intensity in geology, a measurement of the observed effects of an earthquake

intertidal zone the region on the beach between the high-tide level and the low-tide level

intrusive igneous rock rock that forms when magma solidifies below earth's surface

inversion aloft a temperature inversion that forms in areas of high pressure

ion an atom that has unequal numbers of protons and electrons, and is therefore positively or negatively charged

ionic bond a bond that forms between a positively charged ion and a negatively charged ion

ionosphere the part of earth's upper atmosphere consisting primarily of ionic particles

irregular galaxy a galaxy with an irregular shape; a galaxy that is neither spiral nor elliptical

isotope (IY-suh-tohp) a form of an element that is defined by the number of neutrons in the nucleus; different isotopes of the same element have different numbers of neutrons and different atomic masses

Kelvin the SI scale of temperature; the temperature scale on which water freezes at 273 and boils at 373

kilogram the SI base unit of mass

latitude a measure of distance north or south of the equator

lava magma on the surface of the earth

law of crosscutting relationships in geology, the principle that states that a feature that cuts across layers of rock is younger than the rock layers it cuts across

lignite the lowest grade of coal; contains about 70 percent carbon by mass

lithosphere the rigid, outer physical layer of earth that includes the crust and part of the mantle

Local Group a cluster of galaxies that includes the Milky Way

longitude a measure of distance east or west of the prime meridian

low cloud a cloud at an altitude of less than 2,000 m

lunar eclipse the darkening of a full moon that occurs when the moon moves into earth's shadow

luster a description of the quality of light reflected by a mineral

magma molten rock below earth's surface

magnetic field an area where magnetic force is detected

magnitude a measure of the amount of energy released during an earthquake

main sequence star a star that lies along a roughly diagonal line on a Hertzsprung-Russell diagram; most stars spend most of their lives on the main sequence

main shock the most vigorous shaking during an earthquake

mantle the compositional layer of earth that lies between the core and the crust

mantle plume a column of hot mantle rock that rises to the surface

maritime air mass an air mass that forms over an ocean

mass spectrometer an instrument that measures the ratio of a particle's mass to its charge; used to determine the mass of an ion and to measure the concentrations of different isotopes in a sample

mass wasting a downhill movement of rock and soil due to gravity

mechanical weathering the process in which a rock or mineral is broken into smaller pieces, but with little to no change in chemical composition

meridian a line of longitude

mesosphere 1. the physical layer of earth between the asthenosphere and the inner core; 2. the layer of earth's atmosphere between the stratosphere and the thermosphere

metamorphic rock rock that has experienced a change in chemical composition due to heat, pressure, or both

meter the SI base unit of length

microfossil a fossil that is too small to be seen with the naked eye

microwave electromagnetic radiation with wavelengths between 0.3 cm and 300 cm

mid-altitude cloud a cloud at an altitude of between 2,000 m and 6,000 m

mid-ocean ridge a chain of underwater volcanoes on the ocean floor that forms where two tectonic plates are moving apart

Milky Way the galaxy that includes our solar system

mineral a naturally occurring, inorganic solid with a definite chemical composition and crystalline structure

Modified Mercalli Intensity Scale a scale used to rate the observed damage of an earthquake

molecule two or more atoms that are covalently bonded

mollisol a very fertile soil found in grasslands in humid continental regions

monocline a steplike fold in rock

monsoon a regional wind system that blows in different directions during different seasons

neap tide a tide in which the difference in height between high tide and low tide is smaller than usual

neritic (nuh-RIH-tihk) zone the ocean water that covers the sublittoral zone

neutrino (noo-TREE-noh) a particle with no electric charge and almost no mass

neutron a particle in an atom with the same weight as a proton, but no electric charge

nitrogen cycle the reservoirs of nitrogen and the processes by which nitrogen moves between reservoirs

nitrogen-fixing bacteria bacteria that convert nitrogen gas into compounds that plants can use

nonfoliated a texture in a metamorphic rock in which there are no visible layers or sheets of minerals

normal fault a fault in which the hanging wall moves downward compared to the footwall

nuclear (NOO-klee-uhr) fusion a reaction in which the nuclei of two elements combine to produce a nucleus of a heavier element and a great deal of energy

nucleus (NOO-klee-uhs) the center of an atom, containing the protons and neutrons

occluded front a front in which two or more cold air masses surround a warmer air mass

ocean trench a very deep valley on the ocean floor, generally formed at convergent plate boundaries

oceanic zone the area of open ocean water covering the bathyal and abyssal zones

oceanographer a scientist who specializes in the study of the oceans

ore a body of rock that contains a high enough concentration of a useful material that it is profitable to mine

outer core the liquid, metallic physical layer of earth located between the mesosphere and the inner core

oxisol a highly oxidized, nutrient-poor soil found in warm, humid areas that receive a great deal of rainfall

P wave a seismic wave that causes tension and compression in the matter it moves through

parallel a line of latitude

parent isotope a radioactive isotope that decays to form a daughter isotope

peat a mass of partially decomposed plants that is about 60 percent carbon by mass

pelagic (puh-LA-jihk) zone the open water of the ocean, above the ocean floor

percolation the downward movement of water through soil and rock

permafrost a layer of soil that remains frozen all year

petroleum a mixture of hydrocarbons that can be refined into products like gasoline, kerosene, and tar; also known as crude oil

phosphorus cycle the reservoirs of phosphorus and the processes by which phosphorus moves between reservoirs

photosphere the very thin layer of gas that makes up the visible surface of the sun

photosynthesis the process in which a living organism uses solar energy to convert carbon dioxide and water into sugar and oxygen

plate boundary an area where tectonic plates meet

plate tectonics the theory that states that earth's lithosphere is broken into plates that move slowly over earth's surface

polar air mass an air mass that forms at high latitudes

positron a particle with the same mass as an electron, but with a positive charge

precipitation 1. any form of water that falls from the atmosphere to Earth's surface; 2. the process in which solid materials come out of a solution

prime meridian an imaginary line running north to south through Greenwich, England; the line of 0° longitude

principle of superposition the principle that states that, in an undisturbed rock body, the oldest rock layers are on the bottom and the youngest are on the top

prominence a large cloud of glowing gas that reaches out from the sun's photosphere

proton a positively charged particle in an atom

pyroclastic material ash, rock, and dust produced during an explosive volcanic eruption

radiative zone the zone of the sun through which energy travels as radiation

radio wave electromagnetic radiation with wavelengths between about 10^{-4} m and 10^{6} m

radioactive decay the process in which an unstable isotope breaks down into an isotope of the same or a different element

radiometric dating a technique used to determine the age of a material by measuring the ratio of parent to daughter isotopes in a sample

rain gauge an instrument used to measure the amount of rain that falls in an area

rain shadow the side of a mountain that has a dry climate because precipitation falls mainly on the other side

recumbent fold a syncline or anticline turned on its side

red dwarf a star that is relatively cool, dim, small, and old

reflecting telescope a telescope that uses mirrors to collect and focus light from distant objects

refracting telescope a telescope that uses lenses to collect and focus light from distant objects

respiration the process by which an organism obtains energy from chemical compounds

reverse fault a fault in which the hanging wall moves upward compared to the footwall

Richter Scale a scale that is used to describe the magnitude of an earthquake

rock cycle the series of processes by which rocks form and change from one type to another

rock formation a layer or body of rock with similar age and characteristics throughout

runoff water that moves over the land's surface and into oceans, either directly or through other bodies of water

S wave a seismic wave that causes particles to move perpendicular to the direction that the wave is traveling

salinity a measure of the amount of dissolved salts in water

savanna a region that receives a moderate amount of rainfall and has tall grasses and shrubs

seafloor spreading the movement of two oceanic plates away from each other as new ocean floor forms between them

seamount a underwater volcano that is not located at a mid-ocean ridge

second the SI base unit used to measure time

sediment particles of rock or organic matter

sedimentary rock rock that forms by the cementation of sediment or by the crystallization of minerals from a water solution

seismogram a recording of movement of earth at a specific location caused by an earthquake

seismograph an instrument that measures and records ground motion

shield volcano a low, broad volcano consisting of multiple layers of lava deposited by repeated, nonexplosive eruptions

SI (Système International d'Unités, or International System of Units) a scientific system of measurement that includes seven base units

silicate mineral a mineral containing compounds of silicon and oxygen

sill a sheetlike body of igneous rock that forms parallel to other, preexisting layers of rock

silt rock particles that are larger than clay but smaller than sand

soil a mixture of rock pieces, air, water, organic matter, and living things

solar eclipse the darkening of the sun that occurs when the moon moves directly between earth and the sun

solar flare a violent ejection of charged particles from the sun's photosphere

solar system a star and all the objects that orbit it, including planets, their moons, comets, and asteroids

solar wind the continuous stream of particles coming from the sun

spiral galaxy a galaxy with a bright, central mass of stars and long thin arms extending away from the center

spodosol a soil found in coniferous forests in cool, humid climates

spring tide a tide in which the difference between high tide and low tide is larger than usual

star a large, gaseous body in space that emits light

stratosphere the layer of earth's atmosphere located below the mesosphere and above the troposphere

stratus cloud a horizontal, layered cloud

streak the color of a mineral powder that forms when the mineral is rubbed on a unglazed porcelain plate

strike-slip fault a fault in which the fault blocks move horizontally in opposite directions

stromatolite (stroh-MAH-tuh-liyt) a structure made of mats of algae and bacteria layered with sediment

subduction the process in which oceanic lithosphere sinks into the mantle at a convergent boundary

sublittoral (suhb-LIH-tuh-ruhl) zone the region of the ocean floor that extends from the low-tide level to about 200 m from shore

summer solstice the day on which the Northern or Southern Hemisphere is most tilted toward the sun

sunspot a dark, relatively cool area of the sun's photosphere that forms where strong magnetic field lines intersect the sun's surface

supernova a large explosion caused by the collapse of a massive star

surface inversion a kind of temperature inversion in which a cold land surface absorbs heat from the air near the ground

surface waves seismic waves that can cause particles to move in circular paths

syncline a fold in rock in which the youngest layer is in the interior of the fold; most synclines are U-shaped

taiga (TIY-guh) a subarctic coniferous forest

tectonic plate a section of earth's lithosphere that can move slowly over earth's surface

temperate forest a deciduous forest that experiences widely varying temperatures

temperate grassland an area that has four distinct seasons and receives relatively little rainfall

temperature a measure of how hot or cold something is; a measure of the average kinetic energy of the particles of a substance

temperature inversion a condition in which a warm air mass is located above a cold air mass, causing the cold air mass to be trapped near the ground

theory in science, a well-supported explanation for many independent observations

thermosphere the outermost layer of earth's atmosphere

tides the rise and fall of water levels due to gravitational forces

topographic map a map that shows the physical features of earth's surface

topography the physical features of the land

tornado a violent and potentially destructive windstorm that takes on the form of a rotating, funnel-shaped cloud

transform boundary a plate boundary at which two tectonic plates slide past each other horizontally

transform fault a fault along which two large rock bodies slide past each other horizontally

transverse wave a wave that moves particles in a direction perpendicular to the direction the wave is traveling

transpiration the process in which plants lose water to the atmosphere through their leaves

tropical air mass an air mass that forms at low latitudes

tropical forest a forest that receives large amounts of precipitation and is warm all year

tropics the area between the Tropic of Cancer and the Tropic of Capricorn, characterized by a warm, humid climate

troposphere a layer of the atmosphere closest to earth's surface

trough the lowest point in a transverse wave

tundra an area that is cold year-round, receives little rainfall, and has little vegetation

ultisol a generally nutrient-poor soil that contains a significant amount of clay

ultraviolet radiation electromagnetic radiation with wavelengths between about 10^{-10} m and 10^{-7} m

valence electron an electron in the outermost electron shell of an atom

ventifact a rock shaped by windblown sediment

vernal equinox the day on which the lengths of day and night are equal, and after which days in a hemisphere begin to get longer; also called the spring equinox

vertisol a soil that contains extremely high concentrations of clay minerals

visible light electromagnetic radiation with wavelengths between about 4×10^{-6} m and 7×10^{-6} m; light that people can see

volcano a crack in earth's surface through which magma, ash, and gases erupt

volume a measure of the amount of space occupied by a substance

warm front a front that forms where a warm air mass replaces a cold air mass

wavelength in transverse waves, the horizontal distance between one crest or trough and the next

weathering the process in which a rock or mineral is broken down

well a hole drilled down to an aquifer to retrieve the water stored in the aquifer

white dwarf the remains of a star that has lost much of its mass

winter solstice the day on which the Northern or Southern Hemisphere is least tilted toward the sun; shortest day of the year

X ray electromagnetic radiation with wavelengths between about 10^{-12} m and 10^{-10} m

Index

SUBJECTS are in all capital letters.

A

absolute magnitude, 175
abyssal plains, 126, 127
abyssal zone, 143, 145
acidic soil, 45
aeration tank, 209
Africa
  microfossils, 87
  ocean, 124
  soils, 44–45
  wind erosion, 47
African plate
  fundamentals, 58–59, 60
  ocean floor topography, 128
aftershocks, 13
Ag. *See* silver (Ag)
Age of Fishes, 75
Age of Mammals, 75
Age of Reptiles, 75
agricultural land, 45
agriculture, air pollution, 90
Aguinas Current, 137
air chambers, 114
AIR MASSES AND FRONTS, 116–117
AIR MASSES IN THE UNITED STATES AND CANADA, 112–113
AIR POLLUTION, 90–91
air pressure. *See also* pressures
  atmospheric, 89
  barometer, 114
  Mars, 95
  precipitation and climate, 102–103
  Venus, 94, 168
Al. *See* aluminum (Al)
Alaska, 50, 62
Alaskan Current, 136
Aldebaran, 181
Aldrin, Edwin, 2
alfisols, 44
algae, 85, 86
Alioth, 174, 175
Alkaid, 174, 175
alkali metals, 22
alkaline-earth metals, 22
Al_2O_3. *See* corundum (Al_2O_3)
Alpha Centauri, 181
alpha emission, 76
altering the hypothesis, 15
altitude, 88–89, 120–121
altocumulus clouds, 120
altostratus clouds, 120
aluminum (Al)
  atomic structure, 25
  crust, 54
  mantle, 54
  minerals, 29
  mineral resources, 196–197
Alvin, 2
ammonia, 84, 146
ampere, 6
amphibole, 31
amplitude, 138
analog seismographs, 12
analyzing data, 15
Andes Mountains, 65
andisols, 44
aneroid barometer, 114
anhydrite ($CaSO_4$), 29
animals
  carbon dioxide, 148
  phosphorus cycle, 150
  representative, 108
Antarctic Circle, 19
Antarctic plate, 58–59, 60–61
Antarctica, 124
Antares, 181
anthracite, 199
anticline fold, 42
apatite, 32
Apollo 11, 2
Appalachian Mountains, 17
apparent magnitude, 175
AQUIFERS, 206–207
aquifers, 152
Arabian plate, 59, 61
Arabic numeral system, 3
Archaeopteryx, 75
Archean eon, 72
Arctic Circle, 19
area measurements, 7
aridisols, 44
Aristarchus, 4
Arizona, 80
Armenia, 51
arms, galaxy, 184–185, 190
Armstrong, Neil, 2
arthropods, 75
Asian plate, 57, 64–65
asteroids, 188
asthenosphere, 55
Atlantic Ocean
  air masses, 113
  coast, weather impact, 105
  hurricanes, 123
  plate motions effect, 57
Atlantis (research vessel), 67
atmosphere
  air pollution, 90–91
  circulation, 100–101
  composition, 94–95
  freshwater storage, 206
  fundamentals, 21
  history, 84–85
  Jupiter, 170
  layers, 88–89
  nitrogen cycle, 146–147
  Precambrian period, 86–87
  solar energy, 92–93
  Venus, 168
  water cycle, 152–153
ATMOSPHERIC CIRCULATION, 100–101. *See also* WIND PATTERNS
atmospheric pressure, 89
ATOMS, 24–25
Au. *See* gold (Au)
Australia
  Bitter Springs Formation, 87
  El Niño–Southern Oscillation, 110
  electricity consumption, 205
  mineral resources, 197
  ocean, 124
  population, 211
Australian-Indian plate, 59, 61
automobiles, air pollution, 90
autumnal equinox, 164
axis of rotation, seasonal impact, 164–165

B

B. *See* boron (B)
Babylonians, 3
bacteria
  denitrifying, 147
  fossilized stromatolites, 86
  nitrogen-fixing bacteria, 146
  phosphorus cycle, 150
  wastewater treatment, 208–209

bands of minerals. *See* foliated metamorphic rock; minerals
barometer, 114
barred spiral galaxies, 184
basalt, 34, 35
bathyal zone, 143, 144
bauxite, 196–197
Be. *See* beryllium (Be)
benthic zone, 143
Berkshires, 8
beryllium (Be), 22, 182, 183
beta emission, 76
Betelgeuse, 181
Bhuj, India, 51
big bang theory, 5, 192–193, 194
Big Dipper, 174–175
Big Island, Hawaii, 62
binary star system, 191
biological sedimentary rocks, 37
BIOMES, 106–107
biomes, 106–107, 108–109
biosphere, 21
biotite, 154
birch trees, 107
birds
  biomes, 107, 108
  earth's history, 75
Bitter Springs Formation, 87
bituminous coal, 199
Bjerknes, Jacob, 111
black holes, 177
blue stars, 180
blue-shifted light, 195
boreal forest, 106
bornite, 157
boron (B), 23, 182, 183
Br. *See* bromine (Br)
Brazil, 204, 211
Brazil Current, 136
breakers, 139
Bright Angel Shale, 80–81
brightness, star, 174, 175
bromine (Br), 23, 28
Buffalo, New York, 105
Burgess Shale formation, 86

C

C. *See* carbon (C) and carbon atoms
Ca. *See* calcium (Ca)
$CaAl_2Si_2O_8$. *See* plagioclase ($CaAl_2Si_2O_8$)
$CaCO_3$. *See* calcite ($CaCO_3$)
CaF_2. *See* fluorite (CaF_2)
calcite ($CaCO_3$)
  carbonate mineral group, 28
  formation, 155
  hardness, 32
  marble, 39
  ore-derived materials, 157
  properties, 33
calcium (Ca), 22, 25
California Current, 136
Cambrian period
  geologic maps, 11
  geologic time scale, 73
  Grand Canyon, 80–81
  life history, 74, 75
$(Ca,Mg)CO_3$. *See* dolomite ($(Ca,Mg)CO_3$)
Canada
  air masses, 112–113
  Appalachian Mountains, 17
  Burgess Shale formation, 86
  Ediacaran fossils, 87
Canary Current, 136
candela, 6
carbon (C) and carbon atoms
  anthracite, 199
  bituminous coal, 199
  cycle, 148–149
  lignite, 198
  mineral groups, 28
  peat, 198
  CARBON CYCLE, THE, 148–149
carbon dioxide (CO_2), 84, 94, 168
carbon monoxide (CO), 90
carbonates mineral group, 28, 33
Carboniferous period, 73, 74, 75
Caribbean plate, 58, 60
Cascades range, 16
$CaSO_4$. *See* anhydrite ($CaSO_4$)
$CaSO_4 \bullet 2H_2O$. *See* gypsum ($CaSO_4 \bullet 2H_2O$)
Celsius conversions, 7
Cenozoic era, 72, 73, 74, 75
centi- prefix, 6
centimeter length, 7
Central America, 47, 112
Cerro Azul (Quizapu), central Chile, 62
chalcopyrite, 157
Challenger Deep (Mariana Trench), 129
CHANGES IN EARTH'S SURFACE, 46–47
CHARACTERISTICS OF BIOMES, 108–109
chemical bonds, 26–27
chemical plants, air pollution, 90
chemical sedimentary rocks, 37
chemical weathering, 46
Chile, 50, 62
China
  astronomers, 4, 191
  electricity consumption, 205
  population, 211
  scientific contributions, 3
chlorination, 209
chlorine (Cl), 23, 26, 28
chlorite, 154
chromosphere, 158
cinder cone volcanoes, 71
circulation. *See also* WIND PATTERNS
  atmosphere, 100–101
  sea-level temperature, 96–97
  wind patterns, 98–99
cirrocumulus clouds, 120
cirrus clouds, 120
Cl. *See* chlorine (Cl)
clastic sedimentary rocks, 36
clay and clay soil, 36, 45, 157
clay mineral soil, 45
cleavage, 32
CLOUDS, 120–121
clouds and cloud cover, 93, 118–119
CMB. *See* cosmic microwave background (CMB)
CO. *See* carbon monoxide (CO)
CO_2. *See* carbon dioxide (CO_2)
coagulation tank, 208
coal
  air pollution, 90
  deposits, 75
  formation, 198–199
  United States resources, 200–201
  world resources, 196–197
COAL RESOURCES, 200–201
Coconino Sandstone, 80–81
Cocos plate, 58, 60
cold fronts, 116, 119
Collins, Michael, 2
color
  hematite streak, 33
  lunar eclipse, 163
  metamorphic rock, 38
  mineral properties, 33
  red shift, 194–195
  refracting telescopes, 178
Colorado, 104
Colorado Plateau, 16, 80
Colorado River, 16, 80
comets, 84, 188, 191
communication disruption, 159
compasses, 3
composite volcanoes, 71
COMPOSITION OF ATMOSPHERES, 94–95
compositional layers, earth, 54
conclusions, drawing, 15
conglomerate, 36
conifers and coniferous trees, 106, 108
construction, air pollution, 90
consumers, 106, 107
continental drift, 66–67

continental lithosphere, 58–59
continental margin, 126
continental polar (cP) air masses, 113
continental rise, 126
continental shelf, 126
continental slope, 126
continental tropical (cT) air masses, 112
continents, locations, 56–57
contour lines, topographic maps, 9
CONTRIBUTIONS TO SCIENCE, 2–3
convection cells, 101
convective zone, 158
convergent plate boundary, 43, 61, 68
cool air, 91, 100, 116
cool water, 134
cooling, rock cycle, 40–41
Copernicus, 4
copper (Cu), 23, 28, 157
copper penny, hardness, 32
coquina, 37
corals, 37, 149
core
  Earth, 54
  Jupiter, 171
  sun, 158
  Venus, 169
corona, 158
coronal mass, 159
corundum (Al_2O_3), 29, 32
cosmic microwave background (CMB) radiation, 5
Costa Rica, 62
covalent bonds, 27
cP. *See* continental polar (cP) air masses
Crab Nebula, 191
crest, 138
Cretaceous period, 73, 74, 75
crosscutting relationships, 82–83
crust, 54, 169
cryosphere, 21
crystalline forms. *See* minerals
crystallization
  halides, 28
  igneous rocks, 34–35
  mineral formation, 154, 155
  ores, 156
  sedimentary rocks, 36–37
cT. *See* continental tropical (cT) air masses
Cu. *See* copper (Cu)
cumulonimbus clouds, 121
cumulus clouds, 121
Curie, Marie, 172
current, electric, 6
currents, ocean, 136–137
cyanobacteria, 85

D

data, gathering and analyzing, 15
daughter isotopes, 76–77, 78–79
decay, 76–77
deci- prefix, 6
decomposers, 147, 150, 198
deep-sea hypothermal vents, 2
deep submergence vehicle (*Alvin*), 2
deka- prefix, 6
deltas, 45
denitrifying bacteria, 147
dense air, 101
density
  gas planets, 170
  Jupiter, 170
  planets in solar system, 167
  Venus, 168
DENSITY OF OCEAN WATER, 134–135
deposition, rock cycle, 41
depositional basin, 16
depression contour lines, 9
deserts
  biomes, 107, 108
  precipitation and climate, 102
deuterium, 182
Devonian period, 11, 73, 74, 75
dew point, 118–119
diameter, planets, 167, 170
diamonds
  hardness, 32
  mineral groups, 28
  mineral resources, 196–197
  ore-derived materials, 157
digital seismographs, 12
dinosaurs, 75
directions
  front movements, 116–117
  wave movements, 139
  winds, weather maps, 119
divergent plate boundary, 61
divided highways, 9
dolomite ($(Ca,Mg)CO_3$), 28, 155
double-chain silicates, 31
drawing conclusions, 15
drizzle, 119
Dubhe, 174, 175
dull luster, 33
dust, wind erosion, 47
dwarf galaxies, 185

E

early Archean era, 72
early Cambrian period, 81
early Jurassic period, 56
early middle Permian period, 81
early Pennsylvanian period, 81
early Permian period, 81
early Proterozoic era, 72
Earth
  characteristics, 167
  latitude and longitude, 18–19
  orbit around sun, 4
  pressures, 94
  rotation, 98–99, 101
  solar system location, 167
  spheres, 20–21
  temperature, 94
  Venus similarity, 168–169
  wind patterns, 98–99
earth, history
  geologic cross sections, 82–83
  geologic time scale, 72–73
  Grand Canyon, 80–81
  life, 74–75
  radioactive decay, 76–77
  radiometric dating, 78–79
earthquakes
  intensity, 48–49
  magnitude, 48–49
  major historical earthquakes, 50–51
  mass wasting, 46
  seismic waves, 52–53
  seismographs, 12–13, 48–49
EARTH'S ELEMENTS, 22–23
earth's history, 72–75
EARTH'S LAYERS, 54–55
EARTH'S OCEANS, 124–125
EARTH'S PLATES, 58–59
earthy luster, 33
East Australian Current, 137
East Greenland Current, 136
East Wind Drift, 136–137
Eastern Hemisphere, 18–19
ECLIPSES, 162–163
eclipses, 4, 162–163
Ediacaran fossils, 87
EL NIÑO–SOUTHERN OSCILLATION, 110–111
electric current, 6
electricity use, world comparison, 205–206
electrolysis, 203
electromagnetic radiation wavelengths, 93
ELECTROMAGNETIC SPECTRUM, THE 172–173
electron capture, 76
electron clouds, 24
electrons
  atomic structure, 24
  stars, element formation, 182
  valence, 26–27

ELEMENT FORMATION IN STARS, 182–183
elements, 22–23, 182–183
elements, native. *See* native elements
elevation, topographic maps, 8
elliptical galaxies, 185, 190
energy, 92–93, 136
enriched uranium, 197
ENSO. *See* EL NIÑO-SOUTHERN OSCILLATION
entisols, 44
environmental damage, coal mining, 201
Eocene epoch, 57, 73, 74
Eohippus, 75
equator, 18, 19, 100–101
erosion, 41, 47
Ethiopia, fossils, 3
Eurasian plate
  ocean floor topography, 128
Europe, 44, 197
evaporation
  halides, 28
  mineral formation, 154
  sulfates, 29
  water cycle, 153
evaporite, 37, 156, 157
Everglades, 17
expansion, universe, 192–193, 194–195
experiments, performing, 14
extinction, 75
extrusive igneous rock, 34. *See also* intrusive igneous rock

F

F. *See* fluorine (F)
Fahrenheit conversions, 7
fall equinox, 164
fault blocks, 43
faults, 42–43
Fe. *See* iron (Fe)
FEATURES OF THE OCEAN FLOOR, 126–127
feldspar, 32, 155
Fe_2O_3. *See* hematite (Fe_2O_3)
fertile soil, 45
fertilizers, 147, 150
FeS_2. *See* pyrite (FeS_2)
filtration, 208
filtration tank, 208
fingernails, hardness, 32
first-quarter moon, tides, 140
fish, 75
floodplains, 45
flowering plants, 75
fluorescence, 33
fluorine (F), 23, 28
fluorite (CaF_2), 28, 32
fog, 104, 119
folds, 42–43
FOLDS AND FAULTS, 42–43
foliated metamorphic rock, 38, 39
footwall, faults, 43
forest fires, 90
FORMATION OF COAL, THE, 198–99
FORMATION OF THE HIMALAYAS, 64–65
forming a hypothesis, 14
fossil fuels
  air pollution, 90–91
  carbon cycle, 148–149
  coal formation, 198–199
  mineral resources, 196
  nitrogen cycle, 147
fossils
  atmospheric conditions, 85
  discovery, 3
  Precambrian period, 86–87
fracture, mineral property, 32
France, 205, 211
Franklin, Benjamin, 2
freezing, 153
freezing rain, 119
fresh water, 152, 153, 206
fronts, weather, 116–117, 118–119
frozen water, 21, 46, 102. *See also* water
full moon, 140, 163
fusion, 182–183

G

galaxies. *See also* solar systems; universe
  scale of the universe, 186–187
  structure of the universe, 188–189
  types, 184–185
galaxy clusters, 189
galena (PbS), 29, 33, 157
Galilei, Galileo, 3, 5
gamma rays, 172
Gamow, George, 5
garnet, 154
gases. *See also specific gas*
  atmosphere, 21, 89
  comets, 191
  igneous rock, 34
  volcanoes, 62
gelisols, 44
gemstones, 35
GEOLOGIC MAPS, 10–11
GEOLOGIC TIME SCALE, 72–73
geologists, 201
geosphere, 20
giga- prefix, 6
glaciers, 153. *See also* ice; water
  freshwater storage, 206
  Great Lakes formation, 17
  landscape shaped by, 8
  surface changes, 47
glass, hardness, 32
glass, volcanic, 34
glassy (vitreous) luster, 33
Glomar Challenger (research vessel), 67
glucose, 148
gneiss, 38
gold (Au)
  formation, 155, 156
  native elements, 28
  ore-derived materials, 157
  suspended sediment, 157
gram, 7
GRAND CANYON, THE, 80–81
Grand Canyon, 16, 36, 80–81
granite, 35
graphite, 28, 39, 157
grasslands, soil, 45
gravity and gravitational force
  atmosphere, 21, 84, 88
  explanation, 5
  Jupiter, 170
  mass wasting, 46
  planets in solar system, 167
  tides, 140–141
  Venus, 168, 169
Great Central Valley, 16
Great Lakes, 17, 105, 113
Great Plains, 16
Great Red Spot (Jupiter), 171
greenhouse gases
  atmospheric composition, 94, 95
  carbon cycle, 149
  Venus, 168
Griggs, David, 66
ground motion, sensors, 12–13
groundwater, 152
GROUPS OF MINERALS, 28–29
Gulf of Mexico, 17, 105, 122
Gulf Stream, 2, 105, 136
Gunflint Chert, 86
gypsum ($CaSO_4 \cdot 2H_2O$)
  evaporation, 156
  evaporite, 37
  hardness, 32
  ore-derived materials, 157
  sulfates, 29

H

H. *See* hydrogen (H)
Hadean eon, 72
hail. *See* precipitation
half-life, 77
halides, 28, 29, 156
halite (NaCl)
  defined, 28
  evaporite, 37
  formation, 154
  ore-derived materials, 157
halogen elements. *See* halides
halogens, 23
hanging wall, faults, 43
hardness, mineral properties, 32
Hawaii
  formation, 128
  igneous rock, 34
  volcanoes, 62, 70
haze, 119
He. *See* helium (He)
heat, 38–39, 40–41. *See also* temperatures
heat (weather), 113
hectare, 7
hecto- prefix, 6
heliocentric model, 4
helium (He)
  big bang theory, 192
  early atmosphere, 84
  Jupiter, 170
  stars, element formation, 182
hematite (Fe_2O_3)
  atmospheric history, 85
  fundamentals, 203
  ore-derived materials, 157
  oxides, 29
  streak, 33
Hermit Shale, 80–81
HERTZSPRUNG-RUSSELL DIAGRAM, THE, 180–181
high clouds, 120
Himalayas
  formation, 64–65
  wind pattern effects, 98
HISTORY OF EARTH'S ATMOSPHERE, 84–85
HISTORY OF LIFE ON EARTH, 74–75
HISTORY OF THE UNIVERSE, 192–193
histosols, 45
Holocene epoch, 73, 74
Homo sapiens, 75. *See also* humans
Hooker telescope, 5
Horsehead Nebula, 191
hot water, mineral formation, 156
Hubble, Edwin, 5
HUMAN POPULATION, 210–211
humans
  air pollution, 90
  electricity consumption, 204–205
  history, 75
  water treatment, 208–209
humidity, 113
hurricanes, 122–123
Hutton, James, 3
hydrocarbons, 149
hydrogen (H)
  atoms and atomic structure, 24
  big bang theory, 192
  early atmosphere, 84
  exhaustion, 176, 181
  Jupiter, 170, 171
  periodic table, 22
  stars, element formation, 182
hydrogen bonds, 27
hydrosphere, 20
hydrothermal veins, 157
hydrothermal vents, deep-sea, 2
hypotheses, 14, 15

I. *See* iodine (I)
ice, 153. *See also* glaciers; water
  freshwater storage, 206
  ocean water density, 134
  precipitation and climate, 102
  surface changes, 47
  weather maps, 119
Iceland, 34, 129
IGNEOUS ROCKS, 34–35
igneous rocks
  mineral formation, 155, 156
  Pennsylvania, 11
  rock cycle, 40
inceptisols, 45
India
  earthquakes, 51
  electricity consumption, 205
  mineral resources, 197
  population, 211
  scientific contributions, 3
  soils, 45
  wind patterns, 98–99
Indian plate, 57, 64–65
Indonesia, 51, 63
industry, air pollution, 90
inflation, universe, 192
infrared light, 173
inner core, 55
instruments, weather, 114–115
Interior Craton, 17
International Date Line, 18
International System of Units (Système International d'Unités) (SI), 6–7
INTERPRETING GEOLOGIC CROSS SECTIONS, 82–83
intertidal zone, 142, 144
intrusive igneous rock, 35. *See also* extrusive igneous rock
inversion aloft, 91
iodine (I), 23, 28
ionic bonds, 26
ionosphere, 88
ions, 26
iron (Fe)
  atmospheric history, 85
  atomic structure, 25
  core, 54
  inner core, 55
  mantle, 54
  mineral groups, 29
  ore-derived materials, 157
  outer core, 55
  world resources, 202–203
iron nails, hardness, 32
iron ore resources, 196–197
IRON RESOURCES, 202–203
irregular galaxies, 185, 190
irrigation methods, 3, 44
isotopes, 76–77, 78–79

Japan
  earthquakes, 51
  electricity consumption, 205
  population, 211
  volcanoes, 63
Juan de Fuca plate, 16, 58, 60
Jupiter, 167
JUPITER: A GAS GIANT, 170–171
Jurassic period
  geologic maps, 11
  geologic time scale, 73

K. *See* potassium (K)
Kaibab Limestone, 80–81
Kanto, Japan, 51
kaolinite, 157
Kelvin (measurement), 6–7
Kenya, 205, 211
Kepler, Johannes, 3, 5
Kilauea, Hawaii, 62
kilo- prefix, 6
kilogram, 6, 7
kiloliter, 7

kilometer, 7
kilowatt-hours (kWh) consumption, 204–205
kimberlite rock, 157
KINDS OF VOLCANOES, 70–71
Krakatau volcano, Indonesia, 63
Kuroshio Current, 137

L

Labrador Current, 136
lake-effect snow, 105
Lake Erie, 105. *See also* Great Lakes
lakes, 152, 206
land, 92–93, 211
LANDFORMS OF THE CONTIGUOUS UNITED STATES, 16–17
late Archean era, 72
late Cambrian period, 81
late Jurassic period, 56
late middle Permian period, 81
LATITUDE AND LONGITUDE, 18–19
latitude
- biomes, 106
- effects on precipitation and climate, 102–103
- fundamentals, 18
- ocean temperature, 130–131
- ocean water density, 135
- weather, local influences, 104–105

lava. *See* extrusive igneous rock
LAYERS OF THE EARTH'S ATMOSPHERE, THE, 88–89
Le Pichon, Xavier, 67
lead (Pb), 23, 157
Leeuwin Current, 137
Lemaître, Georges-Henri, 5
length measurements, 6–7
Li. *See* lithium (Li)
lichens, 108
life, solar system, 166
light, electromagnetic spectrum, 172–173
light elements, big bang theory, 192
light observation, 194–195
lignite, 198
limestone
- coquina, 37
- formation, 155, 156
- Kaibab Limestone, 80–81
- marble, 39
- ore-derived materials, 157
- Unkar Group, 81

liquid volume measurements, 7
liter, 7
lithium (Li), 22, 182, 192
lithosphere
- ocean floor, 127, 128
- physical layer, 55
- tectonic plates, 58–59

living things, 152
Local Group, 187, 189
LOCAL INFLUENCES ON THE WEATHER, 104–105
locations on earth. *See* LATITUDE AND LONGITUDE
longitude lines, 18
low clouds, 120
low-mass stars, 177
Lucy, fossil, 3
luminous intensity, 6
lunar eclipses, 163. *See also* ECLIPSES
luster, mineral property, 33
Lyell, Charles, 3

M

magma
- crystallization, 154, 155
- formation, 156
- intrusive igneous rock, 35
- mineral formation, 154, 155
- ocean floor, 127, 143
- volcanoes, 62, 68
- Zoroaster Granite, 81

magnesium (Mg), 22, 24, 54
magnetic fields, sunspots, 159
magnetic rocks, 3
magnetism, mineral properties, 33
magnetite, 85, 157, 203
magnitude, 175
main-sequence stars, 180–181
main shock, earthquakes, 13
MAJOR EARTHQUAKES, 50–51
making observations, 14
malachite, 157
manganese (Mn), 22, 156
mantle, 54, 68, 169
mantle plumes, 128–129
maps, weather, 118–119
marble, 39, 157
Mariana Arc, 61
Mariana Trench, 129
MARINE ORGANISMS, 144–145
marine organisms, 75
marine sediment, 151
maritime polar (mP) air masses, 112–113
maritime tropical (mT) air masses, 112–113
Mars, 167
marshlands, 17
Martinique, West Indies, 62
Masaya volcano, Nicaragua, 62
mass
- stars, 176–177
- Jupiter, 170
- measurements, 7

mass (measurements), 6
mass spectrometer, 78
mass wasting, 46
massive stars, 177
Matthews, Drummond, 67
Matuyama, Motonori, 66
Mauna Loa volcano, Hawaii, 70
Maya civilization, 2
MEASUREMENTS, 6–7
MEASURING EARTHQUAKE SEVERITY, 48–49
mechanical weathering, 46
medium-sized stars, 176
mega- prefix, 6
Megrez, 174, 175
melting, rock cycle, 40–41
Merak, 174, 175
Mercury (planet), 167
meridians, 18
Mesopotamia, 3
mesosphere, 55, 88, 89
Mesozoic era, 72, 73, 74, 75
metal ions, 28
metals, 22–23
metamorphic rock
- ore-derived materials, 157
- Pennsylvania, 11
- rock cycle, 41
- Vishnu Schist, 80–81

METAMORPHIC ROCKS, 38–39
metamorphism, mineral formation, 154
meteor showers, 4
meteorites, 78
meter, 6–7
methane, 84, 94, 149
Mexico
- earthquakes, 50
- electricity consumption, 204
- population, 211
- volcanoes, 68–69, 71

Mexico City, Mexico, 50
Mg. *See* magnesium (Mg)
$(Mg,Fe)_2SiO_4$. *See* olivine ($(Mg,Fe)_2SiO_4$)
mica (muscovite)
- cleavage, 32
- foliated metamorphic rock, 38 (*See also* minerals)
- mineral formation, 154, 155
- sheet silicates, 31

mice, 106, 107
micro- prefix, 6
microfossils, 87

microgram, 7
micrometer, 7
microwaves, 173
Mid-Atlantic Ridge, 67, 128
mid-ocean ridges, 67, 127, 129, 143
middle Archean era, 72
middle clouds, 120
Middle East, 196
middle Proterozoic era, 72
Milky Way galaxy
  historical events, 5
  location, 189
  scale of the universe, 187
  spiral galaxy, 190
milli- prefix, 6
milligram, 7
milliliter, 7
millimeter, 7
MINERAL FORMATION, 154–155
MINERAL RESOURCES, 196–197
minerals. *See also specific type of mineral*
  carbon cycle, 148
  chemical sedimentary rock, 37
  foliated metamorphic rock, 38, 39
  formation, 154–155
  groups of, 28–29
  ores, 156–157
  properties, 32–33
  resources, 196–197
  silicate, 30–31
mining, phosphorus cycle, 150
mining engineers, 201
Miocene epoch, 73, 74
Mississippi River, 17
Mississippian period, 11, 73, 74, 81
Missouri River, 17
Mizar, 174, 175
mL (measurement), 7
Mn. *See* manganese (Mn)
Modified Mercalli Intensity Scale, 49
Mohs hardness scale, 32
mole (measurement), 6
mollisols, 45
molten rock. *See* igneous rocks
monocline, 42
monsoons, 98, 99
Montana, 17
Montserrat, West Indies, 62
moon
  eclipse, 162–163
  formation, 84
  Jupiter, 170
  phases, 160–161
  planets, 167
  scale of the universe, 186–187
  solar systems, 167, 188
  tides, 140–141
  Venus, 168
Morgan, W. Jason, 67
mostly cloudy, weather maps, 119
motion, laws of, 4, 5
Mount Cleveland, Chuginadak Island, 62
Mount Etna, Sicily, 63
Mount Everest, 129
Mount Marsili, Naples, 63
Mount Myoko, southwestern Niigata Prefecture, 63
Mount Nyamuragira, Democratic Republic of the Congo, 63
Mount Pelée, Martinique, 62
Mount Pinatubo, 34
Mount Saint Helens, Washington, 16, 62
Mount Tambora, Sambawa, 63
Mount Vesuvius, Naples, 63
mountains
  granite, 35
  tectonic plate movement, 56
  thick lithosphere, 64–65
mP. *See* maritime polar (mP) air masses
mT. *See* maritime tropical (mT) air masses
Muav Limestone, 80–81
mud, 36
muscovite (mica)
  cleavage, 32
  foliated metamorphic rock, 38
  mineral formation, 154, 155
  sheet silicates, 31

N

N. *See* nitrogen (N)
Na. *See* sodium (Na)
NaCl. *See* halite (NaCl)
nano- prefix, 6
native elements, 28
Nazca plate, 58, 60
neap tides, 140
negatively charged ions, 26
Neptune, 167, 170
neritic zone, 142
network silicates, 31
neutrino, 182, 183
neutron stars, 177
neutrons, 192
Nevada, 104
New England, 8
New Mexico, 16
new moon, solar eclipse, 162
new moon, tides, 140
New York, 8, 105
New Zealand, 59
Newton, Isaac, 5
Ni. *See* nickel (Ni)
nickel (Ni), 23, 54, 55
nimbo- prefix, 121
nimbostratus clouds, 120
nimbus- suffix, 121
nitrates, 146
nitrogen (N)
  atmosphere, 95
  atomic structure, 24
  cycle, 146–147
  Venus, 94
NITROGEN CYCLE, THE, 146–147
nitrogen-fixing bacteria, 146
nitrogen oxides (NO_x), 90
noble gases, 23
nonmetals, 23, 28
nonreflective luster, 33
normal faults, 43
North America
  air masses, 112–113
  coal deposits, 196
  historic hurricane paths, 123
  mineral resources, 196
  surface changes, 47
  weather, local influences, 105
  wind patterns, 98–99
North American plate, 58
North Dakota, 105
North Equatorial Current, 136–137
North Pacific Current, 136
North Pole, 125. *See also* poles
Northern Africa (Sahara), 44
Northern Hemisphere
  equator, 19
  seasons, 164–165
  surface temperature fluctuations, 96–97
Northern Lights, 158
Norwegian Current, 137
novas, 176
NO_x. *See* nitrogen oxides (NO_x)
nuclear fusion, 182–183
numeral system, Arabic, 3

O

O. *See* oxygen (O)
observations, making, 14
obsidian, 34
occluded fronts, 117, 119
OCEAN SALINITY, 132–133
OCEAN TEMPERATURE, 130–131
OCEAN WAVES, 138–139
oceanic lithosphere, 34, 58–59
oceanic zone, 143, 145
oceans
  carbon reservoir, 149
  density, 134–135
  dissolved chemicals, 154

floor, 126–127, 128–129
formation, 84
mineral formation, 154
phosphorus reservoir, 151
salinity, 132–133
surface currents, 136–137
temperature, 130–131
tides, 140–141
water reservoir, 153
waves, 138–139
weather, local influences, 104
zones, 142–143
Oligocene epoch, 73, 74
olivine ($(Mg,Fe)_2SiO_4$), 29, 30
openpit mining, 203. *See also* strip mining
Ordovician period, 11, 73, 74, 75
ORES, 156–157
organic sedimentary rocks, 37
organisms, biosphere, 21
Orion, constellation, 191
OUR PLACE IN THE UNIVERSE, 4–5
OUR SOLAR SYSTEM, 166–167
OUR SUN, 158–159
outer core, 55
oxides, 29
oxisols, 45
oxygen (O)
abundance, 30
atmosphere, 85, 95
atomic structure, 24
covalent bonds, 27
oxides, 29
Oyashio Current, 137
ozone layer, 85, 88

P. *See* phosphorus (P)
P seismic waves, 52
P waves, 52
Pacific Ocean
air masses, 112
El Niño–Southern Oscillation, 110–111
historic hurricane paths, 123
Pacific plate
Mariana Arc, 61
Mariana Trench, 129
ocean floor topography, 128
paints, air pollution, 90
Paleocene epoch, 73, 74
Paleozoic era, 72, 73, 74
PANGAEA AND TECTONIC ACTIVITY, 56–57
paper mills, air pollution, 90
parallels, lines of latitude, 18
parent isotopes, 76–77, 78–79
Paricutín volcano, Mexico, 68–69, 71
partial lunar eclipse, 163
partial solar eclipse, 162
particulate matter, 90
partly cloudy, weather maps, 119
Pb. *See* lead (Pb)
PbS. *See* galena (PbS)
pearly luster, 33
peat, 198
pegmatite, 35
pelagic zone, 143
Pennsylvania, geologic map, 10–11
Pennsylvanian period, 11, 73, 74
penny, hardness, 32
percolation, 152
performing experiments, 14
Periodic Table of the Elements, 22–23
periods, earth's history
geologic time scale, 72–73
life, 74–75
permafrost, 44, 106
Permian period, 11, 73, 74, 75, 81
Peru Current, 136
petroleum, world resources, 196–197
Phanerozoic eon, 72
PHASES OF THE MOON, 160–161
Phecda, 174, 175
Philippines, 34
Philippine plate, 59, 61, 129
phosphorus (P), 23, 150–151
PHOSPHORUS CYCLE, THE, 150–151
photon, 182
photosphere, 158
photosynthesis, 85, 148
photosynthetic organisms, 86, 148
physical layers, earth, 54
phytoplankton, 147
pico- prefix, 6
placer deposits, 157
plagioclase ($CaAl_2Si_2O_8$), 29
plains. *See* abyssal plains
planets. *See also* solar systems
characteristics, 167
Jupiter, 170–171
solar system, 166–167, 188
Venus, 168–169
plants, 75, 148, 150
plasticlike luster, 33
plastics, burning, 90
plate movements and motions. *See* PANGAEA AND TECTONIC ACTIVITY
PLATE TECTONICS: DEVELOPMENT OF A THEORY, 66–67
Pleistocene epoch, 16, 73, 74
Pliocene epoch, 73, 74
Pluto, 167
Poas volcano, Lake Poas, 62
polar air masses, 112
polar ice, 106
poles
atmospheric circulation, 100–101
ocean water density, 134–135
oceans, 124–125
ponds, 152
population, human, 210–211
positively charged ions, 26
positrons, 182
potassium (K), 22, 25, 76
power plants, air pollution, 90
Precambrian eon, 11, 72
PRECAMBRIAN LIFE AND ROCKS, 86–87
Precambrian period, 73, 74
Precambrian shield rocks, 86
precipitation
biomes, 106–107, 108
clouds, 120–121
mineral formation, 154, 155
water cycle, 153
weather maps, 118–119
PRECIPITATION AND CLIMATE, 102–103
prefixes, International System of Units, 6
pressures. *See also* air pressure
anthracite, 199
bituminous coal, 199
inflation, universe, 192
metamorphic rock, 38–39
rock cycle, 40–41
prevailing winds, 102
prime meridian, 18–19
Prince William Sound, Alaska, 50
producers, 106
prominences, solar, 159
PROPERTIES OF MINERALS, 32–33
Proterozoic eon, 72
protons, 192
protostars, 176, 177
Proxima Centauri, 181
Ptolemy, 4
pumice, 34
pyrite (FeS_2), 29, 156
pyroxene, 30

quartz
fracture, 32
hardness, 32
luster, 33
mineral formation, 154, 155
network silicates, 31
ore-derived materials, 157
sandstone, 36
quartz sand, 44

quartzite, 81
Quaternary period, 11, 73, 74
Quizapu (Cerro Azul), central Chile, 62

R

Ra. *See* radium (Ra)
radiation
atmosphere, 88
big bang theory, 193
cosmic microwave background, 5
gamma rays, 172
radiative zone, 158
radio waves, 173
RADIOACTIVE DECAY, 76–77
radioactive isotopes, 78–79
radioactivity, mineral properties, 33
RADIOMETRIC DATING, 78–79
radium (Ra), 22, 172
rain, weather maps, 119
rain gauge, 115
rainfall
biomes, 107
climate, 102–103
mass wasting, 46
wind patterns, 98–99
rain forests, 103, 107
Rasht, Iran, 51
reactions, mineral formation, 154
reclamation, coal mining, 201
recrystallization, 39, 41
recumbent fold, 42
RED SHIFT, 194–195
reddish color
dwarf stars, 181
giant stars, 176
lunar eclipse, 163
stars, 180
supergiant stars, 177
Redwall Limestone, 80–81
redwood forests, 104
refineries, air pollution, 90
reflective luster, 33
reflecting telescopes, 178
refracted light, lunar eclipse, 163
refracting telescopes, 178, 179
Regulus, 180
reservoirs, freshwater storage, 206–207
resinous luster, 33
respiration, 148, 152
reverse faults, 43
revolution, planets, 167, 170
Richter scale, 49, 50–51
Rio Grande, 16
rivers
erosion, 47
freshwater reservoir, 206
placer deposits, 157
water cycle, 152
roads, 9
ROCK CYCLE, THE, 40–41
rocks and rock bodies. *See also specific type of rock*
carbon cycle, 148
cycle, 40–41
earth's layers, 54–55
formations, 10
geologic maps, 10–11
geosphere, 20
igneous, 34–35
magnetic, 3
melting, 68
metamorphic, 38–39
phosphorus cycle, 150
Precambrian shield, 86
sedimentary, 36–37
silicate minerals, 29
superposition, 82–83
ventifacts, 47
Rocky Mountains, 16, 104
rotation
Earth, 98–99, 101
Jupiter, 170
planets in solar system, 167
Venus, 168
running water, 8, 47
runoff, 146, 150, 152
Russia
electricity consumption, 205
mineral resources, 197
population, 211
soil, 44, 45

S

S. *See* sulfur (S)
S waves, 53
Sahara desert, 44
salinity, 134
saltwater, 134, 156
San Andreas fault zone, 16
San Francisco, California, 50
sand, ore-derived materials, 157
sandstone
aquifers, 207
clastic sedimentary rock, 36
Coconino Sandstone, 80–81
ore-derived materials, 157
Unkar Group, 81
Saturn, 167, 170
savanna, 107, 108
SCALE OF THE UNIVERSE, THE, 186–187
scattered clouds, 119
schist, 38, 154
scientific contributions, 2–3
SCIENTIFIC METHOD, 14–15
scientific procedures and tools
geologic maps, 10–11
latitude and longitude, 18–19
measurements, 6–7
scientific method, 14–15
seismographs, 12–13
topographic maps, 8–9
Scotia plate, 58, 60
seafloor, 2
sea ice, 153
sea level, 8
SEA LEVEL TEMPERATURE, 96–97
seamounts, 127
seashell fragments, 37
SEASONS, 164–165
seasons, 96–97, 130–131
second (measurement), 6
sediment, marine, 47, 127, 151
SEDIMENTARY ROCKS, 36–37
sedimentary rocks
calcite, 28
defined, 36–37
dolomite, 28
ore-derived materials, 157
Pennsylvania, 11
rock cycle, 41
SEISMIC WAVES, 52–53
seismogram, 52
SEISMOGRAPHS, 12–13
seismographs, 48–49
seismometers, 66
severe weather
clouds, 120–121
tornadoes and hurricanes, 122–123
sewage treatment plants, 151
shale, 36, 80–81
sheet silicates, 31
shell fragments, 37
shellfish, 149
shield volcanoes, 70
showers (rain or snow), 119
Si. *See* silicon (Si)
SI *See* Système International d'Unités (International System of Units)
Sierra Nevada, 16, 104
SILICATE MINERALS, 30–31
silicates, 29
silicon-oxygen tetrahedron, 30
silicon (Si), 29
atomic structure, 25
crust, 54
mantle, 54
Silurian period
geologic maps, 11
geologic time scale, 73
life, 74, 75

silver (Ag), 23, 28, 157
single-celled organisms, 85
single-chain silicates, 30
single-tetrahedron silicates, 30
slate, 38
sleet. *See* precipitation
smelting, 203
smithsonite, 157
snow. *See also* precipitation
  air masses, 113
  water cycle, 153
  weather, local influences, 105
  weather maps, 119
SO_2. *See* sulfur dioxide (SO_2)
sodium (Na)
  atomic structure, 24
  chemical bonds, 26
  ions, 26
Soufrière Hills, Montserrat, 62
soil
  biomes, 107, 108
  mineral resources, 196
  phosphorus cycle, 150
  types, 44–45
SOILS OF THE WORLD, 44–45
solar eclipses, 4, 162. *See also* ECLIPSES
SOLAR ENERGY, 92–93
solar flare, 158, 159
solar systems. *See also* galaxies; planets
  scale of the universe, 186–187
  structure of the universe, 188–189
solar wind, 158, 159
solid iron, 55
solvents, air pollution, 90
Somali Current, 137
South America
  El Niño-Southern Oscillation, 110
  mineral resources, 196
South American plate, 58, 60, 128
South Equatorial Current, 136–137
South Pole, 125. *See also* poles
Southeast Asia, 45
Southwest Indian Ridge, 61
Southern Hemisphere
  equator, 19
  ocean, 124
  seasons, 164–165
  surface temperature fluctuations, 96–97
Soviet Union, 3
space missions, 2–3
sphalerite, 157
SPHERES OF EARTH, THE, 20–21
spiral galaxies, 184, 190
spodosols, 45
spring equinox, 165
spring tides, 140
Sputnik, 3
square centimeter, 7
square kilometer, 7
square meter, 7
STAR LIFE CYCLES, 176–177
STAR MAGNITUDES, 174–175
stars. *See also* solar systems
  element formation, 182–183
  Hertzsprung-Russell Diagram, 180–181
  life cycles, 176–177
  magnitudes, 174–175
stationary fronts, 117, 119
steel, 196
steel file, 32
storms, Jupiter, 171
stratosphere, 88, 89
stratus clouds, 120
streak, mineral properties, 32, 33
streak plate, 32
streams
  erosion, 47
  placer deposits, 157
  topographic maps, 8
  water cycle, 152
stress, 55
strike-slip faults, 43
strip mining, 201. *See also* openpit mining
stromatolites, 86
STRUCTURE OF THE UNIVERSE, THE, 188–189
subduction, 68, 127
sublittoral zone, 142, 143, 144
submetallic luster, 33
sulfates, 29
sulfide materials, 157
sulfides, 29
sulfur dioxide (SO_2), 90
sulfur (S)
  ores, 157
  sulfides, 29
summer solstice, 164, 165
sun and sunlight
  distance, planets in solar system, 167
  fundamentals, 158–159
  gravitational force, 140
  hydrogen exhaustion, 176
  main-sequence star, 180, 181
  solar energy, 93–94
  solar system center, 4, 5
  surface temperature fluctuations, 96–97
  water cycle, 153
  wind patterns, 98
sunspots, 159
Supai Group, 80–81
supercontinent (Pangaea), 56–57
supernova, 177, 191
superposition, 82
SURFACE CURRENTS OF THE OCEAN, 136–137
surface inversion, 91
surface waves, 53
surface water, 152
swamps, 75, 152
syncline, 42
Système International d'Unités (International System of Units) (SI), 6–7

T

tables
  air pollution, 90
  elements, 22–23
  biome characteristics, 108
  isotopes, 76
  Jupiter characteristics, 170
  measurements, 6–7
  Modified Mercalli Intensity Scale, 49
  ore-derived materials, 157
  planet characteristics, 167
  population, 211
  radiometric dating, 79
  Richter scale, 49
  solar system characteristics, 167
  tides, 141
  Venus characteristics, 168
taiga, 106, 108
talc, 32, 33
Tapeats Sandstone, 80–81
Tau Ceti, 181
tectonic collisions, 16, 64
tectonic plates
  boundary types, 60–61
  collisions, 16
  current locations, 58–59
tectonics
  activity, 56–57
  Himalayas formation, 64–65
  ocean floor topography, 128–129
  Pangaea, 56–57
  plate boundaries, 60–61
  plates, 58–59
  theory development, 66–67
  volcanoes, 62–63
telescope, 5
TELESCOPES, 178–179
temperate forest, 107, 108
temperate grasslands, 107
temperatures. *See also* heat
  air masses and fronts, 116–117
  air pollution, 91
  anthracite, 199
  atmosphere, 88–89, 94–95
  biomes, 106–107

bituminous coal, 199
chromosphere, 158
clouds, 120
convective zone, 158
core of sun, 158
corona, 158
deserts, 107
greenhouse gases, 149
Gulf Stream, 2
inflation, universe, 192
International System of Units, 6
inversions, 91
Jupiter, 170
lignite, 198
Mars, 95
mineral formation, 155
ocean water density, 134–135
planets in solar system, 167
precipitation and climate, 102–103
rock cycle, 40–41
solar energy, 93
stars, 180–181
sun core, 158
surface fluctuation, 96–97
Venus, 94, 168
weather maps, 118–119
wind patterns, 98–99
TERRESTRIAL BIOMES, 106–107
Tertiary period, 11, 73, 74
testing the hypothesis, 14
tetrahedrons
double-chain silicates, 31
network, 31
sheets, 31
single-chain silicates, 30
Th. *See* thorium (Th)
theory, scientific, 15
thermosphere, 88, 89
third-quarter moon, tides, 140
thorium (Th), 22, 76
TIDES, 140–141
tides, intertidal zone, 142, 144
Tiktaalik, 75
time scale, geologic, 72–73
topaz, 32, 156
TOPOGRAPHIC MAPS, 8–9
topography, 102, 106
TOPOGRAPHY OF THE OCEAN FLOOR, 128–129
Tornado Alley, 105, 122
tornadoes, 105, 122
TORNADOES AND HURRICANES, 122–123
Toroweap Formation, 80–81
total lunar eclipse, 163
total solar eclipse, 162
trade winds, 110–111
transform faults, 129
transform plate boundary, 60
transition metals, 22–23
transpiration, 152
transportation, world resources, 201, 203
trenches, ocean floor, 127
Triassic period, 11, 73, 74
trilobites, 75, 81
Tropic of Cancer, 19
Tropic of Capricorn, 19
tropical air masses, 112
tropical rain forest, 107, 108
tropics, 19, 102–103
troposphere, 88
trough, 138
tsunami, 51
tundra, 106, 108
TYPES OF CHEMICAL BONDS, 26–27
TYPES OF GALAXIES, 184–185
TYPES OF PLATE BOUNDARIES, 60–61
Tyrannosaurus rex, 75

U. *See* uranium (U)
ultisols, 45
ultraviolet light, 33
ultraviolet radiation, 88, 172
uniformitarianism, 3
UNIQUE OBJECTS IN THE UNIVERSE, 190–191
United States
air masses, 112–113
aquifers, 206
coal resources, 200–201
earthquakes, 50
electricity consumption, 204
fossilized stromatolites, 86
population, 211
soil, 44
tornado frequency, 122
volcanoes, 62
weather, local influences, 104
universe. *See also* galaxies
blue-shifted light, 195
history, 192–193
red shift, 194–195
scale, 186–187
structure, 188–189
unique objects, 190–191
Unkar Group, 80–81
uranium (U), 22, 76, 196–197
Uranus
characteristics, 167
density comparison, 170
solar system location, 167
Ursa Major, constellation, 174

valence electrons, 26–27
Vailulu'u (underwater), Samoa Islands, 62
veins, metamorphic rock, 157
ventifacts, 47
Venus, 94, 167
VENUS: A TERRESTRIAL PLANET, 168–169
vernal equinox, 165
vertisols, 45
Vine, Fred, 67
Vishnu Schist, 80–81
visible light, 173, 178
vitreous (glassy) luster, 33
volatile organic compounds, 90
volcanic ash, soils, 44
volcanic chains
convergent boundaries, 61, 68
volcanic islands, 34
VOLCANIC PROCESSES, 68–69
VOLCANOES, 62–63
volcanoes. *See also specific volcano*
air pollution, 90
atmosphere, 84
fundamentals, 62–63
locations, 62–63
mass wasting, 46
ocean floor, 127, 128–129
processes, 68–69
types, 70–71

warm air
air masses and fronts, 116–117
air pollution, 91
atmospheric circulation, 100
ocean water density, 134
warm fronts, 116, 119
warm water, 134
wastewater, 151, 208
water
aquifers, 206–207
clouds, 120–121
cryosphere, 21
cycle, 152–153
fog, 104
freshwater, 152, 153
hydrosphere, 20
ice, 153
landscape shaped by, 8
mechanical weathering, 46
molecule, 134
ocean water density, 134–135
precipitation and climate, 102

solar energy, 92–93
tectonic plates, 68
temperatures, 130–131
WATER CYCLE, THE, 152–153
WATER TREATMENT, 208–209
water vapor, 84, 120–121
wavelength (light), 173
wavelength (ocean), 138–139
waves, ocean, 138–139
waxy luster, 33
weather
air masses, 112–113, 116–117
atmospheric circulation, 100–101
Canada, 112–113
clouds, 120–121
fronts, 116–117
hurricanes, 122–123
instruments, 114–115
local influences, 104–105
maps, 118–119
ocean impact, 124
precipitation and climate, 102–103
tornadoes, 122–123
troposphere, 88
United States, 112–113
wind patterns, 98–99
WEATHER INSTRUMENTS, 114–115
WEATHER MAPS, 118–119
weathering, phosphorus cycle, 150
Wegener, Alfred, 66
well, 207
West Australian Current, 137
West Indies, 62
West Wind Drift, 136–137
Western Hemisphere, 18–19
white dwarf stars, 176, 177, 180
wind
atmospheric circulation, 100–101
ocean currents, 136–137
patterns, 98–99
precipitation and climate, 102
solar, 158, 159
surface changes, 47
weather maps, 118–119
WIND PATTERNS, 98–99. *See also* ATMOSPHERIC CIRCULATION
winter solstice, 164, 165
winter storms, 113. *See also* snow
WORLD ELECTRICITY USE, 204–205

X rays, 158, 172

zinc (Zn), 23, 157
zincite, 157
Zn. *See* zinc (Zn)
ZONES OF THE OCEAN, 142–143
Zoroaster Granite, 81